T0280339

Lecture Notes in Computer Science 14504

The series Lecture Notes in Computer Science (LNCS), including its subseries Lecture Notes in Artificial Intelligence (LNAI) and Lecture Notes in Bioinformatics (LNBI), has established itself as a medium for the publication of new developments in computer science and information technology research, teaching, and education.

LNCS enjoys close cooperation with the computer science R & D community, the series counts many renowned academics among its volume editors and paper authors, and collaborates with prestigious societies. Its mission is to serve this international community by providing an invaluable service, mainly focused on the publication of conference and workshop proceedings and postproceedings. LNCS commenced publication in 1973.

Hai Jin · Zhiwen Yu · Chen Yu · Xiaokang Zhou ·
Zeguang Lu · Xianhua Song
Editors

Green, Pervasive, and Cloud Computing

18th International Conference, GPC 2023
Harbin, China, September 22–24, 2023
Proceedings, Part II

Editors
Hai Jin
Huazhong University of Science
and Technology
Wuhan, China

Chen Yu
Huazhong University of Science
and Technology
Wuhan, China

Zeguang Lu
National Academy of Guo Ding Institute
of Data Science
Beijing, China

Zhiwen Yu
Harbin Engineering University
Harbin, China

Xiaokang Zhou
Shiga University
Shiga, Japan

Xianhua Song
Harbin University of Science and Technology
Harbin, China

ISSN 0302-9743 ISSN 1611-3349 (electronic)
Lecture Notes in Computer Science
ISBN 978-981-99-9895-1 ISBN 978-981-99-9896-8 (eBook)
https://doi.org/10.1007/978-981-99-9896-8

This Springer imprint is published by the registered company Springer Nature Singapore Pte Ltd.
The registered company address is: 152 Beach Road, #21-01/04 Gateway East, Singapore 189721, Singapore

Preface

As the program chairs of the 18th International Conference on Green, Pervasive, and Cloud Computing (GPC 2023), it is our great pleasure to welcome you to the proceedings of the conference, which was held in Harbin, China, 22–24 September 2023, hosted by Huazhong University of Science and Technology, Harbin Engineering University, Harbin Institute of Technology, Northeast Forestry University, Harbin University of Science and Technology, Heilongjiang Computer Federation and National Academy of Guo Ding Institute of Data Sciences. The goal of this conference was to provide a forum for computer scientists, engineers and educators.

This conference attracted 111 paper submissions. After the hard work of the Program Committee, 38 papers were accepted to be presented at the conference, with an acceptance rate of 34.23%. There were at least 3 reviewers for each article, and each reviewer reviewed no more than 5 articles. The major topic of GPC 2023 was Green, Pervasive, and Cloud Computing. The accepted papers cover a wide range of areas related to Edge Intelligence and Mobile Sensing and Computing, Cyber-Physical-Social Systems and Industrial Digitalization and Applications.

We would like to thank all the program committee members of both conferences, for their hard work in completing the review tasks. Their collective efforts made it possible to attain quality reviews for all the submissions within a few weeks. Their expertise in diverse research areas helped us to create an exciting program for the conference. Their comments and advice helped the authors to improve the quality of their papers and gain deeper insights.

We thank Xuemin Lin, Nabil Abdennadher and Laurence T. Yang, whose professional assistance was invaluable in the production of the proceedings. Great thanks should also go to the authors and participants for their tremendous support in making the conference a success. Besides the technical program, this year GPC offered different experiences to the participants. We hope you enjoyed the conference.

September 2023

Zhiwen Yu
Christian Becker
Chen Yu
Xiaokang Zhou
Chengtao Cai

Preface

Organization

Honorary Chair

Rajkumar Buyya	University of Melbourne, Australia

General Chairs

Zhiwen Yu	Harbin Engineering University, China
Christian Becker	University of Stuttgart, Germany

Program Chairs

Chen Yu	Huazhong University of Science and Technology, China
Xiaokang Zhou	Shiga University, Japan
Chengtao Cai	Harbin Engineering University, China

Organization Chairs

Wei Wang	Harbin Engineering University, China
Haiwei Pan	Harbin Engineering University, China
Hongtao Song	Harbin Engineering University, China
Yuhua Wang	Harbin Engineering University, China
Jiquan Ma	Heilongjiang University, China
Xingyu Feng	NSFOCUS, China

Publication Chairs

Weixi Gu	University of California, Berkeley, USA
Xianhua Song	Harbin University of Science and Technology, China
Dan Lu	Harbin Engineering University, China

Sponsorship Chair

Bingyang Li	Harbin Engineering University, China

Session Chairs

Chen Yu	Huazhong University of Science and Technology, China
Xianwei Lv	Huazhong University of Science and Technology, China

Web Chair

Wang Chen	Huazhong University of Science and Technology, China

Registration/Financial Chair

Fa Yue	National Academy of Guo Ding Institute of Data Science, China

Steering Committee

Hai Jin (Chair)	Huazhong University of Science and Technology, China
Nabil Abdennadher (Executive Member)	University of Applied Sciences, Switzerland
Christophe Cerin	University of Paris XIII, France
Sajal K. Das	Missouri University of Science and Technology, USA
Jean-Luc Gaudiot	University of California-Irvine, USA
Kuan-Ching Li	Providence University, Taiwan
Cho-Li Wang	University of Hong Kong, China
Chao-Tung Yang	Tunghai University, Taiwan
Laurence T. Yang	St. Francis Xavier University, Canada/Hainan University, China
Zhiwen Yu	Harbin Engineering University, China

Program Committee Members

Alex Vieira	Federal University of Juiz de Fora, Brazil
Alfredo Navarra	Università degli Studi di Perugia, Italy
Andrea De Salve	Institute of Applied Sciences and Intelligent Systems, Italy
Arcangelo Castiglione	University of Salerno, Italy
Bo Wu	Tokyo University of Technology, Japan
Carson Leung	University of Manitoba, Canada
Feng Tian	Chinese Academy of Sciences, China
Florin Pop	University Politehnica of Bucharest, Romania
Jianquan Ouyang	Xiangtan University, China
Ke Yan	National University of Singapore, Singapore
Kuo-chan Huang	National Taichung University of Education, Taiwan
Liang Wang	Northwestern Polytechnical University, China
Marek Ogiela	AGH University of Science and Technology, Poland
Mario Donato Marino	Leeds Beckett University, England
Meili Wang	Northwest A&F University, China
Pan Wang	Nanjing University of Posts and Telecommunications, China
Shingo Yamaguchi	Yamaguchi University, Japan
Su Yang	Fudan University, China
Tianzhang Xing	Northwest University, China
Wasim Ahmad	University of Glasgow, UK
Weixi Gu	China Academy of Industrial Internet, China
Xiaofeng Chen	Xidian University, China
Xiaokang Zhou	Shiga University, Japan
Xiaotong Wu	Nanjing Normal University, China
Xin Liu	China University of Petroleum, China
Xujun Ma	Télécom SudParis, France
Yang Gu	Chinese Academy of Sciences, China
Yanmin Zhu	Shanghai Jiao Tong University, China
Yaokai Feng	Kyushu University, Japan
Yirui Wu	Hohai University, China
Yuan Rao	Anhui Agricultural University, China
Yuezhi Zhou	Tsinghua University, China
Yufeng Wang	Nanjing University of Posts and Telecommunications, China
Yunlong Zhao	Nanjing University of Aeronautics and Astronautics, China

Zengwei Zheng	Hangzhou City University, China
Zeyar Aung	Khalifa University, United Arab Emirates
Zhijie Wang	Hong Kong Polytechnic University, China

Contents – Part II

Wireless and Ubiquitous Networking

Contents – Part I

Industrial Digitization and Applications

Edge Intelligence

OPECE: Optimal Placement of Edge Servers in Cloud Environment

Tao Huang[1], Fengmei Chen[1](✉), Shengjun Xue[1], Zheng Li[2], Yachong Tian[1], and Xianyi Cheng[1]

[1] School of Computer Science and Technology, Silicon Lake College, Suzhou 215332, China
30038378@qq.com, xycheng@ntu.edu.cn

[2] School of Computer Secience, Nanjing University of Information Science and Technology, Nanjing 210044, China

Abstract. Cloud computing offloads user tasks to remote cloud servers, which can effectively enhance the user's network experience, but in recent years, as the number of offloaded tasks increases and users' real-time requirements improve, cloud services are becoming increasingly challenging to meet users' needs. Edge computing deploys multiple edge servers around the users. The shorter distance from users can significantly reduce the transmission time of task data and avoid unpredictable network latency, which is especially suitable for normal users whose tasks are mainly data-intensive tasks. However, the large variability in the density of users in different areas and the type of computing tasks (i.e., compute-intensive and data-intensive) in the same area leads to the challenge of optimally deploying multiple edge servers. To address this challenge, we design a method named optimal placement of edge servers in the cloud environment (OPECE). First, this optimal placement problem is modeled as a constrained multi-objective optimization model with task time and server utilization as the two optimization objectives. Then, this multi-objective optimization model is optimized using the Non-dominated Sorting Differential Evolution (NSDE) algorithm. Finally, the effectiveness and superiority of OPECE are demonstrated by comparing it with the currently used methods.

Keywords: Cloud computing · edge server · optimal placement · NSDE · task time · server utilization

1 Introduction

With the growth of cloud computing, many network users are offloading local computing tasks to remote cloud servers with powerful computing capabilities for execution [1], which not only reduces the computing time of these tasks but also can help local devices to handle more tasks concurrently [2]. However, the increasing complexity and real-time requirements of various computing tasks, as

H. Jin et al. (Eds.): GPC 2023, LNCS 14504, pp. 3–16, 2024.
https://doi.org/10.1007/978-981-99-9896-8_1

well as the increasing number of network users and applications, and even enterprise users are gradually deploying a large number of compute-intensive applications to cloud servers, resulting in an increased load on these cloud servers [3]. In addition, compared to enterprise users, the tasks of normal users are mainly some data-intensive tasks, such as video transmission, etc. The long transmission distance between normal users and cloud servers often leads to higher data transmission time [4] and unpredictable network latency, affecting the experience of normal users. In short, the traditional cloud-based service model is increasingly unable to meet the needs of normal users.

Edge computing can effectively relieve the pressure on cloud servers by deploying edge servers in the user's area, which can offload the computing tasks within their service area to the corresponding edge servers for execution [5]. Although the computational performance and resources of edge servers are slightly less than those of cloud servers, the shorter transmission distance between users and edge servers can significantly reduce the transmission time of task data, which in turn can significantly improve the response time of these tasks [6], especially for those data-intensive tasks. Therefore, benefiting from edge computing, we need to deploy multiple edge servers in the user area to offload more computational tasks to the edge servers for execution, thereby maximizing the experience of the normal user in the target area. However, the density of users in different areas and the type of computing tasks (i.e., compute-intensive and data-intensive) in the same area vary greatly [7]. For example, the density of users in the suburbs is much lower than that in the cities, while even in the user-intensive cities, there are a large number of both normal and enterprise users distributed. Therefore, it has become a challenge to optimally place edge servers to improve users' experience in the target area.

1.1 Motivation

Cloud servers typically have high computational performance and resources that help reduce task computation time. Still, longer distance from users often results in higher data transmission time and unpredictable network latency. On the contrary, although edge servers' computing performance and resources are inferior to cloud servers, the distance between edge servers and users is often shorter, which helps shorten the data transmission time of tasks and avoid network latency. Therefore, the compute-intensive tasks of enterprise users are better suited to be offloaded to cloud servers, while data-intensive tasks of normal users are better suited to be offloaded to edge servers, which can maximize the experience for all network users. In addition, to maximize the utilization of all edge servers, we want each edge server to serve as many users as possible, so it is often ideal for deploying each edge server in the area with the highest density of users. However, deploying multiple edge servers often results in overlapping service coverage and underutilization for some edge servers. In summary, the optimal placement of edge servers typically requires the following three requirements to be met:

1. Edge servers perform as many data-intensive tasks as possible rather than compute-intensive tasks;

2. Edge servers are deployed in user-intensive areas as much as possible, but their service areas should not overlap as much as possible;
3. The load on each edge server must not exceed its maximum load.

To address these requirements, we model the optimal placement of edge servers as a constrained multi-objective optimization problem and propose a method for optimal placement based on the Non-dominated Sorting Differential Evolution (NSDE) algorithm.

1.2 Our Contribution

In this paper, the main contributions include the following.

1. We model the optimal placement of edge servers in the cloud environment as a constrained multi-objective optimization model, with user task time and server utilization as the two critical optimization objectives;
2. The NSDE algorithm is used to optimize the fitness function of the model, and an efficient method for optimal placement of edge servers is proposed.
3. The effectiveness and superiority of OPECE proposed in this paper is demonstrated by comparing it with the currently used methods.

The rest of this paper is organized as follows. In Sect. 2, the basic concepts edge server placement model is presented, including the time computation model, server utilization model, objective function, and constraint. Section 3 elaborates on the OPECE method. Section 4 compares and analyzes the experimental results of different methods. Section 5 summarizes the related work, while Sect. 6 draws a conclusion of the paper and presents the future work.

2 Problem Formulation

In this section, firstly, the placement of all nodes and edge servers in the resource request area is modeled. Secondly, the optimization placement model of edge servers is further modeled, and the user time model and the utilization model of the edge servers in the optimization model are analyzed in detail. Finally, the optimal placement problem of edge servers is modeled as a constrained multi-objective optimization problem.

2.1 Problem Modeling

We make a formal representation of all nodes and servers in the resource request area.

The resource request area is modeled as a right-angle coordinate model, as shown in Fig. 1. The positions of all resource request nodes can be represented as a set $\mathcal{A} = \{A_0, A_1, ..., A_n, ..., A_{N-1}\}$, where $A_n = [a_n, b_n](0 \leq a_n \leq x_{max}, 0 \leq b_n \leq y_{max})$ denotes the position of the $n-$th resource request node A_n, and a_n, b_n denote the position of this node in the X-axis and Y-axis directions, respectively. The amount of data transmitted by all nodes can be denoted as $\Gamma = \{\gamma_0, \gamma_1, ..., \gamma_n, ..., \gamma_{N-1}\}$, where γ_n represents the amount of data transmitted by A_n. The computation amount of all nodes can be denoted as $\Theta = \{\theta_0, \theta_1, ..., \theta_n, ..., \theta_{N-1}\}$, where θ_0 represents the computation amount required by A_n.

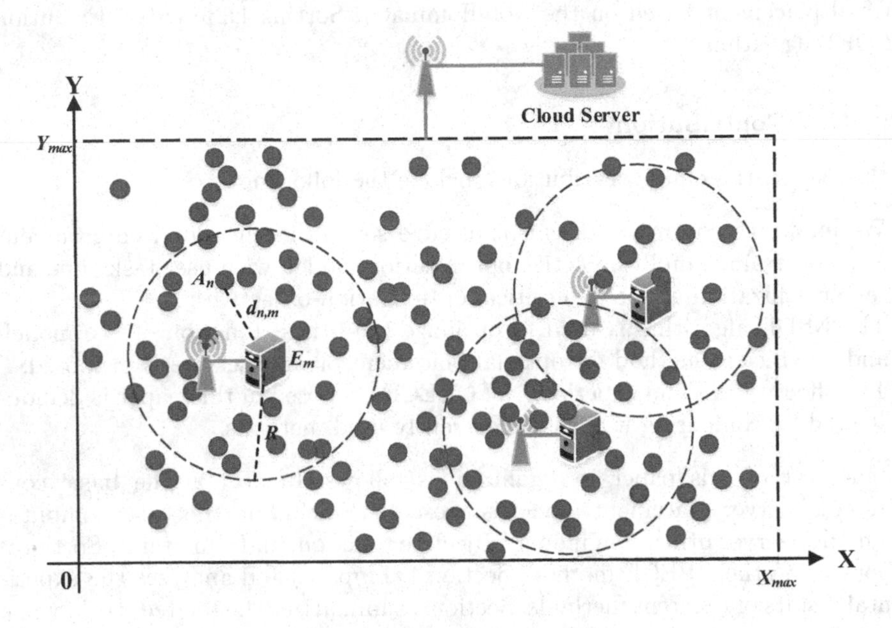

Fig. 1. Problem model

Assume M edge servers are deployed in this request area and the maximum service distance of each edge server is R. Then their deployment locations can be denoted as $\mathcal{E} = \{E_0, E_1, ..., E_m, ..., E_{M-1}\}$, where $E_m = [x_m, y_m](0 \leq x_m \leq x_{max}, 0 \leq y_m \leq y_{max})$ denotes the deployment location of the $m-$th edge server Em location. Then the distance $d_{n,m}$ between A_n and E_m can be expressed as Eq. (1):

$$d_{n,m} = \sqrt{(x_m - a_n)^2 + (y_m - b_n)^2}. \tag{1}$$

If there exist one or more edge servers E_m around A_n whose distance $d_{n,m}$ from A_n is less than R, then A_n will always request resources from the edge server E_m that is closest to it, i.e., $\mathsf{MIN}(d_{n,m})$, otherwise it will request resources from the remote cloud server. Therefore, the server selection strategies of all nodes can be expressed as $\Psi = \{\psi_0, \psi_1, ..., \psi_n, ..., \psi_{N-1}\}$, Where ψ_n denotes the server selection strategy of A_n, as shown in Eq. (2):

$$\psi_n = \begin{cases} m; \mathsf{MIN}(d_{n,m}) \leq R \\ -1; \mathsf{MIN}(d_{n,m}) > R \end{cases}. \tag{2}$$

$\psi_n = m$ and $\psi_n = -1$ indicate that A_n will request resources from edge server E_m and remote cloud server, respectively.

It can be clearly seen that ψ_n depends mainly on all edge servers' deployment locations $\mathcal{E}$. Generally, deploying edge servers in resource-request-intensive areas is more beneficial to improving the service efficiency for user nodes and the utilization of some servers, but the overlapping service area of multiple edge servers also tends to cause redundancy of computing resources. Therefore, further improving server utilization becomes a challenge while ensuring service efficiency for user nodes.

2.2 Time Computation Model

User nodes offload local computational tasks to remote cloud servers or edge servers for computation, so the time computation model mainly includes data transmission time and task computation time. Note that the data transmission time already includes all data transmission time in both "*node to server*" and "*server to node*" directions.

Data Transmission Time. Since the distance between A_n and remote cloud servers is usually long, the distance between them can uniformly default to D_{ac}. Assume that S_{ac} and S_{ae} denote the bandwidth of the remote cloud server and the edge server, respectively. Then the data transmission time $Tran_n$ of node A_n can be expressed as Eq. (3):

$$Tran_n = \begin{cases} d_{n,m} * \gamma_n / S_{ae}; \psi_n \in [0, M) \\ D_{ac} * \gamma_n / S_{ac}; \psi_n = -1 \end{cases}. \quad (3)$$

Typically, S_{ac} is lower than S_{ae}, while D_{ac} is much higher than any of the $d_{n,m}$. From Eq. 3, it is obvious that offloading computational tasks to the edge server is beneficial to reducing the data transmission time of user nodes.

Task Computation Time. Assuming that the computational performance of remote cloud servers and edge servers are p_c and p_e, respectively, then the task computation time $Comp_n$ of A_n can be expressed as Eq. (4):

$$Comp_n = \begin{cases} \theta_n / p_e; \psi_n \in [0, M) \\ \theta_n / p_c; \psi_n = -1 \end{cases}. \quad (4)$$

Generally, p_c is significantly higher than p_e, then from Eq. 4, it is obvious that offloading the computational tasks to the cloud servers is beneficial to reducing the task computation time of user nodes.

The total time T_n required for A_n can be expressed as Eq. (5):

$$T_n = Tran_n + Comp_n. \quad (5)$$

Eventually, the average task time T_{avg} of all nodes is expressed as Eq. (6):

$$T_{avg} = \sum_{n=0}^{N-1} T_n / N. \tag{6}$$

2.3 Server Utilization Model

Based on the server selection strategy ψ_n for each node A_n, the total computational resources U_m used by each edge server E_m can be calculated as Eq. (7):

$$U_m = \sum_{n=0}^{N-1} \theta_n; \psi_n = m. \tag{7}$$

Assuming that the maximum capacity of computing resource pool of each edge server is H, the average utilization U_{avg} of all edge servers can be expressed as Eq. (8):

$$U_{avg} = \sum_{m=0}^{M-1} U_m / H * M. \tag{8}$$

2.4 Objective Function and Constraint

Our objective is to improve the service efficiency for user nodes and the utilization of edge servers, i.e., to reduce the average task time T_{avg} of all nodes and to increase the average utilization U_{avg} of all edge servers. In addition, a constraint still needs to be satisfied that the actual load on each edge server cannot exceed its maximum capacity H, as Eq. (9):

$$\forall U_m \leq H, m \in [0, M). \tag{9}$$

Thus, the optimal deployment problem for edge servers can be modeled as a Constrained Multi-objective optimization model:

$$\begin{cases} \mathsf{MIN}(T_{avg}); \\ \mathsf{MAX}(U_{avg}); \\ s.t. \forall U_m \leq H, m \in [0, M); \end{cases} \tag{10}$$

3 OPECE

In Sect. 2, the optimal placement problem of edge servers in a cloud environment is modeled as a constrained multi-objective optimization model. In this section, we use the NSDE Algorithm to solve this optimization model. NSDE consists of the following four main phases: initialization, mutation, crossover, and selection, where mutation, crossover, and selection are executed iteratively until the termination condition is satisfied.

3.1 Initialization

NSDE is a multi-objective optimization algorithm based on population optimization, where the initial population is usually generated by randomization.

Population: A population consists of multiple individuals, and each individual includes multiple genes. In this optimal placement model of edge servers, the placement location of each edge server $E_m = [x_m, y_m]$ is treated as a gene, and placement locations of all edge servers constitute an individual, i.e.,$\mathcal{X}_i = (E_{i,0}, E_{i,1}, ..., E_{i,m}, ..., E_{i,M-1})$, where $E_{i,m}$ denotes the $m-$th gene in the $i-$th individual X_i. Eventually, the $g-$th generation population can be expressed as $\mathcal{X}^g = (\mathcal{X}_0, \mathcal{X}_1, ..., \mathcal{X}_i, ..., \mathcal{X}_{NP-1})$, where $\mathcal{X}^0$ denotes the initial population, and NP represents the size of the population. The larger the NP, the higher the population diversity, but the slower the convergence of NSDE, and vice versa. Therefore, NP is usually set between 5M and 10M.

Population Initialization: The initial population is generated by taking random values for each gene in the population within a specified range of values, as Eq. (11):

$$\begin{cases} x_{i,m} = \mathsf{Rand}(0, x_{max}), \\ y_{i,m} = \mathsf{Rand}(0, y_{max}), \\ \forall m \in [0, M), \end{cases} \Rightarrow E_{i,m} = [x_{i,m}, y_{i,m}] \tag{11}$$

where $\mathsf{Rand}(a, b)$ denotes a randomly selected real number from a to b.

Eventually, the initial population $\mathcal{X}^0 = (\mathcal{X}_0, \mathcal{X}_1, ..., \mathcal{X}_i, ..., \mathcal{X}_{NP-1})$ is generated, where $\mathcal{X}_i$ represents the i-th individual in $\mathcal{X}^0$.

3.2 Mutation

Three individuals $\mathcal{X}_a$, $\mathcal{X}_b$, and $\mathcal{X}_c$ are randomly selected from the parent population $\mathcal{X}^g$, and the mutation operation is performed according to Eq. (12) to generate the corresponding mutation individual $\mathcal{H}_i = (E'_{i,0}, E'_{i,1}, ..., E'_{i,m}, ..., E'_{i,M-1})$.

$$\mathcal{H}_i = \mathcal{X}_a + F * (\mathcal{X}_b - \mathcal{X}_c) \tag{12}$$

More specifically, each mutation gene $E'_{i,m}$ in $\mathcal{H}_i$ can be calculated according to Eq. (13).

$$\begin{cases} x'_{i,m} = x_{a,m} - F * (x_{b,m} - x_{c,m}), \\ y'_{i,m} = y_{a,m} - F * (y_{b,m} - y_{c,m}), \\ \forall m \in [0, M), \end{cases} \Rightarrow E'_{i,m} = \left[x'_{i,m}, y'_{i,m}\right] \tag{13}$$

where F is the mutation factor that determines the mutation degree of each mutation individual. Usually, the larger the F, the higher the population diversity, but the slower the convergence of NSDE, and vice versa. Therefore, F is usually set between 0.5 and 2.

Eventually, the mutation population $\mathcal{H}^g = (\mathcal{H}_0, \mathcal{H}_1, ..., \mathcal{H}_i, ..., \mathcal{H}_{NP-1})$ is generated, where $\mathcal{H}_i$ represents the i-th mutation individual in $\mathcal{H}^g$.

3.3 Crossover

Based on each $\mathcal{X}_i$ and its corresponding $\mathcal{H}_i$, a gene $E^{'}_{i,m}$ randomly selected from $\mathcal{H}_i$ is first retained in the crossover individual $\mathcal{R}_i = (E^{''}_{i,0}, E^{''}_{i,1}, ..., E^{''}_{i,m}, ..., E^{''}_{i,M-1})$ to ensure that at least one gene in $\mathcal{R}_i$ is from the mutation individual $\mathcal{H}_i$. Then the remaining genes are selected from $\mathcal{X}_i$ and $\mathcal{H}_i$ according to a certain probability, i.e., the crossover factor CR, which can be expressed as Eq. (14):

$$\begin{cases} x^{''}_{i,m} = x^{'}_{i,m} \&\& y^{''}_{i,m} = y^{'}_{i,m}, \mathsf{Rand}(0,1) < CR \parallel m = Rand(0,M) \\ x^{''}_{i,m} = x_{i,m} \&\& y^{''}_{i,m} = y_{i,m}, otherwise \\ \forall m \in [0, M) \end{cases}$$
$$\Rightarrow E^{''}_{i,m} = \left[x^{''}_{i,m}, y^{''}_{i,m}\right] \tag{14}$$

where CR determines the crossover degree of crossover individuals $\mathcal{R}_i$. The larger CR ensures that more mutation genes are retained in $\mathcal{R}_i$. Therefore, the larger the CR, the higher the population diversity, but the slower the convergence of NSDE, and vice versa. Usually, CR is set at around 0.5.

Eventually, the crossover population $\mathcal{R}^g = (\mathcal{R}_0, \mathcal{R}_1, ..., \mathcal{R}_i, ..., \mathcal{R}_{NP-1})$ is generated, where $\mathcal{R}_i$ represents the i-th crossover individual in $\mathcal{R}^g$.

3.4 Selection

NSDE uses the non-dominated sorting algorithm to select suitable individuals which retained in the next generation population $\mathcal{X}^{g+1}$.

First, the parent population $\mathcal{X}^g$ and the crossover population $\mathcal{R}^g$ are merged into a population $\mathcal{Y}^g$ whose population size is 2NP. Then, for each individual $E_{i,m}$, the objective function values (i.e., average task time T_{avg} and average utilization U_{avg}) and the constraint values (i.e., total used computational resources U_m per edge server) are calculated separately according to Eqs. (6),(7),(8).

Assuming that $\mathcal{X}_i$ and $\mathcal{X}_j$ are two different individuals, the average task time of $\mathcal{X}_i$ is less than that of $\mathcal{X}_j$ while the average utilization of $\mathcal{X}_i$ is greater than that of $\mathcal{X}_j$, then $\mathcal{X}_i$ can be said to dominate $\mathcal{X}_j$, expressed as $\mathcal{X}_i \succ \mathcal{X}_j$. Their relationship can be expressed as Eq. (15):

$$\begin{cases} T_{avg}(\mathcal{X}_i) < T_{avg}(\mathcal{X}_j) \\ U_{avg}(\mathcal{X}_i) > U_{avg}(\mathcal{X}_j) \end{cases} \Rightarrow \mathcal{X}_i \succ \mathcal{X}_j, i \neq j. \tag{15}$$

However, if only some objective function values of $\mathcal{X}_i$ are better than that of $\mathcal{X}_j$, while all other objective function values of $\mathcal{X}_i$ are worse than that of $\mathcal{X}_j$, then $\mathcal{X}_i$ and $\mathcal{X}_j$ are said to be mutually non-dominated.

The population $\mathcal{Y}^g$ is divided into multiple non-dominated layers $\mathcal{L}_n$ (e.g., $\mathcal{L}_0$, $\mathcal{L}_1$,$\mathcal{L}_2$, etc.) by the non-dominated sorting algorithm for all individuals in $\mathcal{Y}^g$. They obey the following two rules:

- **Rule 1:** All individuals in the same non-dominated layer do not dominate each other.

- **Rule 2:** All individuals in the lower non-dominated layer are able to dominate all individuals in the higher non-dominated layer, which can be expressed as:

$$\begin{cases} \mathcal{X}_i \in \mathcal{L}_N, \mathcal{X}_j \in \mathcal{L}_M; \\ N < M; \end{cases} \Rightarrow \forall \mathcal{X}_i \succ \forall \mathcal{X}_j. \tag{16}$$

Eventually, individuals satisfying the constraint (i.e., Eq. (9)) are sequentially extracted from the lower non-dominated layer (e.g., $\mathcal{L}_0$) and continuously added to the next generation parent population $\mathcal{X}^{g+1}$ until the population size of $\mathcal{X}^{g+1}$ reaches NP.

3.5 Iteration

Based on the next generation parent population $\mathcal{X}^{g+1}$, the mutation, crossover and selection operations are performed iteratively until the iteration number g of NSDE reaches the predefined maximum number G_{max} or the output optimal solution reaches a predefined error precision.

4 Evaluation

4.1 Experimental Settings

To evaluate the performance of OPECE, we conducted a series of experiments to compare it with two common task offloading methods in cloud environments: traditional Offloading based on Cloud (OBC) and Optimal Placement based on K-Means (OPBKM):

- **OBC:** The most original task offloading strategy, i.e., all task nodes offload tasks to the remote cloud servers for execution before the introduction of edge servers;
- **OPBKM:** After introducing edge servers, the placement location of each edge server is determined based on K-Means. That is, based on the distribution location of all task nodes, the K-Means clustering algorithm is used to find out the centroids of K clusters, i.e., the final placement location of K edge servers.
- **OPECE:** After introducing edge servers, the placement locations of all edge servers are globally optimized by NSDE.

In order to detect the performance impact of the same load demand on different numbers of edge servers, we conduct experiments based on simulated scenarios. In this case, the number of task nodes is always set to 100, which is fixed because the number of users in each region is relatively fixed even in real scenarios.

4.2 Performance Evaluation

Within a constant area, the number of edge servers directly impacts the placement of all edge servers. Therefore, in this experiment, we deploy 2, 6 and 10 edge servers in the area respectively, and use the average task time of all task nodes in the area and the average utilization of all edge servers as evaluation metrics for testing the performance of several placement methods, as shown in Fig. 2, Fig. 3 and Fig. 4. Generally, all edge servers in the area cover as many task nodes as possible, or some computationally intensive tasks are offloaded to remote cloud servers with stronger computational performance for execution, which can be beneficial to improve the comprehensive performance of the approach.

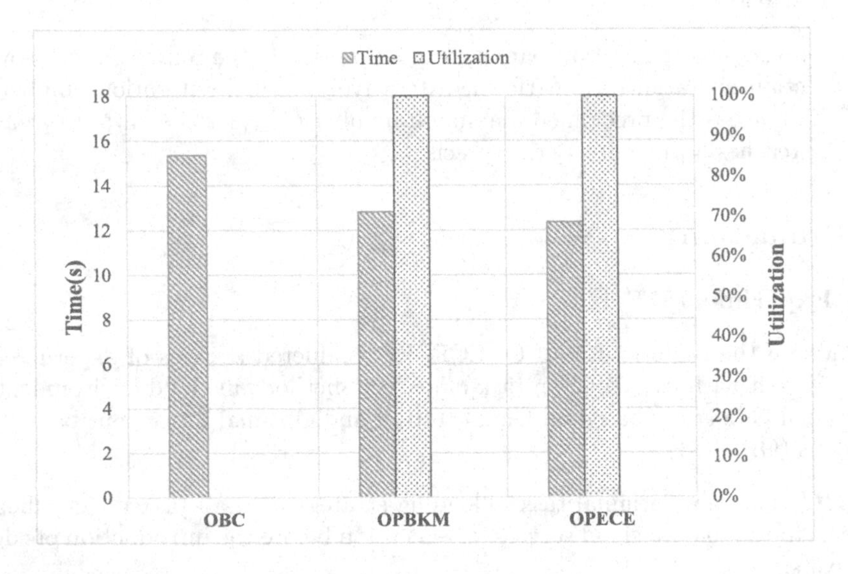

Fig. 2. Performance comparison of three methods when number of edge servers is 2

Figure 2 shows the performance comparison of the three approaches when only two edge servers are deployed in the area. Firstly, it can be seen that OBC does not introduce edge servers, resulting in it having the highest average task time among the three methods. Secondly, both OPBKM and OPECE achieve 100% average utilization because when the number of edge servers is small, these servers are usually deployed in areas with dense nodes and far apart from each other, which is sufficient to ensure that each edge server is maximally utilized, i.e., 100% utilization. However, OPECE slightly outperforms OPBKM in the average task time metric because OPBKM only considers the distance relationship between edge servers and task nodes, while OPECE can directly consider and optimize both the utilization of edge servers and the task time of each node rather than the indirect distance relationship.

Figure 3 shows the performance comparison of the three approaches when only six edge servers are deployed in the area. Firstly, OBC still has the highest average task time among the three methods. Secondly, both performance metrics of OPECE are significantly better than OPBKM, this is because when more edge servers are deployed, the distances between the centroids of multiple clusters extracted by the K-Means clustering algorithm are closer, and even some edge servers have overlapping service coverage, resulting in some edge servers' resources not being maximally utilized and more tasks being offloaded to the remote cloud for execution. OPECE directly optimizes the utilization of edge servers and the task time of each node, which can effectively reduce or even circumvent such service coverage overlap and ensure that more tasks are offloaded to the edge servers for execution.

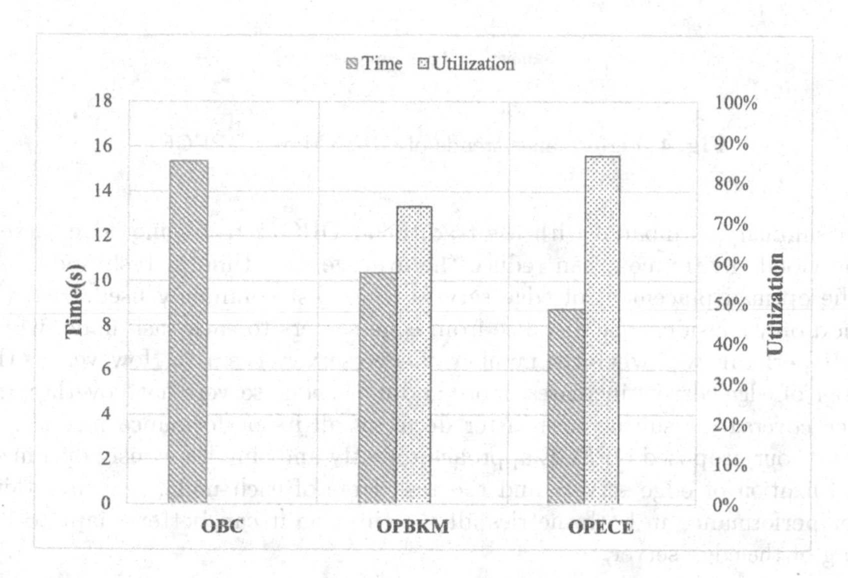

Fig. 3. Performance comparison of three methods when number of edge servers is 6

To further investigate the impact of increasing the number of edge servers on the performance of OPBKM and OPECE, we increase the number of edge servers to 10. The performance trends of OPBKM and OPECE are shown in Fig. 4.

It is obvious from Fig. 4 that OPECE decreases significantly faster than OPBKM in terms of the average task time metric and lower than OPBKM in terms of the average utilization metric. In other words, this indicates that as the number of edge servers increases, OPECE is more likely to guarantee the full utilization of each edge server and is more conducive to reducing the average task execution time of each task node. In conclusion, OPECE is more adaptable than OPBKM to expand the edge server scale.

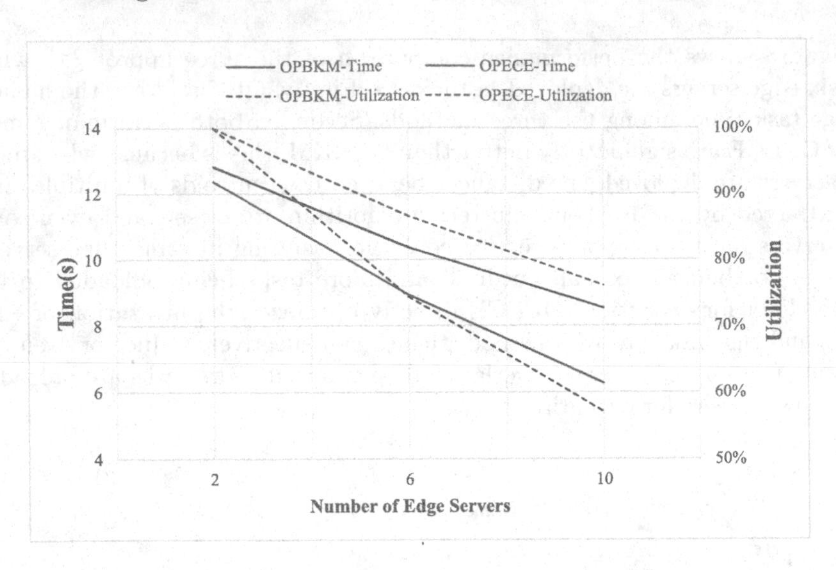

Fig. 4. Performance trends of OPBKM and OPECE

In summary, compared with the traditional OBC, introducing edge servers in the cloud environment can reduce the average task time in task nodes. As for the optimal placement of edge servers, the most commonly used OPBKM method only considers the distance from edge servers to each task node, which usually performs well when the number of edge servers is small. However, as the number of edge servers increases, more and more edge servers have overlapping service coverage, resulting in a faster decay of all its performance metrics. In contrast, our proposed OPECE approach directly and simultaneously optimizes the utilization of edge servers and the task time of each node, and maintains better performance in both metrics all the time, so it can better adapt to the scaling of the edge server.

5 Related Work

Currently, many researchers have been studying the task offloading problem in the cloud environment, mainly in terms of task execution efficiency, energy efficiency, and overhead of servers.

Guan et al. [8] proposed a novel mobility-aware offloading model to optimize the offloading execution efficiency and energy efficiency. Azzedine et al. [9] proposed a novel task-centric energy-aware Cloudlet-based Mobile Cloud model and improved offloading execution efficiency and energy efficiency of user devices. Dai et al. [10] focused on the data transmission delay and computing delay of edge nodes, and proposed an energy-efficient edge offloading scheme to improve the offloading efficiency of UAVs. Yu et al. [6] designed a novel collaborative offloading scheme that formulated the offloading decision problem as a multi-label classification problem, and achieved rapid offloading decisions making, as well as minimized the computation and offloading overhead. Li et al. [11] proposed a vehicular edge computing offloading solution based on deep reinforcement learning, formulated the optimization problem as a Markov decision process model and maximized the total computation rate. Wu et al. [12] proposed a joint computation offloading and task migration optimization algorithm named JCOTM, which can effectively reduce task processing delay and equipment energy consumption, and improve system resource utilization. Fang et al. [7] modeled the offloading decision problem as a two-sided matching game problem, and designed an improved two-sided many-to-one matching game algorithm to reduce the delay and energy cost. Lv et al. [13] researched the task offloading problem in Mobile Edge Computing under the dependency and service caching constraints, and proposed a heuristic algorithm named TBTOA to optimize the makespan and energy consumption. Lu et al. [14] proposed a lightweight offloading framework that can offload the compute-intensive tasks and deploy the server efficiently. Furthermore, they designed a multi-task offloading tactic to optimize the execution time, energy consumption, and CPU utilization rate. In addition, some scholars also consider the constraint of privacy-preserving. Zhao et al. [15] proposed a privacy-preserving computation offloading method based on privacy entropy in Multi-access Edge Computing, which can solve the offloading decisions that satisfy the privacy constraint and the optimal energy consumption target.

6 Conclusion

In this paper, in order to ease the load on cloud servers and enhance the user's network experience, we propose an optimal placement method of edge servers in the cloud environment, namely, OPECE. First, the edge server optimal placement problem is modeled as a constrained multi-objective optimization model with task time and edge server utilization as the two critical optimization objectives. Then, the model is optimized using the NSDE algorithm. Finally, empirical studies demonstrate that our proposed OPECE method has more obvious superiority over other traditional methods, especially for the optimal placement of more edge servers.

In future work, we will consider how to optimize offloading of compute-intensive and data-intensive tasks at the edge server area, which can further enhance the user's network experience.

Acknowledgements. This research is supported by the Natural Science Foundation of the Jiangsu Higher Education Institutions under Grant No.21KJB520001.

References

1. Flores, H., et al.: Evidence-aware mobile computational offloading. IEEE Trans. Mob. Comput. **17**(8), 1834–1850 (2018)

2. Kuang, Z., Shi, Y., Guo, S., Dan, J., Xiao, B.: Multi-user offloading game strategy in OFDMA mobile cloud computing system. IEEE Trans. Veh. Technol. **68**(12), 12190–12201 (2019)
3. Ko, H., Kyung, Y.: Performance analysis and optimization of delayed offloading system with opportunistic fog node. IEEE Trans. Veh. Technol. **71**(9), 10203–10208 (2022)
4. Gao, G., Xiao, M., Jie, W., Han, K., Huang, L., Zhao, Z.: Opportunistic mobile data offloading with deadline constraints. IEEE Trans. Parallel Distrib. Syst. **28**(12), 3584–3599 (2017)
5. Liang, S., Wan, H., Qin, T., Li, J., Chen, W.: Multi-user computation offloading for mobile edge computing: a deep reinforcement learning and game theory approach. In: 20th IEEE International Conference on Communication Technology, ICCT 2020, Nanning, China, October 28–31, 2020, pp. 1534–1539 (2020)
6. Yu, S., Langar, R.: Collaborative computation offloading for multi-access edge computing. In: IFIP/IEEE International Symposium on Integrated Network Management, IM 2019, Washington, DC, USA, April 09–11, 2019, pp. 689–694 (2019)
7. Fang, H., Jia, Y., Wang, Y., Zhao, Y., Gao, Y., Yang, X.: Matching game based task offloading and resource allocation algorithm for satellite edge computing networks. In: International Symposium on Networks, Computers and Communications, ISNCC 2022, Shenzhen, China, July 19–22, 2022, pp. 1–5 (2022)
8. Guan, S., Boukerche, A.: A novel mobility-aware offloading management scheme in sustainable multi-access edge computing. IEEE Trans. Sustain. Comput. **7**(1), 1–13 (2022)
9. Boukerche, A., Guan, S., Grande, R.E.D.: A task-centric mobile cloud-based system to enable energy-aware efficient offloading. IEEE Trans. Sustain. Comput. **3**(4), 248–261 (2018)
10. Dai, M., Su, Z., Li, J., Zhou, J.: An energy-efficient edge offloading scheme for UAV-assisted internet of things. In: 40th IEEE International Conference on Distributed Computing Systems, ICDCS 2020, Singapore, November 29 - December 1, 2020, pp. 1293–1297 (2020)
11. Li, F., Lin, Y., Peng, N., Zhang, Y.: Deep reinforcement learning based computing offloading for MEC-assisted heterogeneous vehicular networks. In: 20th IEEE International Conference on Communication Technology, ICCT 2020, Nanning, China, October 28–31, 2020, pp. 927–932 (2020)
12. Ziying, W., Yan, D.: Deep reinforcement learning-based computation offloading for 5G vehicle-aware multi-access edge computing network. China Commun. **18**(11), 26–41 (2021)
13. Lv, X., Du, H., Ye, Q.: TBTOA: a DAG-based task offloading scheme for mobile edge computing. In IEEE International Conference on Communications, ICC 2022, Seoul, Korea, May 16–20, 2022, pp. 4607–4612 (2022)
14. Junyu, L., et al.: A multi-task oriented framework for mobile computation offloading. IEEE Trans. Cloud Comput. **10**(1), 187–201 (2022)
15. Zhao, X., Peng, J., Li, Y., Li, H.: A privacy-preserving computation offloading method based on privacy entropy in multi-access edge computation. In: IEEE International Conference on Computer and Communications (2020)

Convolutional Neural Network Based QoS Prediction with Dimensional Correlation

Weihao Cao[1(✉)], Yong Cheng[1], Shengjun Xue[1], and Fei Dai[2]

[1] School of Software, Nanjing University of Information Science and Technology, Nanjing 210044, China
941834731@qq.com

[2] College of Big Data and Intelligent Engineering, Southwest Forestry University, Kunming 650224, China

Abstract. In recent years, massive services that provide similar functions continue to emerge. Since services sensitive to latency and throughput are often expected to have high Quality of Service (QoS), how to accurately predict QoS has become a challenging issue. Some current deep learning (DL) based approaches usually simply concatenate the embedding vectors, without considering the correlation between embedding dimensions. Besides, the high-order feature interactions are not sufficiently learned. To this end, this paper proposes a Convolutional Neural Network based QoS prediction model with Dimensional Correlation, named QPCN. First, the two dimensional interaction features is explicitly obtained by modeling the embedding vectors. Then, the convolutional neural network is utilized to perform feature extraction and complete QoS prediction. Compared with the fully connected network, QPCN can build a deeper model and learn high-order features. In addition, the parameters of QPCN are significantly reduced, which will reduce the time and energy consumption of inference. The effectiveness of QPCN is validated by experiments on a real-world dataset.

Keywords: QoS prediction · web service · service recommendation · location-aware

1 Introduction

Currently, cloud computing infrastructures such as Amazon and Google provide a wide range of web services. [1]. As a result, there are many services that provide equal functions for users. When faced with these services, it may be difficult for users to determine which service best meets their requirements. Therefore, recommending high-quality services to users has become an important requirement in service recommendation [2]. In order to make better recommendation, it is necessary to further compare the non-functional properties of equal services, i.e., Quality of Service (QoS), which includes response time, throughput and reliability, etc. [3].

H. Jin et al. (Eds.): GPC 2023, LNCS 14504, pp. 17–31, 2024.
https://doi.org/10.1007/978-981-99-9896-8_2

In order to obtain the QoS of services and recommend them to users, collaborative filtering (CF) based QoS prediction is proposed, which is one of the main QoS prediction method. CF-based QoS prediction methods can be divided into memory-based and model-based [4]. The main idea of memory-based CF is to analyze the QoS matrix and predict missing QoS by collecting historical records of similar users or services. However, the QoS matrix is usually extremely sparse, which will lead to poor performance of CF-based approaches in QoS prediction.

Matrix Factorization (MF) is a model-based method. It maps users and services to the same latent space, obtains their respective latent feature vectors, and then calculates the QoS prediction value by taking the inner product of the two vectors. MF can reduce the impact of data sparsity to some extent and often achieves better performance than CF methods. However, MF can only obtain the low-order linear interactions without considering the higher-order nonlinear interactions between latent features.

Neural network, with its powerful computational capabilities, have been extensively studied by many researchers in the past decades [5]. Now there are many methods using neural network for QoS prediction [6–8]. They perform QoS prediction by extracting deep features of users and services. Compared to traditional CF methods, neural network can automatically learn patterns and regulations within the data. Although these methods improve the accuracy of QoS prediction, most of them are inspired by Neural Collaborative Filtering (NCF) [7], which simply concatenates the embedding vectors and then stacks multi-layer fully connected neural networks for prediction. Compared with CNN, the generalization ability of fully connected NCF is not as good as CNN, and it is not easy to learn deep features [9]. In addition, these approaches do not consider the dimension correlations between embedding vectors. Research has shown that explicitly modeling the interaction of embedding vectors can help improve the generalization of deep learning models on sparse data [10].

To overcome the drawbacks of above methods, we propose a novel dimensional correlation integrated QoS prediction model. The main contributions of this paper include:

- We propose a novel model for QoS prediction. It obtains the two-dimensional interaction features of user and service embedding vectors through outer product. Compared with concatenation of embedding vectors, richer interaction features are obtained.
- The impact of the initial fixed feature sequence is decreased through a layer of shuffle and the feature interaction from local to global is learned through CNN, which help improve the prediction accuracy.
- We conduct experiments on the real-world dataset to demonstrate the effectiveness of our proposed model. Experiments show that the QoS prediction accuracy of it outperforms six baseline models.

The remainder of this paper is organized as follows. In Sect. 2, we review related works on QoS prediction. In Sect. 3, we illustrate the structure and details of QPCN. Experiments are presented and discussed in Sect. 4. A conclusion of this paper is given in Sect. 5.

2 Related Work

The basic idea of memory-based collaborative QoS prediction method is to collect historical records of similar users or services to predict missing QoS values. Shao et al. [11] proposed a QoS prediction method based on user neighborhood, which predicts the QoS value according to similar users of the target user. Chen et al. [12] proposed a service-based QoS prediction method, taking into account the geographical information of service providers. Zheng et al. [13] proposed a hybrid QoS prediction model based on user and service, which improves the prediction accuracy. Wu et al. [14] made an improvement on CF, which uses a ratio-based similarity calculation method to select the neighborhood of user and service.

The model-based CF approach decomposes the QoS matrix to get the latent feature vectors of users and services respectively. QoS prediction results are obtained by the inner product of the two vectors, which decreases the impact of data sparsity to a certain extent. Wu et al. [15] integrated the context information into matrix factorization to realize context-aware QoS prediction. Considering that the QoS values at different moments will be different, Zhang et al. [16] proposed a time-aware tensor decomposition algorithm. Considering the impact of network, Tang et al. [17] combined MF with the underlying network information for QoS prediction. Xu et al. [18] proposed a method named reputation matrix factorization(RMF), which integrates user reputation into matrix factorization, reducing the influence of untrustworthy users on QoS prediction.

In recent years, deep learning has gradually become an effective method for QoS prediction. Wu et al. [6] proposed a DL-based multi-attribute QoS prediction model considering the context. This is the first multi-attribute QoS prediction using DNN. Zhang et al. [8] proposed a location-aware method (LDCF), which uses MLP for feature extraction and combines it with an adaptive corrector (AC) to complete QoS prediction.

3 Proposed Model

To achieve QoS prediction, we propose a location-aware model which integrates outer product and CNN. The structure of it is shown in Fig. 1. It consists of four parts: input layer, embedding layer, interaction layer and prediction layer.

3.1 Input Layer

QoS is influenced by the context of user and service, including network address, network state, subnet, autonomous system, geographic location, etc. Studies [8,15] have demonstrated that the QoS value is greatly affected by location. Furthermore, some service applications are throughput sensitive while others are latency sensitive [19]. Users tend to experience better service with more throughput and faster response time when they invoke services that are close to them. In the input layer, we extract relevant explicit features from the location information, i.e., the region where user and service are located, and the autonomous

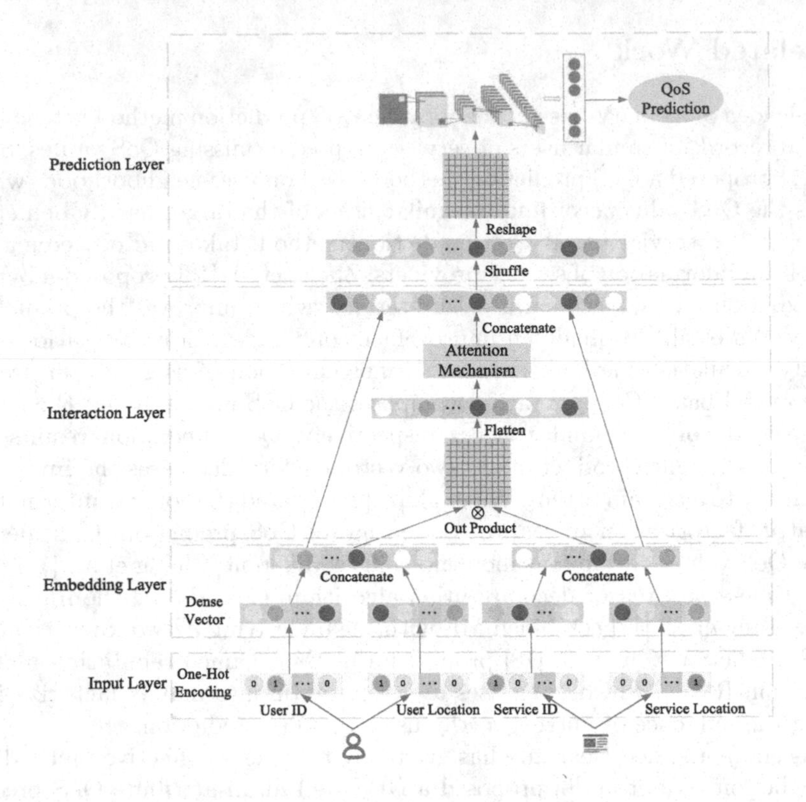

Fig. 1. Structure of the proposed model

system they belong to. Since the number of users, services, autonomous systems and regions is finite, they can be represented separately by non-negative integers. Specifically, we represent the initial features of user and service as shown below:

$$I_u = [u_{id}, u_{rg}, u_{as}] . \tag{1}$$

$$I_s = [s_{id}, s_{rg}, s_{as}] . \tag{2}$$

where u_{id}, u_{rg} and u_{as} represent user's identifier, region, and autonomous system respectively. Similarly, s_{id}, s_{rg}, and s_{as} represent service's identifier, region, and autonomous system. Then every non-negative integer is converted into vector consisting of 1 and 0 by one-hot encoding.

3.2 Embedding Layer

Due to the one-hot vectors are high-dimensional and sparse, inputting them directly into the neural network will not be conducive to the training. Therefore, we add an embedding layer to transform them into low-dimensional dense vectors. Formally, each one-hot vector of user and service is mapped to a latent feature space R^d, where d is the dimensionality of the space. The embedding layer can be regarded as a special fully connected layer, and its formula is as follows:

$$e_j = \sigma \left(W_j X_j + b_j \right). \tag{3}$$

where X_j represents the j-th one-hot vector, W_j represents the corresponding embedding weight matrix, b_j represents the bias term, σ represents the activation function. After that, user's embedding vectors of identifier and location are concatenated as the user's embedding vector, denoted by E_u. Similarly, service's embedding vectors of identifier and location are concatenated as the service's embedding vector, denoted by E_s. After the above processing, we obtain two embedding vectors, represented as follows:

$$E_u = \begin{bmatrix} e_{uid} \\ e_{urg} \\ e_{uas} \end{bmatrix}. \tag{4}$$

$$E_u = \begin{bmatrix} e_{sid} \\ e_{srg} \\ e_{sas} \end{bmatrix}. \tag{5}$$

In this way, we can obtain two low-dimensional dense vectors, which decreases the risk of insufficient expressiveness of categorical features in one-hot encoding.

3.3 Interaction Layer

In the interaction layer, we explicitly model the interaction of two embedding vectors. Although fully-connected MLP is theoretically able to fit any continuous function, it is not easy to capture dimensional correlations [9]. Explicitly modeling the dimensional interactions of the embedding vectors can help improve the generalization ability of deep learning model on sparse data. Therefore, we perform an outer product of the two embedding vectors to explicitly model the dimensional interaction, and the result is denoted as an interaction map. The rich semantics in the interaction map help the subsequent non-linear layers to capture potential complex correlations among dimensions. Moreover, it holds greater significance than a simple concatenation, which only retains the initial information within the embeddings. Formally, the interaction graph $T \in R^{d \times d}$ generated by outer product is defined as follows:

$$T = E_u \otimes E_s. \tag{6}$$

Since T is obtained by the outer product of the embedding vectors of user and service, it includes many redundant features. The importance of these features needs to be distinguished, so we introduce a layer of attention mechanism. Usually, the softmax function is applied in the attention layer. But considering that the softmax function sometimes assigns large weights to certain feature but assigns weights close to 0 to other features, i.e., it will lead to one-hot activation. We use sigmoid to replace softmax which can alleviate this case to some extent. In the attention layer, T is the input of sigmoid function. Then the output of sigmoid function performs element-wise multiplication with T. This process can be defined as follows:

$$V = T \odot sigmoid\,(T)\,. \tag{7}$$

To remain the features of the initial embedding vectors, we flatten V and then concatenate it with the embedding vectors to obtain the feature vector X. X contains the initial embedding features and dimensional interactions. This process is formulated as follows:

$$V_f = Flatten(V). \tag{8}$$

$$X = \Phi\left(V_f, E_u, E_s\right) = \begin{bmatrix} V_f \\ E_u \\ E_s \end{bmatrix}. \tag{9}$$

where Φ is the concatenation operation. After that, the feature vector X is fed to the prediction layer.

3.4 Prediction Layer

In the prediction layer, the choice of hidden layers has a great impact on the QoS prediction performance. A common choice is to use a fully connected network. However, since we generate a large number of dimensional interaction features through outer product in the interaction layer, using a fully connected network will bring a large number of parameters, which will increase the inference energy consumption of the model and will easily cause overfitting. On the other hand, using a fully connected network with fewer layers and fewer neurons per layer to reduce the number of parameters may lead to the model's inability to fully extract high-order features, resulting in a decrease in prediction accuracy.

We consider CNN to be a better choice in this case because of its parameter sharing and local connections, which result in fewer parameters compared to fully connected network. This makes it easier to build a deeper model and allows for the extraction of high-order features. Therefore, to address the shortcomings of fully connected network, we use CNN as the hidden layers.

Prior to this, it should be noted that since the feature vector X is obtained by concatenating the embedding vectors and their dimensional interactions. If X is used directly for model training, the training process will be affected by a fixed feature sequence. Therefore, a layer of Shuffle is used to change the initial feature sequence, the formula is defined as follows:

$$X' = f\left(W_1 X + b\right). \tag{10}$$

where f represents the Relu activation function, $W1$ represents the weight matrix, and b represents the bias item. In this way, the shuffle layer adaptively learns the feature order of X, which is beneficial for the feature extraction process of the CNN-based prediction network.

Next, we perform a Reshape operation on X' to convert it into a $K * K$ matrix which is represented as M_0. Then a multi-layer CNN is used to complete feature extraction on M_0. In CNN, there are N convolutional layers and the i-th layer has P_i convolutional kernels. Then the feature map extracted by the j-th convolutional kernel in the i-th layer can be described as follows:

$$M_i^j = \sigma\left(W_i^j * M_{i-1} + b_i\right). \tag{11}$$

where $*$ represents the convolution operation, W_i^j represents the parameter of the j-th convolution kernel in the i-th layer, and b_i represents the bias item. Then the output feature map of the i-th layer can be described as follows:

$$M_i = \left(M_i^1, M_i^2, \cdots, M_i^{P_i}\right). \tag{12}$$

The size of convolution kernels used in each layer is $2 * 2$ and the stride is set to 2. As a result, the size of the feature map will be reduced to half of its previous size after each layer. Note that we do not apply pooling after convolutional layer in order to preserve more features. In this way, after multiple convolutional layers, the last layer outputs a feature map with a size of $1 * 1 * P_N$, which can be denoted as:

$$M_N = \left(M_N^1, M_N^2, \cdots, M_N^{P_N}\right). \tag{13}$$

Finally, the QoS prediction is completed by fusing the features extracted by CNN through a fully connected layer. The prediction value is defined as follows:

$$Q = \sigma\left(W_m M_N + b_m\right) \tag{14}$$

where σ is the identity function, Q is the prediction value of QoS, and W_m, b_m are the weight matrix and the bias item respectively.

3.5 Model Training

In this paper, the L_1 loss function is used because it is less affected by outliers compared to L_2. Therefore, the objective function of QPCN is defined as follows:

$$\min_{\theta} \sum_{x \in X} |Q(x) - \hat{Q}(x)| + \lambda L_1(\theta) \tag{15}$$

where $Q(x)$ is the true QoS value, $\hat{Q}(x)$ is the predicted QoS value, $L_1(\theta)$ is the regularization term, and λ is corresponding regularization parameter. For model optimization, we choose Adam as the optimizer because of its advantages such as fast convergence and adaptive learning rate. For $L_1(\theta)$ objective function, the updating process of the parameters can be described as follows:

$$\theta = \theta \pm \eta \frac{d\hat{Q}(x)}{d\theta} \tag{16}$$

where θ is the trainable parameters and η is the learning rate.

3.6 Energy Consumption Analysis

We analyze and compare the inference energy consumption when using CNN and DNN in the prediction layer. For a convolutional layer, when the bias term is not considered, the number of Floating Point Operations (FLOPs) can be calculated by:

$$F_C = 2 \cdot C_i \cdot C_o \cdot H_o \cdot W_o \cdot K^2 \tag{17}$$

where Ci and C_o denote the number of input channels and output channels respectively, H_o and W_o denote the height and width of the output feature map respectively, and K is the size of convolution kernel.

For the fully connected layer, when the bias item is not considered, the number of FLOPs can be calculated by:

$$F_D = 2 \cdot D_i \cdot D_o \tag{18}$$

where D_i is the input dimension and D_o is the output dimension. We assume that the number of FLOPs per CPU cycle is f. Then the number of CPU cycles required for forward propagation of a convolutional layer and a fully connected layer are calculated by:

$$T_C = \frac{F_C}{f} \tag{19}$$

$$T_D = \frac{F_D}{f} \tag{20}$$

Then the energy consumption of a convolutional layer and a fully connected layer in the forward propagation are calculated by:

$$E_C = p_c \cdot T_C \tag{21}$$

$$E_D = p_c \cdot T_D \tag{22}$$

where p_c is the computation power of CPU.

As an example, if we set the embedding dimensionality of the identifier, region and autonomous system to 16, the interaction layer will generate a 48*48 feature map after the outer product operation. If we flat it, use fully connected layers and follow the common practice of the half-size tower structure, it is estimated that 3.5 million FLOPs will be generated. But if we set 6 convolutional layers and each layer has 16 channels, the total FLOPs is about 0.75 million. So using CNN in the prediction layer will significantly reduce energy consumption.

Table 1. Dataset Statistics

Statistics	Values
Users	339
Services	5825
Invocations	1974675
User's Regions	31
User's Autonomous Systems	137
Service's Regions	74
Services Autonomous Systems	992

4 Experiment and Analysis

In this section, we conduct experiments on real-world datasets and compare QPCN with current QoS prediction methods to demonstrate the effectiveness of QPCN. In addition, we analyzed the parameters and components of the model.

4.1 Dataset

To evaluate the performance of QoS prediction, we adopted the WS-Dream dataset collected by Zheng et al. [13], which is a widely used dataset for QoS prediction. It includes two QoS attributes, i.e., response time (RT) and throughput (TP). Tang et al. [20] extended it by adding user and service contextual information, such as region and autonomous system. Each entry in the dataset is a seven-tuple, namely (user id, service id, QoS, user region, serivce region, user AS, service AS). The detailed statistics of the dataset are shown in Table. 1.

4.2 Evaluation Metrics

We employ mean absolute error (MAE) and root mean square error (RMSE) to evaluate the QoS prediction accuracy, which are defined as follows:

$$MAE = \frac{1}{|X|} \sum_{x \in X} |Q(x) - \hat{Q}(x)| \tag{23}$$

$$RMSE = \sqrt{\frac{1}{|X|} \sum_{x \in X} (Q(x) - \hat{Q}(x))^2} \tag{24}$$

where $Q(x)$ represents the real QoS value, $\hat{Q}$ represents the predicted QoS value, and $|X|$ is the number of test cases.

4.3 Comparison Method

To illustrate the prediction accuracy of the proposed method, we compare it with 6 baseline methods. They are memory-based CF [11,13], model-based CF and deep learning-based methods [7,8].

UPCC [11]: A user-based collaborative QoS prediction method, which uses PCC to find similar neighbors of the user to predict missing QoS.

IPCC [13]: A service-based collaborative QoS prediction method, which uses PCC similarity to find similar neighbors of the service for missing QoS prediction.

UIPCC [13]: A collaborative QoS prediction method combining user-based and service-based, which performs QoS prediction by finding similar users and similar services.

MF: A model-based approach that maps users and services to the same latent space to obtain their latent feature vectors. Then the QoS prediction value is obtained through inner product of the corresponding vectors.

NCF [7]: A QoS prediction method based on deep learning, which combines GMF model and MLP. It overcomes the limitation that the MF method can only extract low-order linear features.

LDCF [8]: A location-aware and DL-based approach. Feature extraction is completed through MLP, and a similarity adaptive corrector is added to enhance the performance.

DNM [6]: A DL-based QoS prediction model which integrates various contextual information.

For these methods, we adopt the L_1 objective function. UPCC, IPCC, UIPCC, MF and NCF only use the information of QoS matrix as input, i.e., identifiers of user and service. For LDCF and our model, identifiers as well as location information are utilized as input.

4.4 Parameter Setting

Since users only invoke a small number of services, the real-world QoS matrix is extremely sparse. To simulate this situation, we split the QoS dataset into training sets and test sets based on a specific ratio (5%,10%,15%,20%,25%,30%). For example, matrix density (MD) = 5% means that we randomly select 5% of the QoS dataset as the training set and the remaining 95% as the test set. For three memory-based CF methods, the PCC similarity is used to select the top-k neighbors of user or service. We set the neighbor size of user and service as $k = 10$ and $k = 50$, respectively. For MF, we set the size of the latent feature vector to 16. For NCF, LDCF and QPCN, we set the size of the embedding vector to 16 and the size of the mini-batch to 256. For DNM, we use its single-task prediction model and set the size of the embedding vector to 50, while keeping a consistent mini-batch size of 256. We use Adam as the optimizer and set the learning rate to 0.0001. The number of convolutional layers in QPCN is set to 6, as described in Outer product-based neural collaborative filtering (ConvCF) [9]. And the maximum number of iterations is set to 150 for RT and 200 for TP.

4.5 Performance Comparison and Analysis

The RT and TP prediction results of QPCN and baseline methods are shown in Table. 2 and Table. 3, respectively, with the best results marked in bold. As can

Table 2. Performance comparison of response time prediction

Methods	MD=5%		MD=10%		MD=15%		MD=20%		MD=25%		MD=30%	
	MAE	RMSE	MAE	RMSE	MAE	RMSE	MAE	RMSE	MAE	RMSE	MAE	RMSE
UPCC	0.6412	1.4586	0.5577	1.3437	0.5125	1.2729	0.4860	1.2320	0.4685	1.1909	0.4632	1.1812
IPCC	0.6626	1.4276	0.5898	1.4764	0.5020	1.3129	0.4519	1.2565	0.4291	1.2151	0.4120	1.1950
UIPCC	0.6262	1.4196	0.5716	1.3465	0.4895	1.2635	0.4406	1.2062	0.4203	1.1796	0.4053	1.1640
MF	0.4523	1.3836	0.3751	1.2993	0.3457	1.2333	0.3351	1.2086	0.3252	1.1958	0.3209	1.1941
NCF	0.3912	1.3249	0.3413	1.2781	0.3333	1.2324	0.3254	1.2025	0.3199	1.1802	0.3176	1.1859
LDCF	0.3663	1.3233	0.3377	1.2726	0.3214	1.2258	0.3128	1.1924	0.3030	1.1786	0.3048	1.1807
DNM	0.3703	1.3183	0.3402	1.2672	0.3252	1.2482	0.3125	1.2086	0.3032	1.1506	0.2919	1.1274
QPCN	**0.3583**	**1.2685**	**0.3274**	**1.2210**	**0.3081**	**1.1797**	**0.2974**	**1.1483**	**0.2836**	**1.1236**	**0.2763**	**1.1098**

Table 3. Performance comparison of throughput prediction

Methods	MD=5%		MD=10%		MD=15%		MD=20%		MD=25%		MD=30%	
	MAE	RMSE	MAE	RMSE	MAE	RMSE	MAE	RMSE	MAE	RMSE	MAE	RMSE
UPCC	26.6423	62.2416	21.7450	55.3633	20.6025	51.4505	19.2685	49.5425	18.6132	47.6581	17.3216	46.1211
IPCC	26.8092	65.0990	25.2746	62.7262	24.2890	60.0846	23.1043	57.1816	22.1002	54.3921	21.1105	52.1923
UIPCC	26.3189	62.1013	21.6633	54.0820	20.1265	51.2213	18.7669	49.5254	17.6639	47.1032	16.9104	45.1360
MF	18.2153	60.7230	14.5573	51.5910	12.9527	47.0850	12.2309	44.9284	12.1005	43.1323	11.1519	41.7213
NCF	14.8885	48.6459	13.1384	45.9313	12.1308	42.8485	11.6696	41.0652	11.0819	40.4372	10.9025	39.8572
LDCF	13.7035	48.0793	12.3447	44.1337	11.4066	41.5943	10.8795	40.3540	10.9174	39.4807	10.5060	38.4988
DNM	13.6776	47.1867	12.1251	43.2670	11.1598	39.9460	10.6573	38.6675	10.1079	36.8679	9.7542	36.0998
QPCN	**13.0444**	**45.5415**	**11.4640**	**41.5092**	**10.3585**	**37.9071**	**9.9085**	**36.9162**	**9.3886**	**35.3695**	**9.1451**	**34.7018**

be seen from the tables, the MAE and RMSE decrease with increasing matrix density for all methods. A denser matrix provides more information for CF-based methods to find accurate neighbors and more training data for deep learning-based methods to optimize parameters. However, QPCN always achieves the minimum MAE and RMSE at any density, indicating that our proposed method outperforms all baseline methods.

Among the three memory-based methods, UIPCC outperforms UPCC and IPCC in most cases. When the data sparsity is high, the prediction performance of the three methods is very poor. In most cases, the prediction accuracy of MF is better than that of memory-based CF. Through MLP, NCF uses the powerful fitting ability of neural network to learn high-order nonlinear feature interaction from the embedding vectors, which improves the prediction accuracy compared to MF. LDCF further considers the location information and inputs it into the MLP to extract richer interaction features, which results in higher prediction accuracy. DNM considers the contextual information and achieves similar accuracy to LDCF. However, the RMSE values of MF, NCF, LDCF and DNM are sometimes higher than that of memory-based CF. QPCN outperforms LDCF and DNM in response time and throughput prediction tasks, with an average improvement of 5.0% and 5.1% in MAE and RMSE for RT, and an average improvement of 9.5% and 8.2% in MAE and RMSE for TP.

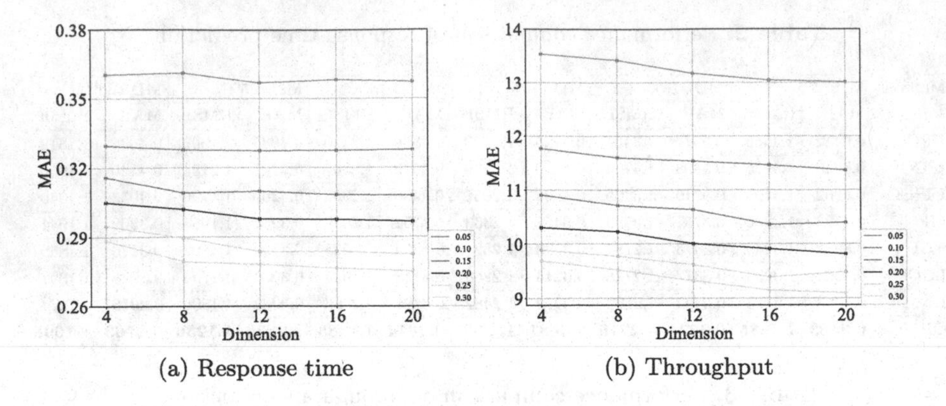

(a) Response time (b) Throughput

Fig. 2. Impact of dimension

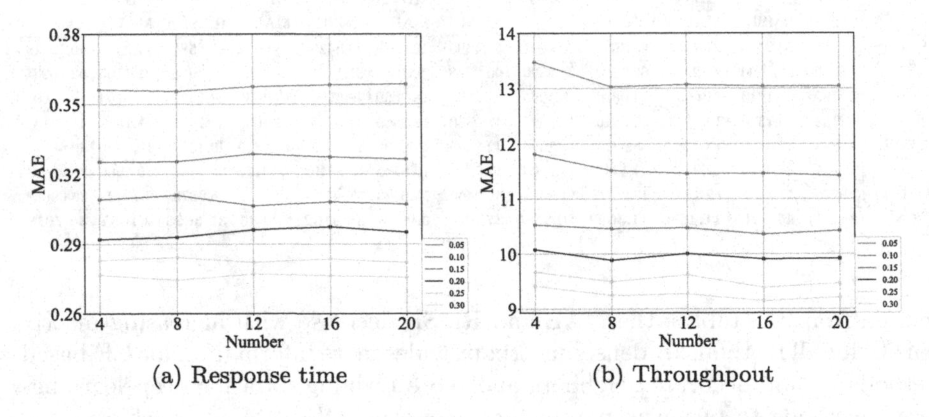

(a) Response time (b) Throughpout

Fig. 3. Impact of convolution kernels number

4.6 Impact of Parameters

The prediction performance of QPCN is influenced by two hyperparameters: the dimensionality of embedding vector and the number of convolution kernels. Accordingly, we conducted experiments to determine the optimal values for these hyperparameters.

Impact of Embedding Dimensionality. The dimensionality d represents the dimension of embedding vector in the embedding layer. The larger d means that the embedding vector contains more latent factors. To test the impact of d, we set it to 4, 8, 12, 16, 20, respectively, and randomly initialize it. Then, we test it under different MDs and the results are illustrated in Fig. 2. As seen in the figure, when the dimensionality d is fixed, the MAE always decreases as the matrix density increases. When the dimensionality d starts to increase, the MAE on RT and TP show a downward trend. Because a larger d enables the embedding vector to capture more latent factors and thus improves the prediction accuracy. However, when the dimensionality d on RT reaches 12, increasing it does not significantly decrease the MAE. When the dimensionality d on TP reaches 16, increasing d will cause MAE to rise under some MDs. Because if the dimension of the embedding vector is too large, it requires more training samples. Otherwise, there is a risk of overfitting. According to the variation trend of MAE under different MDs, to ensure the performance and reduce the number of parameters, we set d to 12 for RT and 16 for TP.

Impact of Convolution Kernels Number. The number of convolution kernels c is also an important hyperparameter in CNN that can impact the prediction performance of QPCN. Increasing the number of convolution kernels can improve the feature extraction ability of the model. To test the impact of it, we set c to 4, 8, 12, 16, 20 respectively and initialized it randomly. Then we test it under different MDs and the results are illustrated in Fig. 3. As seen in the figure, when c is fixed, MAE will always decrease with the increase of MD. When c increases, the change of MAE is not obvious on RT. For TP, when c increased from 4 to 8, the MAE decreased significantly. But when c increases from 8, the MAE no longer changes significantly. This is because using too many convolution kernels will lead to overfitting, especially on sparse datasets. According to the variation trend of RT and TP under different MDs, to obtain better performance and minimize the model parameters, we set c to 4 for RT and c to 8 for TP.

4.7 Ablation Study

QPCN adopts outer product to obtain two-dimensional interaction features and CNN to extract high-order features. In this section, we will conduct comparison experiments to evaluate the impact of these two key components.

Impact of Outer Product. In this section, we evaluate the impact of outer product in the interaction layer. Specifically, to demonstrate its effectiveness, the embedding vectors of user and service obtained in the embedding layer are directly concatenated and fed to the prediction layer. We obtain MAE using outer product and concatenation under six MDs of RT and TP, respectively, as shown in Table. 4. In both RT and TP, the MAE with outer product is smaller than the MAE with concatenation. And the performance on RT and TP is improved by 1.4% and 3.4% on average, respectively. This is because concatenate operation does not consider the dimensional interaction of the embedding vectors compared to outer product.

Table 4. Impact of outer product

QoS	Methods	Matrix Densities					
		MD=5%	MD=10%	MD=15%	MD=20%	MD=25%	MD=30%
RT	Outer Product	**0.3583**	**0.3274**	**0.3081**	**0.2974**	**0.2836**	**0.2763**
	Concatenate	0.3592	0.3308	0.3109	0.3010	0.2838	0.2905
TP	Outer Product	**13.0444**	**11.4640**	**10.3585**	**9.9085**	**9.3886**	**9.1451**
	Concatenate	13.2021	11.8176	10.6707	10.2330	9.9233	9.6092

Table 5. Impact of CNN

QoS	Methods	Matrix Densities					
		MD=5%	MD=10%	MD=15%	MD=20%	MD=25%	MD=30%
RT	CNN	**0.3583**	**0.3274**	**0.3081**	**0.2974**	**0.2836**	**0.2763**
	DNN	0.3662	0.3353	0.3172	0.3076	0.2981	0.2896
TP	CNN	**13.0444**	**11.4640**	**10.3585**	**9.9085**	**9.3886**	**9.1451**
	DNN	13.5368	11.8590	10.8423	10.4771	10.0629	9.7612

Impact of CNN. In this section, we evaluate the impact of CNN in the prediction layer. Specifically, CNN and DNN are respectively used in the prediction layer to realize the prediction of QoS value. We conduct tests under six MDs of RT and TP, and obtain MAE using CNN and DNN, respectively, as shown in Table. 5. The model using CNN obtains smaller MAE on RT and TP, whose performance are improved by 3.4% and 5.0% on average, respectively. The reason is that CNN has fewer parameters, which is easier to build a deeper model than DNN. In the case of DNN, increasing the number of layers will result in a significant increase in the number of parameters, which is not conducive to learning sparse high-order features.

5 Conclusion

In this paper, we propose a new model called QPCN for QoS prediction. We integrate two-dimensional interaction features of user and service embedding vectors in this model. Then, the high-order feature interactions are extracted via CNN. The effectiveness of QPCN is demonstrated by comparing it with CF-based and DL-based QoS prediction methods on real-world datasets. In the future, we will consider the QoS prediction in edge computing environment. Also, considering the dynamic nature of QoS values, we plan to add temporal information to the model for dynamic QoS prediction.

References

1. Zheng, J., Zhang, Z., Ma, Q., Gao, X., Tian, C., Chen, G.: Multi-resource VNF deployment in a heterogeneous cloud. IEEE Trans. Comput. **71**(1), 81–91 (2020)
2. Wang, S., Ma, Y., Cheng, B., Yang, F., Chang, R.N.: Multi-dimensional QoS prediction for service recommendations. IEEE Trans. Serv. Comput. **12**(1), 47–57 (2019)

3. Mistry, S., Bouguettaya, A., Dong, H., Qin, A.K.: Metaheuristic optimization for long-term IaaS service composition. IEEE Trans. Serv. Comput. **11**(1), 131–143 (2016)
4. Ghafouri, S.H., Hashemi, S.M., Hung, P.C.K.: A survey on web service QoS prediction methods. IEEE Trans. Serv. Comput. **15**(4), 2439–2454 (2022)
5. Li, Z., et al.: A knowledge-driven anomaly detection framework for social production system. IEEE Trans. Comput. Soc. Syst., 1–14 (2022)
6. Wu, H., Zhang, Z., Luo, J., Yue, K., Hsu, C.H.: Multiple attributes QoS prediction via deep neural model with contexts. IEEE Trans. Serv. Comput. **14**(4), 1084–1096 (2018)
7. He, X., Liao, L., Zhang, H., Nie, L., Hu, X., Chua, T.S.: Neural collaborative filtering. In: Proceedings of the 26th International Conference on World Wide Web, pp. 173–182 (2017)
8. Zhang, Y., Yin, C., Wu, Q., He, Q., Zhu, H.: Location-aware deep collaborative filtering for service recommendation. IEEE Trans. Syst. Man Cybern. Syst. **51**(6), 3796–3807 (2019)
9. He, X., Du, X., Wang, X., Tian, F., Tang, J., Chua, T.S.: Outer product-based neural collaborative filtering. arXiv preprint arXiv:1808.03912 (2018)
10. Beutel, A., et al.: Latent cross: making use of context in recurrent recommender systems. In: Proceedings of the Eleventh ACM International Conference on Web Search and Data Mining, pp. 46–54 (2018)
11. Shao, L., Zhang, J., Wei, Y., Zhao, J., Xie, B., Mei, H.: Personalized QoS prediction for web services via collaborative filtering. In: IEEE International Conference on Web Services (ICWS), pp. 439–446. IEEE (2007)
12. Chen, Z., Shen, L., Li, F.: Exploiting web service geographical neighborhood for collaborative QoS prediction. Future Gener. Comput. Syst. **68**, 248–259 (2017)
13. Zheng, Z., Ma, H., Lyu, M.R., King, I.: Qos-aware web service recommendation by collaborative filtering. IEEE Trans. Serv. Comput. **4**(2), 140–152 (2010)
14. Wu, X., Cheng, B., Chen, J.: Collaborative filtering service recommendation based on a novel similarity computation method. IEEE Trans. Serv. Comput. **10**(3), 352–365 (2015)
15. Wu, H., Yue, K., Li, B., Zhang, B., Hsu, C.H.: Collaborative QoS prediction with context-sensitive matrix factorization. Future Gener. Comput. Syst. **82**, 669–678 (2018)
16. Hang, W., Sun, H., Liu, X., Guo, X.: Temporal QoS-aware web service recommendation via non-negative tensor factorization. In: Proceedings of the 23rd International Conference on World Wide Web, pp. 585–596 (2014)
17. Tang, M., Zheng, Z., Kang, G., Liu, J., Yang, Y., Zhang, T.: Collaborative web service quality prediction via exploiting matrix factorization and network map. IEEE Trans. Netw. Serv. Manag. **13**(1), 126–137 (2016)
18. Xu, J., Zheng, Z., Lyu, M.R.: Web service personalized quality of service prediction via reputation-based matrix factorization. IEEE Trans. Reliab. **65**(1), 28–37 (2015)
19. Du, Z., Zheng, J., Yu, H., Kong, L., Chen, G.: A unified congestion control framework for diverse application preferences and network conditions. In: Proceedings of the 17th International Conference on emerging Networking EXperiments and Technologies, pp. 282–296 (2021)
20. Tang, M., Zhang, T., Liu, J., Chen, J.: Cloud service QoS prediction via exploiting collaborative filtering and location-based data smoothing. Concurrency Comput. Pract. Exp. **27**(18), 5826–5839 (2015)

Multiple Relays Assisted MEC System for Dynamic Offloading and Resource Scheduling with Energy Harvesting

Jiming Wang and Long Qu(✉)

Faculty of Electrical Engineering and Computer Science, Ningbo University, Ningbo, Zhejiang, China
{2111082176,qulong}@nbu.edu.cn

Abstract. Mobile Edge Computing (MEC) has become an indispensable way to reduce the execution delay of devices. However, for some devices located far away from the MEC server, the transmission delay of communication with MEC is still large. In this case, we consider using multiple relay devices to assist Internet of Things (IoT) devices to communicate with MEC servers. To enhance the energy efficiency of the system, we introduce Energy Harvesting (EH) devices to provide energy for the IoT devices. Our objective is to maximize the utilization of EH devices while minimizing the overall delay in task offloading for the IoT devices. We tackle the problem by formulating it as a Markov Decision Problem (MDP). However, due to the significant expansion of the state space, traditional methods such as relative value iteration and linear iterative reconstruction are ineffective in solving this problem. Hence, we propose an approach called Multi-Relay Assisted Dynamic Computation Offloading (MRADCO) algorithm, which leverages the Lyapunov optimization technique. It is important to note that our proposed algorithm makes decisions solely based on the current state, without requiring the distribution information of the wireless channel and EH process. This characteristic enhances the algorithm's practicality and reduces complexity in real-world implementations. Through rigorous theoretical derivation and comprehensive simulation experiments, we demonstrate that our algorithm is asymptotically optimal. And compared with the benchmark algorithm LODCO, our algorithm reduces the time by 50%.

Keywords: Mobile Edge Computing (MEC) · Lyapunov optimization · Energy Harvesting(EH) · power control · multi-relay · Internet of Things (IoT)

1 Introduction

With the development of wireless communication technology, the delay requirements of compute-intensive tasks for Internet of Things (IoT) devices are becoming more and more stringent. High latency may lead to network congestion and

H. Jin et al. (Eds.): GPC 2023, LNCS 14504, pp. 32–46, 2024.
https://doi.org/10.1007/978-981-99-9896-8_3

affect the Quality of Service (QoS) for users [1]. However, offloading tasks to the core network for processing will cause large transmission delay. Therefore, Mobile Edge Computing (MEC) has been proposed and used in wireless networks. MEC is to place servers closer to IoT devices to provide them with computing, storage and other services [2,3].

Although MEC reduces the transmission delay of communication, it is still valuable to study the delay for MEC systems. Zhang *et al.* [4] studied the benefits of collaboration among multiple MECs for large-scale delay-sensitive tasks. They consider load balancing among MEC servers and study joint parallel offloading of tasks. And they use weights to jointly consider system energy consumption and delay. The problem is formulated as a stochastic programming problem. Authors propose a centralized cost management algorithm (LYPCCMA) based on Lyapunov optimization. And they obtain the optimal average cost of the system in a long time. In [5], authors applied MEC technology to mobile Health (mHealth). Monitoring brain signals using mobile electroencephalogram (EEG) headphones provides an opportunity for seizure detection and prevention. However, due to the limited energy and computational resources of mobile EEG head-mounted devices, they proposed a distributed feature extraction method that depends on the device and the MEC server. Simulation results show the effectiveness of the proposed method for mHealth applications of epilepsy prediction. Xu *et al.* [6] applied MEC to UAV-assisted communication. The MEC server is placed on the UAV to provide computing services for IoT devices and also acts as a relay node to forward tasks to the ground receiving point. By converting non-convex to convex, the problem is decomposed into three low-complexity and easy to solve subproblems.

Energy consumption is also an important issue that cannot be ignored in MEC systems. Due to the limitation of the size of IoT devices, the energy carried by them is limited, which is not enough to support their long-term task processing [7]. In recent years, many scholars have focused on energy harvesting devices. EH devices harvest energy from the natural environment and convert it into electricity for use by the device [8]. It is necessary to jointly consider the energy consumption and delay of the system for MEC systems. In [9] where the authors studied the delay problem of multi-relay assisted single user task offloading. And energy harvesting device (EH) was introduced into the model. The simulation results show that the use of relay devices to assist IoT device offloading reduces the delay of the system. Fu *et al.* [10] studied the model of single relay assisted multiuser. The energy minimization problem is studied by jointly optimizing the transmit power and offloading time. The problem is formulated as a nonconvex optimization problem. Authors propose a continuous convex approximation that achieves lower energy consumption than the baseline scheme. In [11], the Wireless Access Point (WAP) is used as the relay node between the device and the MEC server. Minimizing system energy consumption and meeting device delay requirements were studied by jointly optimizing relay selection and resource allocation. The problem is formulated as a mixed-integer nonconvex optimization problem, which is NP-hard. To this end, the authors propose a heuristic algorithm to find the approximate solution of the problem in two steps. Deng *et al.* [12] proposed a general MEC cooperation mode using the multi-hop task relaying strategy. The goal is to maximize the task throughput. And the authors use matching theory to find a low complexity algorithm. The analysis method presented in the above article provides a comprehensive examination of task offloading in today's MEC system. However, it may not be the most optimal approach for certain special scenarios. For example, when multiple users are far away from the MEC server, large transmission delay and energy consumption will be incurred if tasks are still to be offloaded to the MEC server. However, if only one relay is used to transmit the task, the relay task will be overloaded. For the MEC system of this scenario, this paper gives a detailed solution. Most of the current papers on relay study the task offloading problem in a single relay or a single time slot, and there are few studies on multi-relay assisted and long-term task offloading.

In this paper, we consider MEC systems with multiple relays and multiple IoT devices, where the relay devices can serve multiple IoT devices simultaneously. The model not only used the offloading ability of relay devices to reduce the computing pressure of IoT devices, but also did not cause the overload of relay devices. We equipped each IoT device in the system with EH devices, and established battery energy queues to describe the changes in the available energy of the device. The problem was formulated as a Markov Decision process (MDP) problem. Due to the state space explosion, the traditional relative value iteration method and linear iterative reconstruction method can not solve this problem. Therefore, we designed the MRADCO algorithm using Lyapunov optimization to solve the problem. The asymptotically optimal solution of the system is obtained by jointly optimizing the local computation rate, the local transmission power and the relay selection. Theoretical analysis and simulation results verify the correctness of the results.

2 System Model

In this paper, the relay devices are considered to be located in urban areas with rich network resources, while our IoT devices are located in suburban areas far from the MEC server [9]. We study the average delay of IoT devices to process tasks during a certain busy period of users.

2.1 Multi-relay Assisted Model

We consider a system composed of IoT devices, relay devices, and a Base Station (BS), as shown in Fig. 1. We consider a certain period of time $\mathcal{T}$, which we divide into small time slots of equal duration, each of length τ, denoted by t, where $t \in \{0, 1, 2, \ldots, T-1\}$. We consider $\mathcal{I}$ IoT devices and $\mathcal{J}$ relay devices, denoted by i and j, where $i \in \{1, 2, \ldots, I\}$ and $j \in \{1, 2, \ldots, J\}$, respectively.

We equip each IoT device with an antenna, which is independent of each other. At each time slot, we consider that each IoT device generates a computationally intensive task, defined as $A_i(t) \triangleq (L_i(t), T_d)$, Where $L_i(t)$ (in bits)

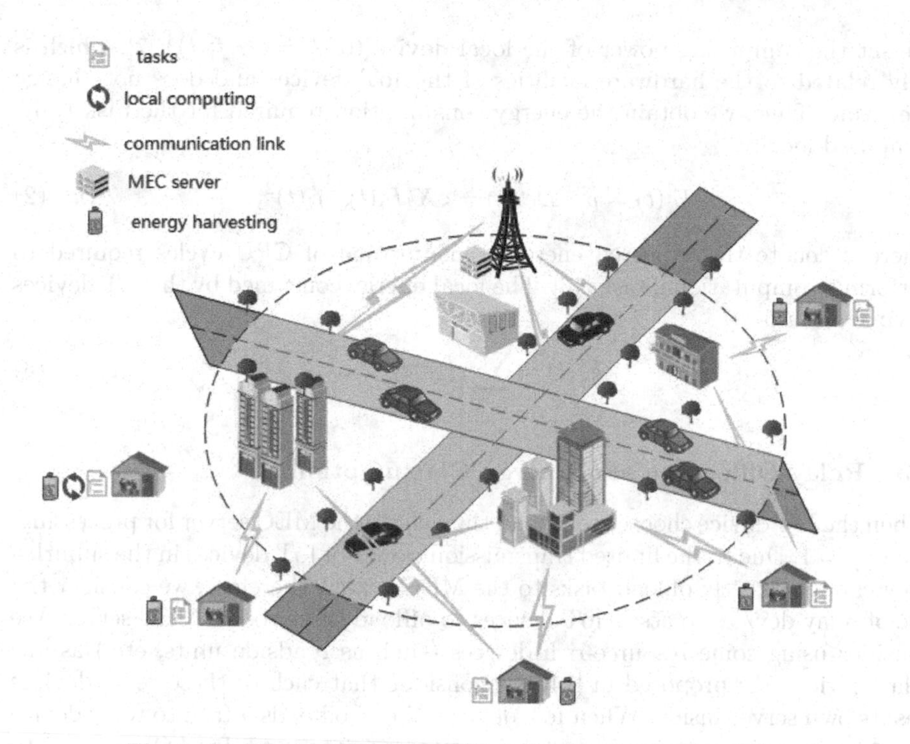

Fig. 1. Multi-relay assisted multi-user offloading tasks model

is the data size of the computation task, and T_d (in second) is the deadline for processing the task (i.e., the maximum time that the task can be kept, $T_d \leq \tau$). We assume that a task is generated at the beginning of a time slot, and this task chooses between two processing modes: local processing or relay offloading. We denote by $\pi(t) \triangleq \{\pi_i^l(t), \pi_i^r(t), \pi_i^d(t)\}$ the selected processing mode: $\pi_i^l(t) = 1$ means local processing is selected, and $\pi_i^r(t) = 1$ means relay offloading is selected. Where $\pi_i^d(t) = 1$ means that the task is discarded. In the following, we discuss the delay and energy consumption required to process tasks from these three aspects.

2.2 Task Generation and Local Computation

In this section, for the tasks generated by IoT devices, we choose to process the tasks locally, that is, $\pi_i^l(t) = 1$. We use X_i to represent the CPU cycle required for an IoT device i to compute 1 bit of data. So we get that the CPU cycle required to compute $L_i(t)$ is $H_i(t) = X_i \cdot L_i(t)$. $f_i(t)$ indicates the computing rate of the local server. From this, we get that the delay of the task processed locally is

$$De_i^l(t) = \frac{H_i(t)}{f_i(t)}, \tag{1}$$

We set the computing power of the local device to $p_i^l = \varsigma * f_i(t)^3$ [7], which is only related to the hardware facilities of the IoT devices and does not change over time. Thus, we obtain the energy consumption required for the task to be computed locally

$$E_i^l(t) = p_i^l \cdot De_i^l(t) = \varsigma X_i L_i(t) \cdot f_i(t)^2, \tag{2}$$

where ς denote the expected energy comsumption of CPU cycles required to perform a computational task [13]. The local energy consumed by the IoT devices in time slot t is

$$E^l(t) = \sum_{i=1}^{I} \pi_i^l(t) E_i^l(t). \tag{3}$$

2.3 Relay Offloading and Energy Consumption

When the IoT device chooses to offload the task to the MEC server for processing, then $\pi_i^r = 1$. Due to the limited transmission power of IoT devices in the suburbs, they cannot directly offload tasks to the MEC server; therefore, we consider the use of relay devices to assist IoT devices to offload tasks to the MEC server. We consider using some resource-rich devices (such as roadside units, etc.) as our relay devices. As proposed in [14], we consider that each of these relay devices has its own server inside. When IoT device $i, i \in \mathcal{I}$ offloads a task to relay device $j, j \in \mathcal{J}$, the server of relay device j creates a container i for IoT device i to process this task. We assume that when relay device j communicates with any IoT device, it will create a container for the IoT device to process tasks.

We optimize communication between IoT devices, relay devices, and MEC servers by employing Orthogonal Frequency Division Multiple Access (OFDMA) technology. This approach guarantees smooth communication without any interference concerns. $\mathcal{M}_i(t)$ represents the relay devices that is idle near IoT device i in time slot t. The IoT device i offloads to the MEC server through $\mathcal{M}_i(t)$ relay devices by dividing the task into $\mathcal{M}_i(t)$ parts. $\xi_{i,j}(t)$ represents the ratio of tasks offloaded to relay device $j, j \in \mathcal{M}_i(t)$. There are many tasks that can be split, such as the convolution operation for DNNS in [16]. As a result, we get the following constraint:

$$\sum_{j=1}^{\mathcal{M}_i(t)} \xi_{i,j}(t) = 1, i \in \mathcal{I}. \tag{4}$$

Then, we consider the transmission delay of this task. Shannon formula is used to obtain the transmission rate of tasks (from IoT device to relay device and relay device to MEC server):

$$R_{i,j}(t) = W \log_2(1 + \frac{p_i(t) \cdot h_{i,j}(t)}{N_0}),$$
$$R_j(t) = W \log_2(1 + \frac{p_j(t) \cdot h_j(t)}{N0}). \tag{5}$$

Among them, the first formula represents the offloading of tasks from IoT devices to relay devices. W denotes the channel bandwidth for task transmission. As mentioned above, each IoT device communicates with the relay device, and the relay device creates a separate container for it to process tasks. Therefore we consider that there is no interference between IoT devices. $p_i(t)$ represents the transmission power of IoT device i, which is an optimization variable. $h_{i,j}(t)$ represents the channel gain between IoT device i and relay device j, which is obtained by $h_{i,j}(t) = g_0(\frac{d_0}{d_{i,j}(t)})^{\nu}$. Where g_0 represents the path loss constant, $d_{i,j}(t)$ is the actual distance from the ith IoT device to the j th relay device, and d_0 is the reference distance. N_0 represents the white Gaussian noise during transmission. $(\frac{d_0}{d_{i,j}(t)})^{\nu}$ represents the large-scale loss between paths. ν is the loss factor, which is only related to the distance of IoT-relay. Similarly, $p_j(t)$ represents the transmission power of the relay device. $h_j(t)$ is the channel gain between the relay device and the MEC server.

Finally, we obtain the delay for the task to be offloaded assisted by the relay. This time is divided into two parts, which are the time $De_{i,j}(t)$ from IoT to the relay and the time $De_j(t)$ from the relay to MEC.

$$De_{i,j}(t) = \frac{\xi_{i,j} L_i(t)}{R_{i,j}(t)}, De_j(t) = \frac{\xi_{i,j} L_i(t)}{R_j(t)}. \tag{6}$$

In this paper, we consider that each IoT device distributes tasks to the relay device simultaneously. Therefore, the total time for IoT device i to offload a task using the relay is

$$De_i^r(t) = \max_{j \in [1, \mathcal{M}_i(t)]} [De_{i,j}(t) + De_j(t)] \tag{7}$$

The energy consumption generated by this process is denoted by

$$E_{i,j}(t) = p_i(t) \cdot T_{i,j}(t), E_j(t) = p_j(t) \cdot T_j(t). \tag{8}$$

Therefore, in time slot t, the energy consumed by IoT device i to offload tasks and the total energy consumption by all IoT devices are

$$E_i^r(t) = \sum_{j=1}^{\mathcal{J}} E_{i,j}(t), E^r(t) = \sum_{i=1}^{\mathcal{I}} E_i^r(t). \tag{9}$$

As mentioned in [15], the MEC server has sufficient computing resources and transmission power, and the amount of data contained in the calculation results is small. Therefore, we ignore the computation energy consumption and the backhaul energy consumption of the MEC server.

2.4 Total Delay and Energy Consumption

In summary, we obtain the total delay and total energy consumption of IoT devices choosing different ways to process tasks as follows:

$$De_i(t) = \pi_i^l(t) De_i^l(t) + \pi_i^r(t) De_i^r(t) + \pi_i^d(t) \varrho, \tag{10}$$

$$E_i(t) = \pi_i^l(t)E_i^l(t) + \pi_i^r(t)E_i^r(t). \tag{11}$$

Here, ϱ represents the cost of task dropping. Without loss of generality, we consider that the system is more inclined to successfully process the task, hence $T_d \leq \varrho$.

It follows that IoT devices process their generated tasks in two ways, namely, local processing or relay offloading. However, if neither of the two methods can meet the deadline, the task will be discarded. And each task of each IoT device can only choose one processing mode. Therefore, we have the following constraint:

$$\pi_i^l(t) + \pi_i^r(t) + \pi_i^d(t) = 1, \tag{12}$$

when the task is successfully processed, the system delay is at least higher than the task deadline,

$$\pi_i^l(t)De_i^l(t) + \pi_i^r(t)De_i^r(t) \leq T_d, i \in \mathcal{I}, t \in \mathcal{T}. \tag{13}$$

2.5 Energy Harvesting and Battery Energy Levels

In this paper, we have configured energy harvesting devices for each IoT device. The device harvest green energy in the environment, such as wind and solar energy, to provide power for IoT devices. We assume that each device collects $EH_i(t)$ unit of energy in each time slot, and the $EH_i(t)$ between time slots is *i.i.d* (independently and identically distributed), and its maximum harvested energy is EH_{max}. At each time slot, a portion of the harvested energy is stored in the battery of the IoT device, denoted by $e_i(t)$, satisfying

$$0 \leq e_i(t) \leq EH_i(t), i \in \mathcal{I}, t \in \mathcal{T}. \tag{14}$$

IoT devices process tasks by invoking energy within their batteries. $B_i(t)$ represents the battery energy of IoT device i at time slot t. At the beginning of each time slot, the energy harvesting device starts to collect green energy and saves a part of the energy (i.e., $e_i(t)$) to the battery to provide energy for the IoT device in the next time slot. Similarly to [13], we assume that $B_i(0) = 0$ and $B_i(t) < \infty$. Thus, we obtain the following battery energy level update process:

$$B_i(t+1) = B_i(t) - E_i(\pi, f, p, t) + e_i(t), \tag{15}$$

where $E_i(\pi, f, p, t) = \pi_i^l(t)E_i^l(t) + \pi_i^r(t)E_i^r(t), i \in \mathcal{I}, t \in \mathcal{T}$.

As a general rule, the energy consumption of an IoT device cannot exceed the energy of its battery

$$E_i(\pi, f, p, t) \leq B_i(t) < \infty. \tag{16}$$

2.6 Problem Formation

According to the above discussion, we introduce in this section the problem of minimizing the delay of the system in the case of long-term stability, the problem is formed as follows:

$$P: \min_{\pi, f, p, \xi, e} \lim_{T \to \infty} \frac{1}{T} \sum_{t=0}^{T-1} \sum_{i=1}^{\mathcal{I}} De_i(t)$$

$$\begin{aligned} s.t.\quad & (4), (12), (13), (14), (16) \\ & E_i(\pi, f, p, t) \in \{0\} \bigcup [E_{min}, E_{max}], i \in \mathcal{I} && (17) \\ & 0 \le p_i(t), p_j(t) \le p_{max} \cdot \pi_i^r(t) && (18) \\ & 0 \le f_i(t) \le f_{max} \cdot \pi_i^l(t) && (19) \\ & \mathcal{M}_i(t) \le J, \pi_i^r(t) = 1, i \in \mathcal{I}, t \in \mathcal{T}. && (20) \end{aligned}$$

Among them, (17) indicates that there are maximum and minimum restrictions on the energy consumption of IoT devices in each time slot. This is reasonable because the batteries of IoT devices can only store a limited amount of energy. (20) indicates that the number of relays used by each IoT device in each time slot cannot be higher than the total number of relays.

In the P problem, the state of the system is composed of the task request, the harvestable energy, the battery energy level and the channel state. Actions are optimal for CPU cycle scheduling and transmit power allocation. In addition, the current action is only related to the current system state, and the objective function is the long-term average delay. Therefore, the problem is a Markov Decision process (MDP) problem. Due to the state space explosion in this paper, the traditional MDP algorithm cannot solve this problem.

3 Algorithm Design

In this paper, we use Lyapunov optimization techniques to solve the problem. Since the energy level of the battery is time-dependent, the system decision is time-dependent. Ordinary Lyapunov optimization techniques cannot be directly applied. We use the weighted perturbation method to solve the problem. A *Multi-Relay Assisted Dynamic Computation Offloading algorithm (MRADCO)* was designed.

3.1 Problem Transfer

First, we define the virtual battery energy queue $\breve{B}_i(t) = B_i(t) - \vartheta$, where ϑ represents the perturbation parameter of the IoT device and its value is as follows

$$\vartheta = \breve{E}_{max} + V\varrho \cdot E_{min}^{-1}, \tag{21}$$

where $\breve{E}_{max} = \min\{\max\{\varsigma H_i(t)(f_i(t))^2, p_{max}\tau\}, E_{max}\}$, V denotes the Lyapunov control parameter, and $0 < V < \infty$.

In the following discussion, the actual battery energy queue is stabilized around ϑ by controlling V, and the optimal delay of the system is obtained. By exploiting the virtual queue, we transform the problem P into P_2.

$$P_2: \min_{\pi,f,p,\xi,e} \breve{B}_i(t)(e_i(t) - E_i(\pi,f,p,t)) + VDe_i(t)$$
$$s.t. \quad (4),(12),(13),(14),(16),(17)-(20).$$

3.2 Problem Decomposition

According to the relationship between the decision variables, we decompose the problem into three sub-problems, which are *the optimal energy harvesting, CPU computing rate and transmission power.*

The Optimal Energy Harvesting: Since $e_i(t)$ is not correlated with other decision variables, we strip it from P_2 and obtain the following problem:

$$\min_{0\le e_i(t)\le EH_i(t)} \breve{B}_i(t)e_i(t), \tag{22}$$

and we solve the optimal solution of this problem as

$$e_i^*(t) = EH_i(t)\cdot \mathbf{1}\{\breve{B}_i(t)\le 0\}. \tag{23}$$

The Optimal CPU Computing Rate: When $\pi_i^l(t) = 1$, the task chooses to be processed locally. At this point, the P_2 problem is transformed into the problem of solving the optimal CPU computing rate.

$$\min_{f_i(t)} -\breve{B}_i(t)\varsigma H_i(t)(f_i(t))^2 + V\frac{H_i(t)}{f_i(t)}$$
$$s.t. \quad 0\le f_i(t)\le f_{max}, \quad De_i^l(t)\le T_d, \tag{24}$$
$$\varsigma H_i(t)(f_i(t))^2 \in [E_{min}, E_{max}]. \tag{25}$$

By the following proposition, we solve the optimal solution of this sub-problem

Proposition 1. *The problem has solution if and only if $f_i^L(t)\le f_i^U(t)$, where $f_i^L(t) = \max\{\sqrt{\frac{E_{min}}{\varsigma H_i(t)}}, \frac{H_i(t)}{T_d}\}$, $f_i^U(t) = \min\{\sqrt{\frac{E_{max}}{\varsigma H_i(t)}}, f_{max}\}$. And its optimal solution is related to the virtual queue $\breve{B}_i(t)$, as follows*

$$f_i^*(t) = \begin{cases} f_i^U(t), & \text{if } \breve{B}_i(t)\ge 0 \text{ or } \breve{B}_i(t)\le 0, f_i^0(t) > f_i^U(t),\\ f_i^0(t), & \text{if } \breve{B}_i(t) < 0, f_i^L(t)\le f_i^0(t)\le f_i^U(t),\\ f_i^L(t), & \text{if } \breve{B}_i(t) < 0, f_i^0(t) < f_i^L(t), \end{cases} \tag{26}$$

where $f_i^0(t) = (\frac{V}{-2\breve{B}_i(t)\varsigma})^{\frac{1}{3}}$.

Proof. We omit the proof due to page limitations.

The Optimal Transmission Power: When $\pi_i^r(t) = 1$, the task is offloaded to the MEC server through the relay device. At this time, the transmission power of IoT and relay devices directly affects the delay of the system. P_2 translates into the following problem

$$
\begin{aligned}
\min_{p_i,p_j,\xi_{i,j}} \quad & -\breve{B}_i(t)\sum_{j=1}^{\mathcal{M}_i(t)}[\frac{p_i(t)\xi_{i,j}(t)L_i(t)}{R_{i,j}(t)}]+ \\
& V\cdot \max_{j\in[1,\mathcal{M}_i(t)]}[\frac{\xi_{i,j}(t)L_i(t)}{R_{i,j}(t)}+\frac{\xi_{i,j}(t)L_i(t)}{R_j(t)}] \\
s.t. \quad & \mathcal{M}_i(t)\le\mathcal{M}, 0\le p_i(t), p_j(t)\le p_{max}, \quad (27) \\
& \max_{j\in[1,\mathcal{M}_i(t)]}[\frac{\xi_{i,j}(t)L_i(t)}{R_{i,j}(t)}+\frac{\xi_{i,j}(t)L_i(t)}{R_j(t)}]\le T_d, \quad (28) \\
& \sum_{j=1}^{\mathcal{M}_i(t)}[\frac{p_i(t)\xi_{i,j}(t)L_i(t)}{R_{i,j}(t)}]\in[E_{min},E_{max}]. \quad (29)
\end{aligned}
$$

Since there are still many variables in this problem, we cannot solve it directly, so we design a heuristic algorithm.

Step 1: When there is only one relay device, which is assumed to be device $j, j \in \mathcal{M}_i(t)$, we offload all the tasks of selecting relay-assisted offloading to device j. We consider that the relay device has sufficient transmission resources and can offload the task to the MEC server with the maximum transmission power. This problem changes as

$$
\begin{aligned}
\min_{p_i(t)} \quad & -\breve{B}_i(t)\frac{p_i(t)L_i(t)}{R_{i,j}(t)}+V\frac{L_i(t)}{R_{i,j}(t)}+G \\
s.t. \quad & 0\le p_i(t)\le p_{max}, \frac{L_i(t)}{R_{i,j}(t)}\le T_d-\frac{G}{V}, \quad (30) \\
& \frac{p_i(t)L_i(t)}{R_{i,j}(t)}\in[E_{min},E_{max}], \quad (31)
\end{aligned}
$$

where $G = VDe_j(t)$, since $p_j(t) = p_{max}$, we just figure out the value of $De_j(t)$. In the following we use a proposition to find the optimal solution of the problem.

Proposition 2. *The problem has a solution if and only if $p_i^L(t) \le p_i^U(t)$. Its optimal solution is given by*

$$
p_i^*(t)=\begin{cases} p_i^U(t), & \text{if } \breve{B}_i(t)\ge 0 \text{ or } \breve{B}_i(t)<0, p_i^U(t)<p_i^0(t), \\ p_i^0(t), & \text{if } \breve{B}_i(t)<0, p_i^L(t)\le p_i^0(t)\le p_i^U(t), \\ p_i^L(t), & \text{if } \breve{B}_i(t)<0, p_i^L(t)>p_i^0(t). \end{cases} \quad (32)
$$

where $p_i^0(t)$ *is the unique solution of equation* $-\breve{B}_i(t)log_2(1+\frac{h_{i,j}(t)p_i(t)}{N_0}) - \frac{h_{i,j}(t)}{(N_0+h_{i,j}(t)p_i(t))ln2} = 0$. *And*

$$p_i^L(t) = \begin{cases} p_i^{L,T_d}(t), & \frac{N_0 L_i(t)ln2}{Wh_{i,j}(t)} \geq E_{min} \\ \max\{p_i^{L,T_d}(t), p_i^{E_{min}}(t)\}, & \frac{N_0 L_i(t)ln2}{Wh_{i,j}(t)} < E_{min}, \end{cases}$$

$$p_i^U(t) = \begin{cases} \min\{p_{max}, p_i^{E_{max}}(t)\}, & \frac{N_0 L_i(t)ln2}{Wh_{i,j}(t)} < E_{max} \\ 0, & \frac{N_0 L_i(t)ln2}{Wh_{i,j}(t)} \geq E_{max}. \end{cases} \quad (33)$$

where $p_i^{L,T_d}(t) \triangleq (2^{\frac{L_i(t)}{WT_d}} - 1)N_0/h_{i,j}(t)$, $p_i^{E_{min}}(t)$ *is the unique solution of* $p_i(t)L_i(t) = R_{i,j}(t)E_{min}$ *given* $N_0L_i(t)ln2(Wh_{i,j}(t))^{-1} < E_{min}$ *and* $p_i^{E_{max}}(t)$ *is the unique solution of* $p_{i,j}(t)L_i(t) = R_{i,j}(t)E_{max}$ *given* $N_0L_i(t)ln2(Wh_{i,j}(t))^{-1} < E_{max}$

Proof. We omit the proof due to page limitations.

Step 2: Repeating step 1, we obtain the optimal transmission power for offloading tasks from IoT devices to relay devices. Next, we require the optimal task ratio of the IoT device offloading to the relay device. According to step 1, we obtain the time $De_{i,j}(t)$ when the task is offloaded through either relay. By Eq. (7), we conclude that to guarantee $De_i^r(t)$ is minimal, we need to guarantee

$$De_{i,j}(t) = De_{i,k}(t), \forall j, k \in [1, \mathcal{M}_i(t)]. \quad (34)$$

Then, according to Eqs. (4) and (34), we find the optimal task allocation ratio $\xi_{i,j}^*(t)$.

4 Algorithm Performance and Simulation Results

Optimality Analysis: First, we analyze the optimality of MRADCO, and we get the following conclusions:

$$P_2^* \leq P^* + \frac{N}{V}, \quad (35)$$

where $N = \frac{1}{2}((EH_{max}^2) + (\breve{E}_{max}^2))$ is a constant. It follows that the performance of the MRADCO algorithm is related to the Lyapunov control parameter V. When $V \to \infty$, the optimal solution of the MRADCO algorithm is infinitely close to the optimal solution of the original problem P. And the MRADCO algorithm's solving time is much less than the traditional algorithm, only needs 20 s to calculate the result.

Simulated Analysis: We set up an MEC system with five IoT devices and four relay devices. The bandwidth is set to 1 MHz [17]. The values of some important parameters in the paper are shown in Table 1. We do the simulation of the battery energy consumption and the average delay of the system.

Table 1. Key Mathematical Symbols

Parameters	
Path-loss constant g_0	-40 dB
Time slot length τ	2 ms
Task deadline T_d	2 ms
Maximum transmission power of the device p_{max}	1 W
Maximum computing rate of the device f_{max}	1.5 GHz

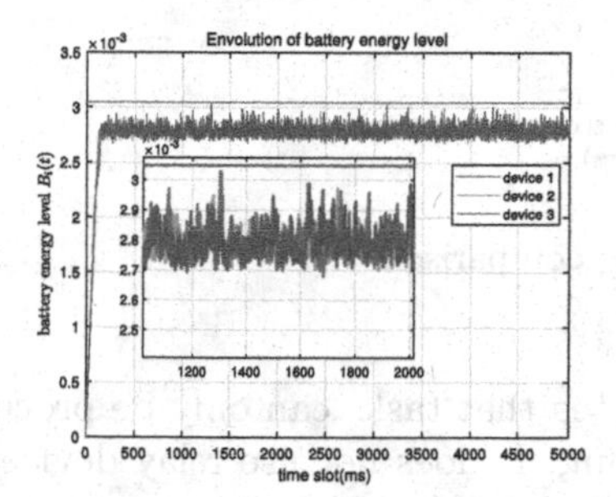

(a) The IoT devices battery energy changes with time.

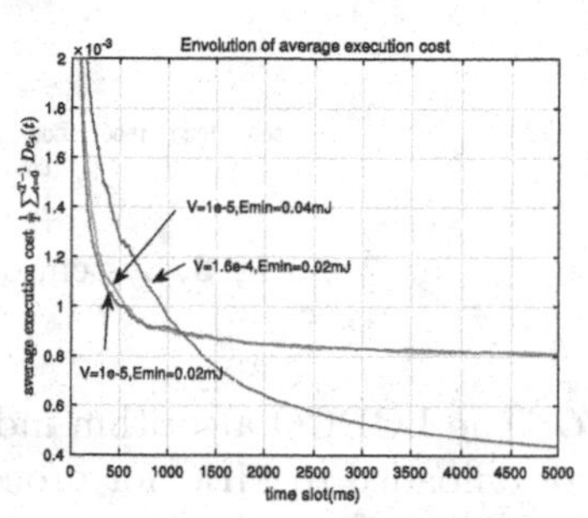

(b) The effect of V and E_{min} on the average delay of the system

Fig. 2. Effect of system parameters on battery energy level and delay.

In Fig. 2(a) we randomly select three IoT devices and show the change of battery energy of each device over time. The number of tasks in each time slot is a random number [1000,3000]. It can be seen from the figure that the battery has been harvesting energy for the first 200 time periods and does not provide energy to the device. The curve is growing. This is because the battery of the IoT device is always in a low power state at the beginning, which cannot provide enough energy to the device, resulting in the processing time of the device being greater than the deadline, and therefore the task is discarded. Subsequently, the battery collects enough energy and stabilizes around the disturbance energy level (i.e.,$\vartheta + E_{max}$). The stabilization process effectively protects the battery and prolongs its service life. It also reduces the number of discarded tasks.

Figure 2(b) depicts the average delay of IoT devices as a function of each parameter. According to the figure, the larger the control parameter V or the smaller the value of E_{min}, the smaller the average delay of the system. Firstly, the larger V is, the closer the solution of P_2 is to the optimal solution, and therefore, the smaller is the average delay of the system. Secondly, a smaller E_{min} leads to a larger perturbation parameter of the system, which leads to a higher stable battery energy level. When the battery has enough battery power, the latency of the system to process the task is also lower.

Figure 3 shows the comparison of the system delay under different methods, where

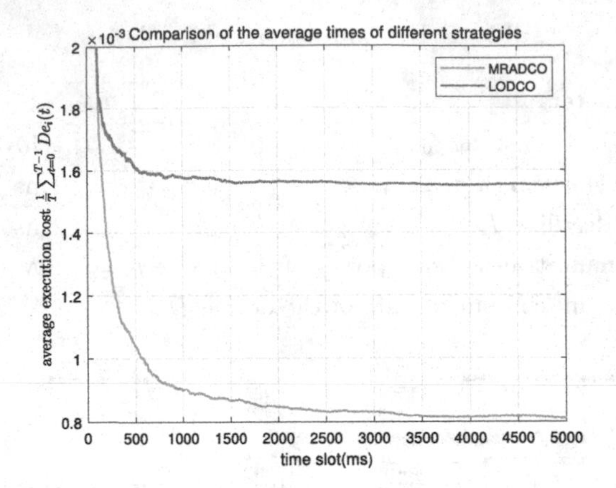

Fig. 3. System delay comparison

- *LODCO*: The LODCO algorithm indicates that tasks can only be processed locally or offloaded to MEC for processing. It does not use relay devices for task offloading [18].

As shown in Fig. 3, we present a comparison between the MRADCO algorithm and the LODCO algorithm. It is evident from the graph that the MRADCO algorithm saves 50% more time compared to the LODCO algorithm when processing the same task. This advantage stems from our MRADCO algorithm's capability to partition tasks into multiple sub-tasks and transmit them to the MEC server via relays. This phased transmission effectively reduces the task load on each transmission path, thereby lowering the required time for task transmission. In contrast, the LODCO algorithm can only handle tasks locally on the device or on the MEC server. Processing intensive tasks locally takes longer due to the limited computational capacity of the local device. When tasks are offloaded directly to the MEC server, without the support of relaying devices, although the IoT transmission power remains unchanged, the need to transmit the entire task at once results in longer transmission times, especially over longer distances. In summary, the MRADCO algorithm optimizes task distribution during transmission through task segmentation and relay transmission strategies, significantly reducing transmission time. In comparison, the LODCO algorithm faces limitations in processing intensive tasks and conducting long-distance transmissions, thereby falling short of achieving the time-saving effectiveness demonstrated by the MRADCO algorithm.

5 Conclusion

In this paper, we study the delay minimization problem during multi-relay assisted multi-user offloading tasks. And energy harvesting devices is added to the model, which collects green energy in the environment to provide energy for

IoT devices. We formulate the problem as an MDP problem. The problem is solved using the weighted perturbation method in the Lyapunov optimization technique. The algorithm named MRADCO is designed. Simulation results and theoretical derivation show that the solution of the algorithm is infinitely close to the optimal solution.

Acknowledgment. This work was supported in part by all of: i) Ningbo Natural Science Foundation (Grant 2021J070), ii) Zhejiang Natural Science Foundation (Grant LY20F010004) and iii) National Natural Science Foundation of China (Grant 61801254).

References

1. Chen, Z., He, Q., Liu, L., Lan, D., Chung, H.-M., Mao, Z.: An artificial intelligence perspective on mobile edge computing. In: 2019 IEEE International Conference on Smart Internet of Things (SmartIoT), Tianjin, China, pp. 100–106 (2019)
2. Mao, Y., You, C., Zhang, J., Huang, K., Letaief, K.B.: A survey on mobile edge computing: the communication perspective. IEEE Commun. Surv. Tutor. **19**(4), 2322–2358, Fourthquarter (2017)
3. Gutierrez, C.A., Caicedo, O., Campos-Delgado, D.U.: 5G and beyond: past, present and future of the mobile communications. IEEE Lat. Am. Trans. **19**(10), 1702–1736 (2021)
4. Zhang, W., Zhang, G., Mao, S.: Joint parallel offloading and load balancing for cooperative-MEC systems with delay constraints. IEEE Trans. Veh. Technol. **71**(4), 4249–4263 (2022)
5. Hashash, O., Sharafeddine, S., Dawy, Z.: MEC-based energy-aware distributed feature extraction for mHealth applications with strict latency requirements. In: 2021 IEEE International Conference on Communications Workshops (ICC Workshops), Montreal, QC, Canada, pp. 1–6 (2021)
6. Xu, Y., Zhang, T., Yang, D., Xiao, L.: UAV-assisted relaying and MEC networks: resource allocation and 3D deployment. In: 2021 IEEE International Conference on Communications Workshops (ICC Workshops), Montreal, QC, Canada, pp. 1–6 (2021)
7. Zhuang, Y., Li, X., Ji, H., Zhang, H.: Optimization of mobile MEC offloading with energy harvesting and dynamic voltage scaling. In: 2019 IEEE Wireless Communications and Networking Conference (WCNC), Marrakesh, Morocco, pp. 1–6. (2019)
8. Li, X., Bi, S., Quan, Z., Wang, H.: Online cognitive data sensing and processing optimization in energy-harvesting edge computing systems. IEEE Trans. Wireless Commun. **21**(8), 6611–6626 (2022)
9. Li, M., Zhou, X., Qiu, T., Zhao, Q., Li, K.: Multi-relay assisted computation offloading for multi-access edge computing systems with energy harvesting. IEEE Trans. Veh. Technol. **70**(10), 10941–10956 (2021)
10. Fu, S., Zhou, F., Hu, R.Q.: Resource allocation in a relay-aided mobile edge computing system. IEEE Internet Things J. **9**(23), 23659–23669 (2022)
11. Zhang, K., Gui, X., Ren, D., Li, J., Wu, J., Ren, D.: Survey on computation offloading and content caching in mobile edge networks. J. Softw. **30**(8), 2491–2516 (2019)

12. Deng, Y., Chen, Z., Chen, X., Fang, Y.: Task offloading in multi-hop relay-aided multi-access edge computing. IEEE Trans. Veh. Technol. **72**(1), 1372–1376 (2023)
13. Zhao, H., Deng, S., Zhang, C., Du, W., He, Q., Yin, J.: A mobility-aware cross-edge computation offloading framework for partitionable applications. In: 2019 IEEE International Conference on Web Services (ICWS) (2019)
14. Wang, X., et al.: Dynamic resource scheduling in mobile edge cloud with cloud radio access network. IEEE Trans. Parallel Distrib. Syst. **29**(11), 2429–2445 (2018)
15. Chen, Y., Zhang, N., Zhang, Y., Chen, X., Wu, W., Shen, X.: Energy efficient dynamic offloading in mobile edge computing for internet of things. IEEE Trans. Cloud Comput. **9**(3), 1050–1060 (2021)
16. Zhang, S.Q., Lin, J., Zhang, Q.: Adaptive distributed convolutional neural network inference at the network edge with ADCNN. In: Proceedings of 49th International Conference on Parallel Process, pp. 1–11 (2020)
17. Tong, Z., Cai, J., Mei, J., Li, K., Li, K.: Dynamic energy-saving offloading strategy guided by lyapunov optimization for IoT devices. IEEE Internet Things J. **9**(20), 19903–19915 (2022)
18. Mao, Y., Zhang, J., Letaief, K.B.: Dynamic computation offloading for mobile-edge computing with energy harvesting devices. IEEE J. Sel. Areas Commun. **34**(12), 3590–3605 (2016)

A Cloud Computing User Experience Focused Load Balancing Method Based on Modified CMA-ES Algorithm

Jihai Luo[1,2], Chen Dong[1,2(✉)], Zhenyi Chen[3], Li Xu[4], and Tianci Chen[4]

[1] College of Computer and Data Science - Fuzhou University, Xueyuan Road, Fuzhou 360108, Fujian, China
dongchen@fzu.edu.cn

[2] Fujian Key Laboratory of Network Computing and Intelligent Information Processing, Fuzhou University, Fuzhou, China

[3] Department of Computer Science and Engineering, University of South Florida, Tampa, FL 33620, USA
zhenyichen@usf.edu

[4] College of Computer and Cyber Security, Fujian Normal University, Fuzhou 350007, China
xuli@fjnu.edu.cn

Abstract. The development of the software and hardware has brought about the abundance and overflow of computing resources. Many Internet companies can lease idle computing resources based on the peak and valley cycles of usage load to provide IaaS service. After the surging development, more and more companies are gradually paying attention to the specific user experience in cloud computing. But there are few works about that aspect in load balance problem. Therefore, we consider the load balancing resource allocation problem, from the perspective of user experience, mainly based on geographical distance and regional Equilibrium, combined with user usage habits, in multiple service periods. A detailed mathematical definition has been established for this problem. This article proposes a mathematical model for the problem, along with an modified CMA-ES (Covariance Matrix Adaptation Evolution Strategy) algorithm, called WS-ESS-CMA-ES algorithm, to allocate computing resources and solve the above problem. The process of using the proposed algorithm to solve the problem of cloud computing resource allocation in real scenarios is simulated, and compared with some other algorithms. The experiment results show that our algorithm performs well.

Keywords: Cloud computing · Optimization algorithm · Evolutionary strategy

This work is supported by the fund of Fujian Province Digital Economy Alliance, the National Natural Science Foundation of China (No. U1905211), and the Natural Science Foundation of Fujian Province (No. 2020J01500).

H. Jin et al. (Eds.): GPC 2023, LNCS 14504, pp. 47–62, 2024.
https://doi.org/10.1007/978-981-99-9896-8_4

1 Introduction

The concept of cloud computing was first proposed in SES San Jose 200 by Eric Emerson Schmidt. Cloud computing service has three levels of types: SaaS, PaaS and IaaS. IaaS (Infrastructure as a service) commonly means that service provider make computing devices as a service and rent it for consumers' various uses. Many companies have developed IaaS services and derive significant benefits from them [1].

The load balance problem of cloud computing has been studied by many researchers. [2] summarizes a large number of load balancing methods and divides them into 7 types, and also concludes the metrics for load balancing.

Many works focus on macro metrics of entire service system, such as, throughput, response time, makespan, migration time, energy consumption etc. These metrics may be somehow not independent on specific software and hardware condition, could not be a pervasive references for various system settings. And much metrics are not directly related to the individual user experience. Besides, few metrics could be embedded with artificial factors that abstractly descript a specific system, and then recalculated. Therefore, this work proposed a original load balancing method, mainly based on two original and pervasive metrics that have a direct and intimate relationship with the specific experience of individual users: usage habit prioritized average access cost, and weighted regional Equilibrium degree.

According to [3], to solve the problem of cloud computing resources allocation with load balancing, there are two types of strategies: dynamic and artificial intelligence. The dynamic strategy predict the nature of consumer's task [4], and the artificial intelligence imitate nature to schedule tasks among resources [5]. ES (Evolution strategy) algorithm is an important branch of artificial intelligence. The original CMA-ES [6] is an evolution strategy that is designed for solving difficult non-linear non-convex optimization problems in continuous domains, and it is still developing till now [7]. Beside, CMA-ES is one of the most famous, widely used, and high-performance ES [8], and has been used for hundreds of applications.

However, the work of load balancing technology based on user experience is still in a relatively vacant state currently. Consequently, this work proposes a novel variant of CMA-ES algorithm, which is extended to be used for discrete problem, and with respect to the specific cloud computing resources allocation problem, could take the real-time conclusional nature of consumers' uses into consideration and make a dynamic solution. The proposed access cost reducing and load balancing method is based on this modified CMA-ES algorithm.

The main contributions of this work are listed as follows:

1. Two pervasive metrics for load balancing of cloud computing, could be used regardless of the limitation of system condition, and customized by artificial factors to adapting to the specific system requirements. These metrics extend IaaS load balance problem issue to a specific user perspective.
2. A detailed, parameterized mathematical model for cloud computing load balancing resources allocation problem, which could consider and apply the

above metrics, and then provide solutions for multiple adjacent service periods.

3. A novel modified CMA-ES algorithm method, to solving resource allocation problem under above mathematical model, and could use prior knowledge and accelerate optimization speed in the later stages of iteration. This algorithm is adaptable and efficient, and the experiment results show that it has a good performance.

2 Related Work

This work proposed a method based on modified CMA-ES algorithm to solve the load balance problem of infrastructure allocation problem in cloud computing service. Therefore, this section mainly consist of two parts: cloud computing load balance problem, and CMA-ES algorithms, to provide horizontal and vertical references.

2.1 IaaS Cloud Computing Load Balance Problem

load balance problem is a popular and difficult problem in IaaS cloud computing scheme. The load balance problem could be defined to that reducing gradients in user tasks and resource utilization to improve the performance and efficiency of computing resources [9].

To evaluate the quality of load balancing, [2] summarized the most popular 10 metrics of load balance problem in cloud computing: throughput, response time, makespan, scalability, fault tolerance, migration time, degree of imbalance, performance, energy consumption, carbon emission.

Evolution strategy is the most commonly used method for this problem [10]. A GWO (Grey Wolf Optimization) algorithm based task allocation method for Load balancing, especially in containerized cloud was proposed in [11]. A load balancing technique by using modified PSO task scheduling (LBMPSO) to achieve optimal utilization of resource was proposed in [12]. [13] proposed a hybridization of meta-heuristic algorithm load balancing method, which integrates modified Particle swarm optimization (MPSO) and improved Q-learning algorithm, called QMPSO. This method had been proven to have better robustness and performs well under a range of metrics.

The load balancing problem in IaaS involves allocating hardware resources to users, which could be regarded as a power-aware resource allocation problem. The core concept of power-aware task or resource scheduling problem is to optimize energy consumption under multi constraint and multi variable conditions [14,15]. These work inspire us that in the IaaS scenario, user experience can be seen as a form of energy consumption. It was mentioned in [3] that the allocation of cloud computing resources is a complex problem due to the presence of heterogeneous application workloads having contentious allocation requirements in terms of resource capacities (resource utilization, execution time, response time, etc.). This work provides an idea for us to a perspective of load balancing based on user usage habits and practical experience.

2.2 CMA-ES Algorithms

CMA-ES algorithm is first proposed by [6], and developed and modified into many varieties.

[16] proposed a modified CMA-ES algorithm embeded with two additional mechanisms, called IPOP-CMA-ES. These two mechanisms are restart criteria and increasing population. This method greatly improves in multimode and composite Class function, and also has the characteristic of strong global search ability. [17] proposed a warm starting variety of CMA-ES algorithm, which could improve the issue of long adaptation phase mainly based on prior knowledge, called WS-CMA-ES. [18] proposed a competitive variable-fidelity surrogate-assisted CMA-ES (CVFS-CMA-ES) algorithm using data mining techniques to improve the computational efficiency and global optimizing ability. [19] proposed HTPC, which is a hybrid method of PSO and CMA-ES, and could complementarily solving the problem of PSO being misled by local information and CMA-ES using poorly utilized global information.

3 Problem Description

This section introduces a formulated cloud computing resource allocation problem, and then proposes two metrics as optimization objective in mathematical description. Our proposed scheme is majorly based on the load balance problem defined in [9], but also revolving the specific user experience perspective, to let it become a underlying optimization objective.

Figure 1 is a schematic diagram of the mathematical model for the above cloud computing resource allocation problem. Our method assumes a framework that uses cloud servers to manage user requests from different geographical locations. The user's location is referred to as RSA (Requirement Sending Place), and the unit device is referred to as a node, taking into account the three dimensions of geographic latitude and longitude and historical service cycle.

Consider a cloud computing service system distributed across g discrete geographic regions, which could be represented by many coordinates $Rg = (x, y)$. The whole system could be indicated as a region set $G = \{Rg_1, Rg_2, ..., Rg_g\}$. And a single region has many cloud computing infrastructures. These infrastructures are abstracted into several resource nodes. Correspondingly, a region with n nodes could be indicated as $S = \{N_1, N_2, ..., N_n\}$. In fact, the terminals in cloud computing services are often not the same. Different terminals mean different response speeds, such as long-distance transmission and download speeds, which are closely related to the user experience. Here, the overall quality of each resource node's terminal is presented as a quality score in a scoring set: $Q = \{Qs_1, Qs_2, ..., Qs_n\}$.

Besides, current resource requests under the same user account are considered as one task. Assuming there are m tasks waiting for cloud computing resource allocation. An the task set could be indicated as $TS = \{T_1, T_2, ..., T_m\}$.

And for each task, a set of requirement sending addresses $R = (x, y)$, is recorded as follows: $RS = \{R_1, R_2, ..., R_p\}$, where p is the count of requirement

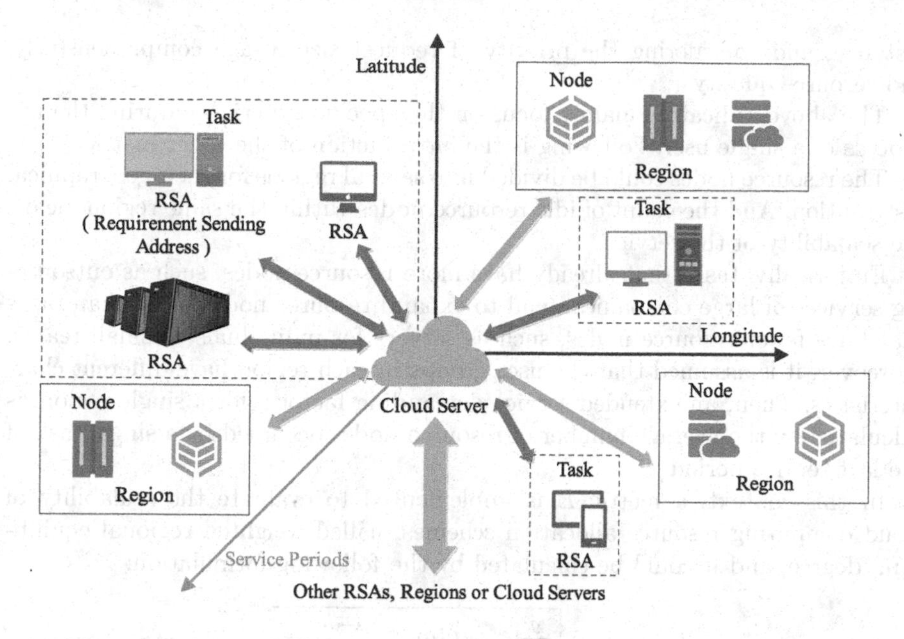

Fig. 1. The framework of cloud computing load balance problem model

sending addresses. Moreover, when a new task is coming, each node of its requirement has a description value pair of usage habit $D = (Af, Sr)$, of which Af: access frequent, and Sr represents for terminal standby time ratio. They could be calculated as total number of network requests in one time round, and the ratio of the time of terminal remains open to the total time of one time round correspondingly. All of these pairs form a description set, which could be recorded as $DS = \{D_1, D_2, ..., D_r\} = \{(Af_1, Sr_1), (Af_2, Sr_2), ..., (Af_r, Sr_r)\}$. These description sets come from the usage information of historical service cycles recorded by cloud servers.

To calculate the cost of an access from a requirement sending place $R = (x_r, y_r)$ to a resource node $P = (x_p, y_p)$, the geolocation distance of them is calculated by Minkowski distance of coordinate points, and used as an intermediate metric:

$$dis(R, P) = \sqrt{(x_r - x_p)^2 + (y_r - y_p)^2} \tag{1}$$

Consider the overall balance of access times for each task, and prioritize it in conjunction with description value pair of usage habit, moreover, the infrastructure quality of allocated nodes, the usage habit prioritized average access cost is introduced as follows:

$$COST = \sum_{t=1}^{m} \sum_{r=1}^{p} dis \cdot \frac{p \cdot Af_r}{m \cdot Sr_r \cdot Qs_r} \tag{2}$$

Above is one of the optimization objectives of our method. The value of $COST$ represents for the relatively average access cost based on geolocation distance, and considering the priority of request size, usage comprehensively, and resource quality.

The above indicators mainly focus on the specific experience during the use process of a single user. Following is the introduction of the other metric:

The resource nodes could be divided into several regions for their geographical distribution. And the count of idle resource nodes within the same region means the scalability of the service.

Empirically, tasks that already have more resource nodes, such as outsourcing services of large companies, tend to expand resource nodes more than tasks that have fewer resource nodes, such as servers for individuals or small teams. Moreover, it is assumed that the user groups in each region have different characteristics. Then, an extended service demanding factor e for a single region, is calculated by the average number of resource nodes occupied by a single task of the last service period.

In this context, a metric E is implemented to evaluate the scalability of cloud computing resource allocation schemes, called weighted regional equilibrium degree, and it could be calculated by the following formulation:

$$EQ = \sqrt[g]{\prod_{l=1}^{g}(1 + \frac{Cnt_l^{\star}}{Cnt_l}) \cdot \frac{\sum_{l=1}^{g} e_l}{g * e_l}} \tag{3}$$

where $Cnt_l^{\star}$ and Cnt_l stand for the number of idle resource nodes and total number of resources nodes of region l respectively.

Metrics $COST$ and EQ represent user satisfaction from the perspectives of tasks and regions, respectively. The optimization goal of the problem is to decrease $COST$ while increasing EQ, which is a dual objective optimization problem.

Following is the final optimization objective of this problem:

$$OBJ = \alpha \cdot q \cdot EQ + (1 - \alpha) \cdot COST \tag{4}$$

where α is a moderating factor with a range of (0,1), used to balance the significance of metric C and E.

4 Methodology

This section introduces a modified CMA-ES algorithm to solve the problem in Sect. 3. First, we briefly introduce the original CMA-ES algorithm, and secondly, introduce the specific implement and improvement of it for the above problem.

4.1 CMA-ES Algorithm

CMA-ES (Covariance Matrix Adaptation Evolution Strategy) is a evolutionary algorithm, and is an optimization algorithm widely used to solve continuous optimization problems. Evolutionary algorithm is effective when we do not know the exact analysis of the objective function or can not directly calculate the gradient.

The calculation process of CMA-ES algorithm is very similar to Gaussian evolution strategy. Define that $f(x)$ as the fitness value function, $\theta = (\mu, \sigma)$ as the parameter of strategy, and $p_\theta(x) \sim N(\mu, \sigma^2 I) = \mu + \sigma N(0, I)$ as the distribution of solution vector x, and $D = \{(x_i, f(x_i))\}, x_i \sim p_\theta(x)$ as the initial population. The evolution steps are mainly summarized as follows:

1. Initialize the parameter θ and the iteration count t.
2. Sampling offspring populations with a population size of n from a Gaussian distribution:

$$D^{t+1} = \{x_i^{t+1} \mid x_i^{t+1} = \mu^t + \sigma^t y_i^{t+1}, i = 1, ..., n\} \tag{5}$$

where y_i^{t+1} is sampled from a distribution:

$$y_i^{t+1} \sim N(x \mid 0, I) \tag{6}$$

3. Sort D^{t+1} in descending order of fitness value. Then select a subset of λ samples with the best fitness value $f(x)$, as the elite set D_{elite}^{t+1}:

$$D_{elite}^{t+1} = \{x_i^{t+1} \mid x_i^{t+1} \epsilon D^{t+1}, i = 1, ..., \lambda, \lambda \leq n\} \tag{7}$$

4. Calculating μ and σ with respect to D_{elite}^{t+1}:

$$\mu^{t+1} = avg(D_{elite}^{t+1}) = \frac{1}{\lambda} \sum_{i=1}^{n} x_i^{t+1} \tag{8}$$

$$\sigma^{2,t+1} = var(D_{elite}^{t+1}) = \frac{1}{\lambda} \sum_{i=1}^{\lambda} (x_i^{t+1} - \mu^t)^2 \tag{9}$$

5. Increase the value of t: $t = t + 1$. Then repeat steps 2–4 until the iteration count t reaches the threshold or the result is good enough.

The uniqueness of CMA-ES is to replace the distribution $p_\theta(x)$ in the above process with the distribution based on Covariance matrix C:

$$\theta = (\mu, \sigma, C), p_\theta(x) \sim N(\mu, \sigma^2 C) = \mu + \sigma N(0, C) \tag{10}$$

The Covariance matrix C is defined as follows:

$$C = \begin{bmatrix} c_{11} & c_{12} & \cdots & c_{1d} \\ c_{21} & c_{22} & \cdots & c_{2d} \\ \vdots & \vdots & \ddots & \vdots \\ c_{d1} & c_{d2} & \cdots & c_{dd} \end{bmatrix}, c_{ij} = Cov(X_i, X_j) \tag{11}$$

where d is the count of dimensions of x, $Cov(a,b) = E[(a - E(a))(b - E(b))]$ means the covariance of variables a, b, and X_i and X_j are the i th and j th dimension of x respectively. All of these dimensions could also be regarded as variables of the optimization problem.

Specifically, if μ_i and σ_i are respectively represented as the mean and variance of the i th variable, and take case of two variables, x and y, as an example, and adaptively set the elite set to the top 25% solution with the highest fitness based on the population size, then the specific calculation process of covariance matrix is as follows:

$$\mu_x^{t+1} = \frac{1}{N_{best}} \sum_{i=1}^{N_{best}} x_i, \mu_y^{t+1} = \frac{1}{N_{best}} \sum_{i=1}^{N_{best}} y_i \tag{12}$$

where N_{best} is set to be the best 25% of solutions.

Then, each covariance could be calculated:

$$\sigma_x^{2,t+1} = \frac{1}{N_{best}} \sum_{i=1}^{N_{best}} (x_i - \mu_x^t)^2 \tag{13}$$

$$\sigma_y^{2,t+1} = \frac{1}{N_{best}} \sum_{i=1}^{N_{best}} (y_i - \mu_y^t)^2 \tag{14}$$

$$\sigma_{xy}^{t+1} = \frac{1}{N_{best}} \sum_{i=1}^{N_{best}} (x_i - \mu_x^t)(y_i - \mu_y^t) \tag{15}$$

In the case of other problem dimensions, the calculation method of C is similar. Based on this calculation, the CMA-ES algorithm is obtained by replacing the sampling process in step 2 with the sampling process based on the mean μ^{t+1} and covariance matrix C^{t+1}.

The covariance matrix describe the relativeness of multiple variables, and it can closely fit the expected distribution of the sample to provide more detailed references for the next iteration. The CMA-ES algorithm can adaptively change the extent of exploration, increase the standard deviation of the search space when far from the target solution, and fine-tune the solution when it is certain that it is close to the optima.

4.2 Modified CMA-ES and Its Implement

The first problem for us is that CMA-ES algorithm could not be directly used for discrete problem. The problem is to allocate every cloud computing task to a single resource node with various quality score, and it obviously is a discrete problem.

CMA-ES output a continuous solution in multiple dimension. Let each dimension stands for the choice of resource nodes for every single requirement sending address, and remind that, in the problem description section, there are resource nodes with different quality scores. Following is the method to use continuous solution to express the choice of this specific problem:

Restrict the value range of solution as $[0, n)^s$, where n and s represent for the number of resource nodes and the number of require sending places respectively. When considering the allocation of resource nodes according to the value of each dimension, the requests with the same integer number are assigned to the same resource node, and the resource quality is determined according to the rank of the specific value in ascending order. For example, if the solution is $(1.154, 1.455, 2.954)$, and there are 2 regions, the first region has 2 nodes with quality scores of $(90, 80)$, and the second region has 1 node. Then the first RSA and second RSA are allocated to region 1 for the values of integer bit of the first and second dimension of the solution, and the third RSA is allocated to region 2 for the same reason. Moreover, the first RSA is allocated to node 1 for its smaller value of corresponding dimension in solution.

Figure 2 shows the resource allocation according to a continuous solution vector and in a simple case of 3 regions and 8 RSAs.

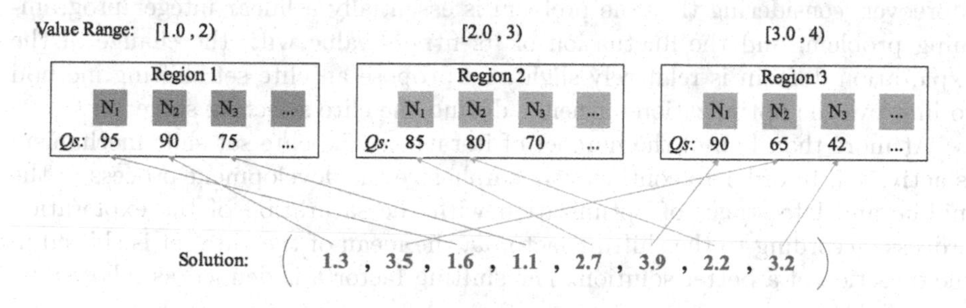

Fig. 2. The framework of cloud computing load balance problem model

4.3 Warm Starting Method

Intuitively, there is the similarity in system circumstance and user characteristics between adjacent service cycles. Motivated by it, we used a population initialization process to achieve warm start of CMA-ES for each service period, which take the information of past service periods as prior knowledge. In other words, we can generate an estimate of the allocation method for the next service period of new tasks as the initial solution based on the resource allocation pattern up to the previous service period.

Following is the detailed description of this process:

At the beginning of an evolutionary, the population is initialized based on the geographical location of tasks, and the usage habit description in the previous round. To achieve this step, a task-node similarity metric is defined as follows:

$$Sim(T, Rg) = -\sum_{i=1}^{p}\sum_{j=1}^{n}\frac{dis(i,j)}{n}\cdot Cos(\bar{D}_{Rg}, D_T) \tag{16}$$

where $Cos(x, y)$ means cosine distance, and $\bar{D}_{Rg}$ represents the average value of usage habits description values for all tasks in the region, and D_T represents the usage habits description values for task users. If it is the first service period, then simply make $Cos(\bar{D}_{Rg}, D_T) = 1$.

Then, generate solutions as pre-elite set that account for 25% of the initial set, and the integer bits of these solutions determine the allocation of each task to the region with the best similarity, with the decimal places randomly and uniformly generated. For each solution, allocation is based on the priority of RSA-resource node similarity. If a node that a RSA tempts to allocate to is already occupied, the RSA is allocated to the node with the next highest similarity, until successful allocation.

Other part of initial set is made of random feasible solutions.

4.4 Elite Set Shifting Mechanism

Moreover, considering that the problem is essentially a linear integer programming problem, and the fluctuation of its fitness value with the change of the exploration domain is relatively slight, we propose an elite set shifting method to improve the optimization efficiency during the elite selecting stage:

At more than half of the number of iterations, the elite set shift mechanism is activated. In order to continuously emphasize the development process in the middle and late stages of optimization with the saturation of the exploration process, according to the shifting factor ω, the mean of the elite set is shifted in the direction of a better solution. The shifting factor a is defined as follows:

$$\omega = 1 - \phi \cdot e^{-(\frac{t}{Tmax} - 0.5)}, t > 0.5 * Tmax \tag{17}$$

where ϕ is a scaling factor, could be set in the range of $[0.9, 0.95]$, and t is the iteration time, while $Tmax$ represents the max iteration time.

After the elite set is derived, an offset mean μ^o is calculated in place of the original mean μ. The calculation equation is as follows:

$$\mu^o = \sum_{i=1}^{N_{best}} [1 + \omega * (\frac{f(x_i)}{\sum_{i=1}^{N_{best}} f(x_i)} - \frac{1}{N_{best}})] \tag{18}$$

After that, the offset mean μ^o and covariance matrix C^{t+1} are used for the next round of sampling generation. This is Equivalent to "shifting" the elite set to a more optimal solution.

4.5 Steps of Proposed WS-ESS-CMA-ES

Now the solution of modified CMA-ES algorithm could be transformed to the arrangement of all resource nodes and tasks. Following is the steps of modified CMA-ES algorithm to solve this problem:

1. Generate pre-elite set using the information from the last service period, and the combined randomly generated solution forms the initial population.

2. Calculate the fitness value for each solution. And then get the elite set from the 25% solutions with best fitness value.
3. Determine whether the number of iterations is more than half of the maximum number of iterations. If not, the mean and variance of elite set are calculated. If so, the mean of the offset variance is calculated for elite set.
4. Based on the mean or cheap mean obtained in the previous step, and the variance, generate the initial population for the next iteration.

5 Experiment Results

The experiment was executed in a personal laptop, utilizing the proposed WS-ESS-CMA-ES algorithm to solve the load balance problem in IaaS cloud computing, with resource nodes in different regions and multiple user tasks. And different problem sizes were considered and tested. All the result values are derived from the average of 100 times of same experiment cases. In our experiment, PSO-GWO [20,21], WOA [22], GOA [23] algorithms have been implemented and applied to the same problem. These algorithms are all excellent swarm intelligence algorithms in recent years, among which the PSO-GWO algorithm combines PSO and GWO, using two-stage calculations to achieve good early optimization speed and late convergence ability. WOA and GOA respectively simulate whale predation and locust migration, with fewer parameters, simple implementation, and trends to jump out of local optima. We compared the proposed algorithms with their comprehensive optimization objectives of $COST$ and EQ, considering only the first round of service period, and conducted repeated comparative experiments on different problem scales. During the experiment, parameters α and ω were set to 0.5 and 0.95 based on comparison of algorithm performance in pre experiments, respectively. In addition, we also conducted ablation experiments on improving warm staring and elite set shifting, which proved their effectiveness. For algorithms that could not directly handle discrete problems, we will use a similar method in Sect. 4.2 for continuation processing.

Among the experiment, $size$ represents the number of regions, and the number of RSAs is set to $5 * size$, belonging to $3 * size$ tasks. The number of resource nodes is set to $10 * size$, belonging to $size$ regions. The quality score of resource nodes is randomly generated by distribution $N(75, 10)$, and the two bits of the description set used by users are randomly generated by distribution $EXP(100)$ and $N(0.5, 0.4)$, respectively. The positions of all resource nodes and request locations are generated in a square with a center at the origin and a side length of $5 * size$. All regions are divided using recursive random region segmentation, and all resource nodes are in a certain region. The max iteration count $Tmax$ is set as $100 * size$, and the population size n is set as $10 * size$.

Table 1 and Table 2 shows the parameter setting of the proposed modified CMA-ES algorithm, where constants are represented by values, and variables are represented by distributions.

Table 1. Constant parameter settings of ES-WSS-CMA-ES algorithm and problem

Parameter	Value
α	0.5
ω	0.95
problem size	$size = 10, 20, 30$
region count	$size$
resource node count	$10 * size$
task count	$size$
RSA count	$5 * size$
side length of map	$5 * size$
max iteration count	$100 * size$

Table 2. Random variable parameter settings of ES-WSS-CMA-ES algorithm and problem

Parameter	Distribution
quality score of resource node	$N(75, 10)$
access frequent of task	$EXP(100)$
stand time ratio of task	$N(0.5, 0.4)$

The algorithm performance comparison of optimization objective value under different problem scales is shown in Fig. 3a, Fig. 3b, and Fig. 3c. Figure 3a is the case of only considering $COST$, Fig. 3b is the case of only considering EQ, and Fig. 3c is the case of considering them both (Because the map size is affected by the problem size $size$, EQ is normalized by multiplied by $size$). Figure 3a illustrates the comparison of the optimization results of $COST$, which represents for the prioritized average access cost. The optimization of $COST$ is a minimization problem. It can be found that WS-ESS-CMA-ES algorithm has a better average $COST$ after $Tmax$ iterations, and it is more pronounced when the problem scale is large. Figure 3b illustrates the comparision of the optimization results of EQ, which represents for the load region Equilibrium of the cloud computing system. The optimization of EQ is a maximization problem. In this figure, it can be found that WS-ESS-CMA-ES algorithm has a better average EQ after $Tmax$ iterations. Figure 3c illustrates the comparision of the optimization results of OBJ, which represents for the hybrid of $COST$ and EQ. The optimization of OBJ is a minimization problem. In conclusion, this figure shows that WS-ESS-CMA-ES algorithm has a better solution quality than other algorithms.

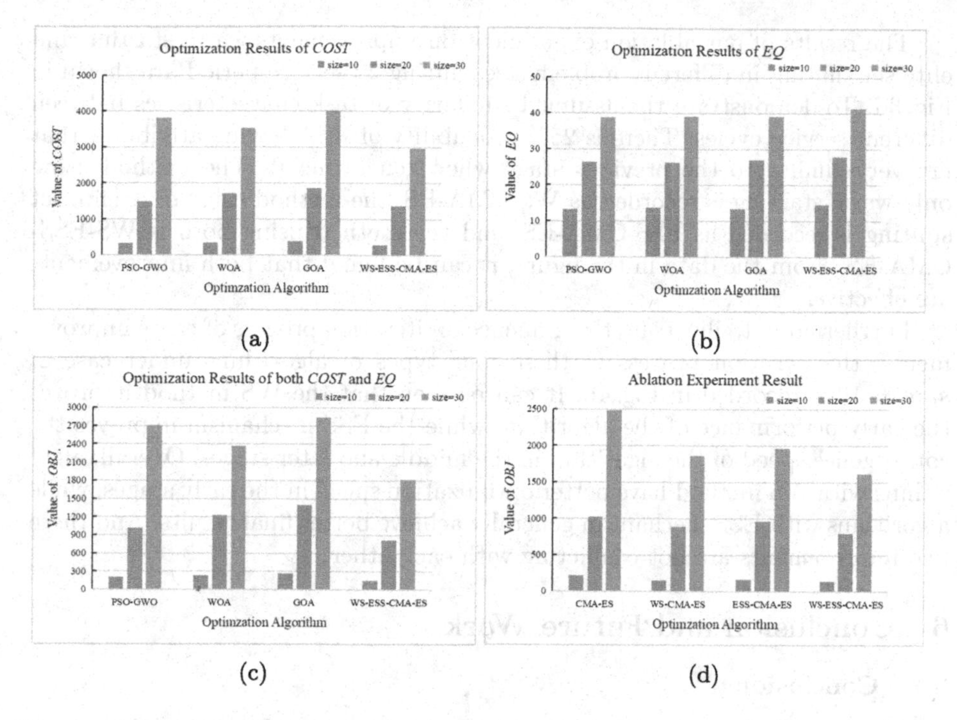

Fig. 3. Algorithm performance comparison on optimization objectives. This set of figures could visually compare the quality of the final solution between the ES-WSS-CMA-ES algorithm cluster and other SOTA algorithms in a limited number of iterations.

Figure 4a visually illustrates the optimization process of the target value of the case of $size = 10$. It can be found that the curve of WS-ESS-CMA-ES algorithm is lower and smoother, which indicates that the WS-ESS-CMA-ES algorithm is more efficiency and less sensitive to iteration count.

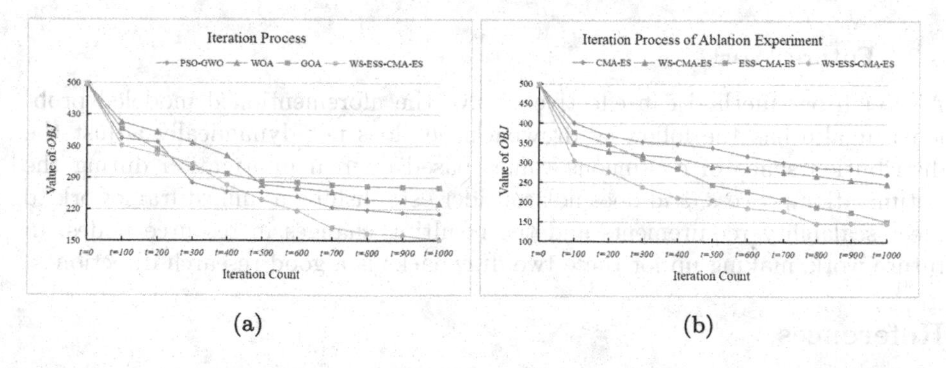

Fig. 4. Algorithm performance comparison on iteration process. This set of figures can demonstrate the influence of the WS method and the ESS mechanism on the iteration process.

The results of our ablation experiment on improvements warm starting and elite set shifting in different problem sizes during 20 service period are shown in Fig. 3d. To demonstrate the assumed similarity of task characteristics between different service cycles, There is 25% probability of RSA having attributes that are very similar to the previous stage when generating it. The method using only warm starting is recorded as WS-CMA-ES, the method using only elite set shifting is recorded as ESS-CMA-ES, and the method using both is WS-ESS-CMA-ES. From the data in the figure, it can be found that both improvements are effective.

Furthermore, to illustrate the influence on iteration process of these improvements, the iteration process of these four types of algorithms under case of $size = 10$ is recorded in Fig. 4b. It can be seen that the WS method improves the early performance of the algorithm, while the ESS mechanism improves the convergence speed of the algorithm in the middle and later stages. Overall, algorithms with WS method have better optimization speed in the early stages, while algorithms with ESS mechanism generally achieve better final results. And these two improvements are not conflicting with each other.

6 Conclusion and Future Work

6.1 Conclusion

In this work, we defined and modeled the load balance problem in cloud computing resource allocation, taking into account information from three dimensions: geographic latitude and longitude, and historical service cycles. The problem resulting from modeling can be seen as a discrete problem of integer programming. The proposed WS-ESS-CMA-ES algorithm can effectively solve the above problems, and experimental results show that compared to other classic algorithms, such as genetic algorithm, particle swarm optimization, gray wolf optimization algorithm, etc., our proposed algorithm has better performance and optimization efficiency and the improvements of warm starting and elite set shifting are effective.

6.2 Future Work

Although our method can effectively solve the aforementioned modeled problems, it also has the following drawbacks: it does not dynamically adjust the distribution space of continuous values based on real information during the optimization process, and does not consider establishing a unified framework to meet scalability requirements and the resulting changes in resource nodes. In future work, making up for these two drawbacks is a good research direction.

References

1. Madni, S.H.H., Latiff, M.S.A., Coulibaly, Y., et al.: Resource scheduling for infrastructure as a service (IaaS) in cloud computing: challenges and opportunities. J. Netw. Comput. Appl. **68**, 173–200 (2016)

2. Ghomi, E.J., Rahmani, A.M., Qader, N.N.: Load-balancing algorithms in cloud computing: a survey. J. Netw. Comput. Appl. **88**, 50–71 (2017)
3. Abid, A., Manzoor, M.F., Farooq, M.S., Farooq, U., Hussain, M.: Challenges and issues of resource allocation techniques in cloud computing. KSII Trans. Internet Inf. Syst. **14**(7) (2020)
4. Dong, C., Luo, J.H., Hong, Q., Chen, Z., Chen, Y.: A dynamic distributed edge-cloud manufacturing with improved ADMM algorithms for mass personalization production. J. King Saud Univ. Comput. Inf. Sci. **35**(8), 101632 (2023)
5. Chen, X., Ye, Y., Dong, C., Chen, Z., Huang, Y.: Grasshopper optimization algorithm combining gaussian and chaos theory for optimization design. In: 2019 3rd International Conference on Electronic Information Technology and Computer Engineering (EITCE) (2019)
6. Hansen, N., Ostermeier, A.: Completely derandomized self-adaptation in evolution strategies. Evol. Comput. **9**(2), 159–195 (2001)
7. Biedrzycki, R.: Handling bound constraints in CMA-ES: an experimental study. Swarm Evol. Comput. **52**, 100627 (2020)
8. Hansen, N.: The CMA evolution strategy: a tutorial. arXiv preprint arXiv:1604.00772 (2016)
9. Afzal, S., Kavitha, G.: Load balancing in cloud computing-a hierarchical taxonomical classification. J. Cloud Comput. **8**(1), 22 (2019)
10. Kumar, P., Kumar, R.: Issues and challenges of load balancing techniques in cloud computing: a survey. ACM Comput. Surv. (CSUR) **51**(6), 1–35 (2019)
11. Patel, D., Patra, M.K., Sahoo, B.: GWO based task allocation for load balancing in containerized cloud. In: 2020 International Conference on Inventive Computation Technologies (ICICT), pp. 655–659. IEEE (2020)
12. Arabinda Pradhan and Sukant Kishoro Bisoy: A novel load balancing technique for cloud computing platform based on PSO. J. King Saud Univ. Comput. Inf. Sci. **34**(7), 3988–3995 (2022)
13. Jena, U.K., Das, P.K., Kabat, M.R.: Hybridization of meta-heuristic algorithm for load balancing in cloud computing environment. J. King Saud Univ. Comput. Inf. Sci. **34**(6), 2332–2342 (2022)
14. Ye, T., Wang, Z.-J., Quan, Z., Guo, S., Li, K., Li, K.: ISAECC: an improved scheduling approach for energy consumption constrained parallel applications on heterogeneous distributed systems. In: 2018 IEEE 24th International Conference on Parallel and Distributed Systems (ICPADS), pp. 267–274. IEEE (2018)
15. Hu, H., et al.: Optimizing the waiting time of sensors in a manet to strike a balance between energy consumption and data timeliness. In: 2019 IEEE 25th International Conference on Parallel and Distributed Systems (ICPADS), pp. 806–813. IEEE (2019)
16. Auger, A., Hansen, N.: A restart CMA evolution strategy with increasing population size. In: 2005 IEEE Congress on Evolutionary Computation, vol. 2, pp. 1769–1776. IEEE (2005)
17. Nomura, M., Watanabe, S., Akimoto, Y., Ozaki, Y., Onishi, M.: Warm starting CMA-ES for hyperparameter optimization. In: Proceedings of the AAAI Conference on Artificial Intelligence, vol. 35, pp. 9188–9196 (2021)
18. Li, Z., Tian, K., Li, H., Shi, Y., Wang, B.: A competitive variable-fidelity surrogate-assisted CMA-ES algorithm using data mining techniques. Aerosp. Sci. Technol. **119**, 107084 (2021)
19. Xu, P., Luo, W., Lin, X., Qiao, Y., Zhu, T.: Hybrid of PSO and CMA-ES for global optimization. In: 2019 IEEE Congress on Evolutionary Computation (CEC), pp. 27–33. IEEE (2019)

20. Teng, Z., Lv, J., Guo, L.: An improved hybrid grey wolf optimization algorithm. Soft. Comput. **23**, 6617–6631 (2019)
21. Fan, X., Ye, Y., Chen, Z., Hong, Z., Qiu, Z., Dong, C.: Combine discussion mechanism and chaos strategy on particle swarm optimization algorithm. In: 2019 IEEE 10th International Conference on Software Engineering and Service Science (ICSESS), pp. 642–645. IEEE (2019)
22. Mirjalili, S., Lewis, A.: The whale optimization algorithm. Adv. Eng. Softw. **95**, 51–67 (2016)
23. Saremi, S., Mirjalili, S., Lewis, A.: Grasshopper optimisation algorithm: theory and application. Adv. Eng. Softw. **105**, 30–47 (2017)

Energy-Efficient Task Offloading in UAV-Enabled MEC via Multi-agent Reinforcement Learning

Jiakun Gao[1], Jie Zhang[2], Xiaolong Xu[1,2(✉)], Lianyong Qi[3], Yuan Yuan[4,5], Zheng Li[6], and Wanchun Dou[2]

[1] School of Software, Nanjing University of Information Science and Technology, Nanjing 210044, China
njuxlxu@gmail.com

[2] State Key Laboratory for Novel Software Technology, Nanjing University, Nanjing 210023, China

[3] College of Computer Science and Technology, China University of Petroleum (East China), QingDao 266580, China

[4] School of Computer Science and Engineering, Beihang University, Beijing 100191, China

[5] State Key Laboratory of Software Development Environment Zhongguancun Laboratory, Beihang University, Beijing 100191, China

[6] School of Computer Science, Nanjing University of Information Science and Technology, Nanjing 210044, China

Abstract. Nowadays, artificial intelligence-based tasks are imposing increasing demands on computation resources and energy consumption. Unmanned aerial vehicles (UAVs) are widely utilized in mobile edge computing (MEC) due to maneuverability and integration of MEC servers, providing computation assistance to ground terminals (GTs). The task offloading process from GTs to UAVs in UAV-enabled MEC faces challenges such as workload imbalance among UAVs due to uneven GT distribution and conflicts arising from the increasing number of GTs and limited communication resources. Additionally, the dynamic nature of communication networks and workload needs to be considered. To address these challenges, this paper proposes a Multi-Agent Deep Deterministic Policy Gradient based distributed offloading method, named DMARL, treating each GT as an independent decision-maker responsible for determining task offloading strategies and transmission power. Furthermore, a UAV-enabled MEC with Non-Orthogonal Multiple Access architecture is introduced, incorporating task computation and transmission queue models. In addition, a differential reward function that considers both system-level rewards and individual rewards for each GT is designed. Simulation experiments conducted in three different scenarios demonstrate that the proposed method exhibits superior performance in balancing latency and energy consumption.

Keywords: Mobile Edge Computing · unmanned aerial vehicles · NOMA · multi-agent deep reinforcement learning

H. Jin et al. (Eds.): GPC 2023, LNCS 14504, pp. 63–80, 2024.
https://doi.org/10.1007/978-981-99-9896-8_5

1 Introduction

The rapid development of artificial intelligence(AI) has led to the emergence of numerous applications with high computational demands and stringent latency requirements. These applications span various domains, including intelligent surveillance, virtual reality, and autonomous driving [1,2]. However, the computational resources and energy provisioning of ground terminals (GTs) are insufficient to meet the increased demands posed by these tasks. Mobile Edge Computing (MEC) has emerged as a new paradigm to tackle this challenge. By deploying MEC servers in close proximity to GTs, tasks can be offloaded from GTs to MEC servers using wireless links [3]. This offloading process helps to reduce the computation latency and energy consumption of resource-constrained GTs [4].

With the advancement of unmanned aerial vehicle (UAV) technology, UAVs can be equipped with MEC servers to provide efficient task offloading services for GTs, leveraging the flexibility of UAVs [5,6]. However, the discrepancy in the number of GTs covered by each UAV leads to an imbalance in workload distribution, resulting in increased computation latency for tasks in densely populated user areas. To mitigate the computational load in high-density user areas, a Fly Ad-hoc Network (FANET) can be utilized to connect multiple UAVs and form an edge service cluster, enabling task offloading among UAVs [7]. Furthermore, existing AI-based tasks can often be partitioned to enable parallel computation across multiple computing nodes, facilitating the parallel execution of tasks [8]. In addition, due to the need for GTs to offload tasks through wireless links, traditional orthogonal multiple access (OMA) techniques allocate mutually exclusive radio spectrum resources to each GT [9,10]. By applying Non-Orthogonal Multiple Access (NOMA) technology to UAV systems, multiple GTs can share the same spectrum resources, thereby improving spectrum efficiency and further enhancing transmission rates [11].

The challenges addressed in the context of UAV-enabled MEC with NOMA as follows. Firstly, the dynamic nature of server computing resources and the divisibility of GTs' tasks need to be considered to determine the optimal offloading strategy based on task size, computational density, and server resource availability [12]. Secondly, the complexity of the communication environment, particularly in the context of NOMA, requires GTs to consider both the distance to the UAV and the interference caused by other GTs during task offloading [13]. Thirdly, there is inherent uncertainty in the environment, where each GT only observes its own information and remains unaware of the information from other GTs [14]. Moreover, employing centralized decision-making based on reinforcement learning for multiple GTs may encounter challenges attributed to the high dimensionality of the action space, resulting in convergence difficulties.

Based on the aforementioned motivations, this paper considers each GT as an independent decision-maker and investigates the decisions regarding transmission power and offloading ratio to jointly optimize latency and energy consumption. The main contributions are summarized as follows:

- In the UAV-enabled MEC with NOMA, the dynamic nature of computing resources and transmission rates is taken into account, targeting tasks that are divisible and exhibit different characteristics (e.g., delay-sensitive or energy-sensitive). The system is modeled based on a queuing model, and a distributed method is employed, treating each GT as an independent decision-maker.
- This paper introduces a distributed method called DMARL based on MAD-DPG. By designing a decision reward function based on different reward schemes, the method achieves more stable convergence and effectively reduces cost expenses for GT.
- We conduct simulation experiments and compared our proposed method with the baseline experiment. The results demonstrate that our method effectively balances latency and energy consumption across different scenarios, showcasing a greater adaptability to diverse scenarios compared to the baseline experiment.

2 Related Works

In this chapter, we will review the research work related to UAV-enabled MEC, as well as the research on the application of NOMA in the MEC environment. Furthermore, we will introduce the relevant research on distributed algorithms with GTs as decision-makers.

UAV-enabled MEC has garnered significant attention in existing research. Zhao et al. [15] investigated the utilization of UAVs as auxiliary edge cloud nodes to facilitate task offloading for GTs. By employing multi-agent reinforcement learning (MARL) techniques to jointly optimize UAV trajectory, task allocation, and UAV transmission power control, their objective is to minimize GT latency and energy consumption. Faraci et al. [7] proposed a reinforcement learning-based system controller that utilizes UAVs equipped with MEC servers to extend 5G network slicing. Their approach manages computation resources and task allocation on UAVs to minimize UAV power consumption and GT latency. Qi et al. [5] focused on addressing the high demand for task processing by clustering multiple UAVs with MEC servers. By introducing a twofold computing offloading mechanism based on the alternating direction method of multipliers (ADMM) and Lyapunov optimization, their goal is to maximize system energy efficiency.

In the context of existing research, the combination of MEC and NOMA has been extensively explored. In [11,16], NOMA is introduced into UAV-enabled edge computing, and both studies utilize convex optimization theory to address different scenarios. A method based on successive convex approximation (SCA) and quadratic approximation was proposed by [11] to decompose the offloading problem into sub-problems of power allocation, resource allocation, and trajectory planning. This approach effectively reduces system energy consumption. the UAV is employed as a relay node between user base stations, and the problem is transformed into sub-problems of trajectory planning, power allocation, and user scheduling using SCA in [16]. This method efficiently reduces UAV energy consumption while improving user quality of service (QoS).

The research on distributed algorithms with GTs as decision-makers has been extensively explored in existing studies. Tang et al. [14] investigated the scenario where each GT only observes local information. By enabling each GT to autonomously make offloading decisions using the conventional Deep Q-Network (DQN) algorithm, the aim was to reduce the average latency for users. Xu et al. [17] focused on achieving a vehicle-mounted networked physical system through collaborative sensing and heterogeneous information fusion by vehicles. The study employed a MARL approach to optimize the decision-making process of information acquisition frequency and priority for each vehicle in an unknown dynamic environment. In Air-Ground Integrated Edge Computing Systems, the task offloading between GTs is modeled as a stochastic game, enabling terminals to optimize their own decisions in unknown dynamic environments, thereby improving the average system performance [18].

However, in the context of UAV-assisted MEC, there is limited research focusing on the perspective of GTs while simultaneously considering the NOMA communication environment and the collaborative offloading scenario involving multiple UAVs. In specific scenarios, the absence of information sharing among GTs, coupled with the complexity posed by the extensive action space dimensions associated with reinforcement learning approaches, makes the application of centralized algorithms for decision-making a challenging endeavor. To address these issues, this paper proposes a distributed method based on MARL, aiming to mitigate the latency and energy consumption of GT tasks.

3 System Architecture

In the scenario illustrated in Fig. 1, UAVs equipped with MEC servers provide services to GTs, without the inclusion of ground-based MEC servers. To maintain brevity, the abbreviation UAV will be used to denote UAV-enabled MEC. The scenario comprises N_{uav} UAVs and N_{gt} GTs, denoted as U and G, respectively. UAVs fly at a fixed altitude, and the total service time T is divided into N_t time slots of duration $\delta = \mathrm{T}/N_t$. Each GT maintains local computation and communication queues and makes real-time decisions to enqueue tasks in these queues. GTs utilize NOMA to offload tasks from the communication queue to the UAVs. The primary notations are summarized in Table 1.

3.1 Communication Model

In this section, we will introduce the communication model between GTs and UAVs. The NOMA-based air-to-ground (A2G) transmission model is employed, allowing GTs to share the same channel and serving multiple GTs simultaneously on the same channel. The channel model for A2G communication adopts a probabilistic path loss model that includes line-of-sight (LoS) and non-line-of-sight (NLoS) links. The path loss between UAV u and GT g on these links can be calculated using the following formula:

$$L_{u,g}^{\xi}(t) = \eta^{\xi}(\frac{4\pi d_{u,g}(t) f_c}{c})^{\alpha}, \xi \in \{LoS, NLoS\} \tag{1}$$

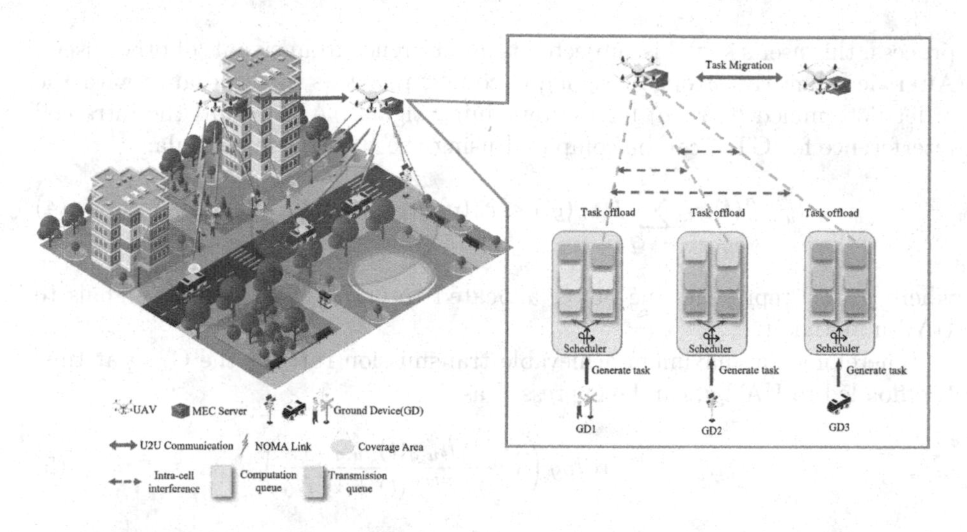

Fig. 1. UAV-enabled MEC with NOMA architecture

where f_c represents the carrier frequency, c represents the speed of light, α represents the path loss exponent, and η^{LoS} and η^{NLoS} represent the average additional path loss for LoS and NLoS links, respectively.

The probability of LoS connection between the UAV u and GT g can be calculated using the following formula:

$$P_{u,g}^{LoS}(t) = \frac{1}{1 + \varpi_1 exp(-\varpi_2(\theta_{u,g} - \varpi_1))} \tag{2}$$

where ϖ_1 and ϖ_2 are constants depending on the actual environment, and $\theta_{u,g}$ is the elevation angle between the UAV u and the GT g. The probability of NLoS is given by $P_{u,g}^{NLoS}(t) = 1 - P_{u,g}^{LoS}(t)$. the signal gain $h_{u,g}(t)$ between the UAV u and the GT g under both LoS and NLoS conditions at time t, is related to the average path loss $L_{u,g}(t)$, and it is obtained using the following formula:

$$h_{u,g}(t) = \frac{1}{L_{u,g}(t)} = \frac{1}{P_{u,g}^{LoS}(t)L_{u,g}^{LoS}(t) + P_{u,g}^{NLoS}(t)L_{u,g}^{NLoS}(t)} \tag{3}$$

Due to potential variations in GT positions and the dynamic nature of UAV positions information, the communication between UAVs and GTs exhibits dynamic characteristics. In the NOMA-based A2G scenario, the set of GTs served by the UAV at time t is denoted as $A_u(t)$. Additionally, it is assumed that UAVs utilize mutually orthogonal frequency bands, ensuring that GTs served by the same UAV are subjected solely to intra-cell interference. When the GT requires task offloading to the UAV, the SIC receiver at the UAV, based on the varying channel gains $h_{u,g}(t)$ from GTs, arranges $h_{u,g}(t)$ in an ordered manner. The term $r_u(g)$ is employed to represent the k-th strongest rank of the channel gain $h_{u,g}(t)$ of GT g within the set $A_u(t)$. The SIC receiver interprets signals based on $r_u(g)$, initially decoding the signal from the strongest user. During the decoding process, this user's signal is subjected to interference from signals of other users. After decoding the strongest signal, decoding proceeds in accordance with the order determined by $r_u(g)$ for the remaining signals. As a result, the intra-cell interference for GT g can be computed using the subsequent formula:

$$I_{u,g}^{intra}(t) = \sum_{g' \in G} \mathbf{1}(r_u(g') < r_u(g)) h_{u,r_u(g')}(t) p_{u,r_u(g')}(t) \tag{4}$$

where $p_{u,g}(t)$ represents the power allocated to GT g when sending signals to UAV u at time t.

Therefore, the maximum achievable transmission rate for the GT g at time t, offloaded to UAV u, can be expressed as:

$$R_{u,g}(t) = W log \left(1 + \frac{h_{u,g}(t) p_{u,g}(t)}{I_{u,g}^{intra}(t) + W N_0}\right) \tag{5}$$

where W represents the bandwidth allocated to each GT, and N_0 denotes the power of the additive white Gaussian noise (AWGN).

Table 1. Summary of primary notations

Notions	Description	Notions	Description
U/G/T	**Set to index difference GTs/UAVs/time slots**	$N_{uav}/N_{gt}/N_t$	**Number of GTs/UAVs/time slots**
δ	**Duration of each time slot**	$L_{u,g}$	**Path loss**
$P_{u,g}^{Los}$	**Probability of LoS connection**	$h_{u,g}$	**Signal gain**
$I_{u,g}^{intra}$	**Intra-cell interference**	$R_{u,g}$	**Transmission rate**
Q^{com}	**Set to index different GT local computation queue**	Φ_g^{loc}	**Task delay**
Q^{tran}	**Set to index different GT local transmission queue**	E_g^{loc}	**Task energy consumption**
$\Phi_g^{tran}(t)$	**Transmission delay of the task generated by GT g at time t**	$p_{u,g}$	**upload power**
λ	**Task generation probability**	s_g	**Task of data size**
K	**Set to index difference tasks generated by GT g**	c_g	**Task of required cycles per bit**
$\phi_g^{fintran}(t)$	**Completion time of task transmission**	E_g^{tran}	**Energy consumption of offloading task**
Q_u^{mec}	**Set to index all GTs computation queue in UAV u**	M_u	**Set to different active GTs in UAV u**
x_g^u	**Proportion of tasks offloaded by GT g to UAV u**	A_u	**Set to different GTs serverd by UAV u**

3.2 Task Computation Model

We assume that GTs generate tasks randomly with a probability $\lambda \in [0,1]$ in each time slot. Specifically, we use $\zeta_g(t) \in \{0,1\}$ to indicate whether GT generates a task at time t. If the GT generates a task at time t, then $\zeta_g(t)=1$; otherwise, $\zeta_g(t)=0$. Each task $k_g(t)=(s_g(t),c_g(t))$ is characterized by a tuple consisting of the task size $s_g(t)$ and the required CPU cycles per bit $c_g(t)$. The set of tasks generated by GTs at time t is denoted as $\mathrm{K}(t) := \{k_1(t), k_2(t), \ldots, k_{|\mathrm{G}|}(t)\}$.

Local Computation Queue Model. Each GT is equipped with a local computation queue, represented by the set $Q^{com}(t)=\{Q_1^{com}(t), Q_2^{com}(t), \ldots, Q_{|\mathrm{G}|}^{com}(t)\}$. At each time slot, GTs make real-time decisions on the local task computation ratio $x_g^{loc}(t)$, resulting in the local computation task size $s_g^{loc}(t)=x_g^{loc}(t)\cdot s_g(t)$. Additionally, the total number of cycles required for computation is given by $c_g^{loc}(t)=s_g^{loc}(t)\cdot c_g(t)$. Consequently, the length of the local computation queue at each time slot can be expressed as:

$$Q^{com}(t)=\{Q_g^{com}(t-1)+c_g^{loc}(t) | g \in G\} \tag{6}$$

We assume a first-in-first-out (FIFO) mode for the local computation queue $Q_g^{com}(t)$, allowing only one task to be processed at a time. GTs have real-time awareness of the length of their local queue. Therefore, the task delay and energy consumption incurred at time slot t can be determined as follows:

$$\Phi_g^{loc}(t)=\frac{|Q_g^{com}(t)|}{\omega_g} \tag{7}$$

$$E_g^{loc}(t)=kc_g^{loc}(t)(\omega_g)^3 \tag{8}$$

where ω_g represents the computational capacity of GT g, and k is a constant. At the end of a time slot, the processed data is removed from the local computation queue, updating $Q_g^{com}(t)=\max\{0, Q_g^{com}(t)-\omega_g\}$.

Transmission Queue Model. Similar to the local computation queue model, the transmission queue is represented by the set $Q^{tran}(t)$, which contains the partial tasks that need to be offloaded from the GT to the UAV for processing. The UAV service provider can send information about the UAVs currently providing services in the service area to the GTs, allowing them to know the set of UAVs available for task offloading. The GT establishes a communication link with the nearest UAV, enabling the offloading of tasks to the MEC through the established communication link with UAV u. We use $Q_g^{pretran}(t)$ to represent the queue status before adding tasks to the transmission queue at time t. The total amount of data that GT g needs to offload at time t is $s_g^{off}(t)=(1-x_g^{loc})\cdot s_g(t)$. These task data will be added to the transmission queue Q_g^{tran}. Therefore, the length of the transmission queue can be expressed as:

$$Q_g^{tran}(t)=Q_g^{tran}(t-1)+s_g^{off}(t) \tag{9}$$

Considering the dynamic communication environment and the varying data requirements of real-time tasks, GTs need to make real-time decisions on the optimal upload power $p_{u,g}(t)$ to meet the task's transmission requirements. However, unlike the local computing queue, the transmission delay of tasks cannot be directly obtained at time t. We use $\phi_g^{fintran}(t)$ to represent the completion time of task transmission and $\phi_g^{tran}(t)$ to represent the time when the front task in the transmission queue is generated at time t, in order to update the transmission delay at each time. Therefore, the transmission delay of the task generated by GT g at time t can be obtained using the following formula:

$$\Phi_g^{tran}(t') = \phi_g^{fintran}(t') - \phi_g^{tran}(t'), t' \in [\phi_g^{tran}, t] \tag{10}$$

when $s_g^{off}(t) \leq R_{u,g}(t)$ and $Q_g^{pretran}(t) = 0$, it indicates that the task at the front of the transmission queue Q_g^{tran} is generated at the current time t, i.e., $\phi_g^{tran}(t) = t$. In this case, the completion time $\phi_g^{fintran}$ of the task's transmission can be computed as $\phi_g^{fintran}(t) = \frac{s_g^{off}(t)}{R_{u,g}(t)} + t$. If $s_g^{off}(t) > R_{u,g}(t)$, it implies that the current transmission rate cannot complete the task within $|t|$ time. Hence, the transmission delay of the task generated at the current time is $\phi_g^{fintran}(t) = |t| + t$. If $Q_g^{pretran}(t) > 0$, the transmission of the task generated at time $\phi_g^{tran}(t)$ must be completed before processing subsequent tasks. For the task generated at time $\phi_g^{tran}(t)$, if the remaining data size $s_g^{off}(\phi_g^{tran}(t)) > R_{u,g}(t)$, it indicates that the tasks generated in the time $t' \in [\phi_g^{tran}(t), t]$ need to wait. Consequently, $\phi_g^{fintran}(t') = \phi_g^{fintran}(t') + |t|$. When $s_g^{off}(\phi_g^{tran}(t)) \leq R_{u,g}(t)$, it signifies that the task generated at time $\phi_g^{tran}(t)$ has already completed its transmission, specifically $\phi_g^{fintran}(t') = \phi_g^{fintran}(t') + \frac{s_g^{off}(t')}{R_{u,g}(t)}$, where $t' = \phi_g^{tran}$. Then, it can be determined whether the remaining transmission rate satisfies the unloading transmission of the task generated at time $\phi_g^{tran}(t) + |t|$.

The energy consumption of offloading tasks can be calculated using the following formula:

$$E_g^{tran}(t) = \begin{cases} p_{u,g}(t), & Q_g^{tran}(t) > 0 \\ 0, & otherwise \end{cases} \tag{11}$$

where $E_g^{tran}(t)$ represents the energy consumed by GT g for offloading tasks at time t. If the communication queue $Q_g^{tran}(t)$ contains tasks, the energy consumption is equal to the transmission power $p_{u,g}(t)$. Otherwise, when the queue is empty, no energy is consumed.

UAV Computation Queue Model. When the GT offloads tasks from the transmission queue to the currently established communication link with UAV u, the tasks can be further relayed to neighboring operational UAVs via UAV u. Each UAV pre-assigns a computation queue for each GT in the region, denoted as $Q_u^{mec}(t) = \{Q_{u,1}^{mec}(t), Q_{u,2}^{mec}(t), \cdots, Q_{u,|\mathrm{G}|}^{mec}(t)\}$, which represents the computation queue set of all GTs in UAV u. The proportion of tasks offloaded by GT g to each UAV u at time t is denoted as $x_g^{off}(t) = \{x_g^1(t), ..., x_g^{|\mathrm{U}|}(t)\}$. Therefore, the task size offloaded to UAV u at time t is $d_{u,g}^{offmec}(t) = x_g^u(t) \cdot s_g(t)$, and the computation size is $c_{u,g}^{offmec}(t) = d_{u,g}^{offmec}(t) \cdot c_g(t)$. $c_{u,g}^{remain}(t)$ represents the remaining computation size of the tasks generated at time t in the computation queue of UAV u, initialized as $c_{u,g}^{offmec}(t)$, which is used to indicate whether the tasks have been completed. The time at which the tasks, after completing the local transmission, arrive at UAV g is $\phi_{u,g}^{arr}(t) = \lceil t + \phi_g^{tran}(t) \rceil$, where $\lceil\rceil$ denotes the ceiling function. In this context, we assume that UAVs communicate with each other using a FANET with extremely high bandwidth, and we neglect the transmission delay between UAVs. Thus, the length of $Q_{u,g}^{mec}(t)$ at time t can be computed using the following formula:

$$Q_{u,g}^{mec}(t) = Q_{u,g}^{mec}(t-1) + \sum_{t' \in [0,t]} (t = \phi_g^{tran}(t'))c_{u,g}^{offmec}(t') \tag{12}$$

UAVs support parallel processing of multiple tasks and allocate computing resources evenly based on the number of active queues in UAV u. If there are pending tasks in $Q_{u,g}^{mec}(t)$, the UAV needs to allocate computing resources for GT g. We denote the set of active GTs at time t as $M_u(t)$. The calculation formula is as follows:

$$M_u(t) = \{g | \left|Q_{u,g}^{mec}(t)\right| > 0 \quad, g \in \mathrm{G} \quad, u \in \mathrm{U}\} \tag{13}$$

Similar to the calculation of transmission delay $\Phi_g^{tran}(t)$, the calculation of $\Phi_{u,g}^{mec}(t)$ takes into account the remaining computation $c_{u,g}^{offmec}(t)$ of the task offloaded to UAV u at time t' and the computation received by the GT at time t. Only when $c_{u,g}^{remain}(t') = 0$, indicating that UAV has completed the task offloaded at time t', can the specific time of offloading to the UAV be determined. In this paper, we focus on the energy consumption of the GT and do not consider the computational energy consumption of the UAV.

3.3 Problem Formulation

The task generated by GT $s_g(t) \in S(t)$ can be divided into three parts: the local computation delay $\Phi_g^{loc}(t)$ at the GT, the time required to migrate the task over the wireless channel to the UAV $\Phi_g^{tran}(t)$, and the time required for task computation on the MEC server $\Phi_{u,g}^{mec}(t)$. Therefore, the total latency of the tasks generated by GT g at each time step can be calculated using the following formula:

$$\Phi_g(t') = \begin{cases} \max\{\Phi_g^{tran}(t') + \max_{u \in U}\{\Phi_{u,g}^{mec}(t')\}, \\ \Phi_g^{loc}(t')\}, \quad |U| = \sum_{u \in U} 1(c_{u,g}^{remain}(t') = 0) \\ \Phi_g(t') + |t|, \quad otherwise \end{cases} \tag{14}$$

where $t' \in [\min_{u \in U} \phi_{u,g}^{mec}(t), t]$, when all UAVs have completed the tasks generated at time t', we can calculate the specific delay of the tasks generated at time t'. We use $N_g(t)$ to denote the number of completed tasks until the current time t, and $N_g(t)$ is updated whenever a task is completed. Otherwise, the delay of the tasks at this moment will increase by one time slot. Therefore, the total delay of the tasks generated by GT g up to the current time t can be calculated using the following formula:

$$\Phi_g^{sum}(t) = \sum_{t' \in [0,t]} \Phi_g(t') \tag{15}$$

In addition, the total energy consumption of the tasks considers the energy consumed by the GT during local computation, denoted as $E_g^{loc}(t)$, and the energy consumed during the task offloading transmission, denoted as $E_g^{tran}(t)$. Since the local computation delay can be directly obtained at time t, the local computation energy consumption $E_g^{loc}(t)$ can be directly calculated. Therefore, the total energy consumption of GT g up to the current time t can be obtained using the following formula:

$$E_g^{sum}(t) = \sum_{t' \in [0,t]} E_g^{loc}(t') + \sum_{t' \in [0,t]} E_g^{tran}(t') \tag{16}$$

Furthermore, the total delay and energy consumption of the tasks generated by GT g are considered as the system cost. It can be represented by the following equation:

$$U_g(t) = \omega_d \Phi_g^{sum}(t) + \omega_e E_g^{sum}(t) \tag{17}$$

where ω_d and ω_e represent the weighting factors of delay and energy consumption in the system cost, respectively.

The objective of this study is to jointly optimize the GT's offloading decisions $X = \{x_g^{loc}(t), x_g^{off}(t) | \forall g \in G, \forall t \in T\}$ and power allocation decisions $P = \{p_g(t) | \forall g \in G, \forall t \in T\}$ on UAV-enabled MEC in the NOMA environment, aiming to minimize the system cost for all GTs. The optimization problem can be formulated as follows:

$$\begin{aligned} &\min_{X,P} \sum_{\forall g \in G} U_g(t) \\ &s.t. C1: p_g \in [0, p^{max}], \\ &\quad C2: x_g^{loc}(t) \in [0,1], \\ &\quad C3: x_g^{u}(t) \in [0,1], \\ &\quad C4: x_g^{loc}(t) + \sum_{u \in U} x_g^{u}(t) = 1 \end{aligned} \tag{18}$$

4 Proposed Solution

In this chapter, we employ a distributed algorithmic approach where each GT is considered as an independent decision maker. Specifically, we reframe the task offloading problem for GTs as a multi-agent extension of Markov Decision Processes (MDPs) and propose a method based on the MADDPG algorithm to solve these MDPs.

4.1 Problem Transformation

We have transformed the aforementioned problem into a partially observable Markov game for all GTs, denoted by $g \in$ G. The system state is defined as s, while o := $\{o_1, o_2, ..., o_{|G|}\}$ represents the observation set for each GT, with each subset o containing a partial observation of the current state s. Similarly, $a := \{a_1, a_2, ..., a_{|G|}\}$ denotes the action set for each GT. Each GT g determines its action based on its observed information o_g using the policy $\pi_g : o_g \rightarrow a_g$.

Observation Space: The local observation information for each GT g at time t can be represented as:

$$o_g^t = \{\left|Q_g^{com}(t)\right|, \left|Q_g^{tran}(t)\right|, h_g(t), s_g(t), M(t-1)\} \tag{19}$$

where $\left|Q_g^{com}(t)\right|$ denotes the length of the local computing queue at GT g at time t, $\left|Q_g^{tran}(t)\right|$ represents the length of the transmission queue at time t, $h_g(t)$ indicates the channel gain of GT g at time t, $s_g(t)$ represents the size of the task generated at time t, and $M(t-1) = \{|M_1(t-1)|, |M_2(t-1)|, ..., \left|M_{|\text{U}|}(t-1)\right|\}$ represents the set of active queue sizes among all UAVs in the previous time slot. We assume that each UAV sends the number of active GTs in its queue to all GTs at the end of each time slot. Therefore, the observation information of each gt includes the information about the number of active GTs in the previous time slot $M(t-1)$.

Action Space: The action space of each GT encompasses the decisions regarding task offloading and power allocation and can be represented by the following equation:

$$a_g^t = \{x_g^{loc}(t), x_g^{off}(t), p_g(t)\} \tag{20}$$

where, $x_g^{loc}(t)$ and $x_g^{off}(t)$ respectively denote the proportions of tasks generated at time t that are executed locally at the GT and offloaded for computation on unmanned aerial vehicles (UAVs). $p_g(t)$ represents the power allocation used by the GT for offloading tasks. All of these actions belong to continuous sets.

Reward: In each time slot t, based on the observation information $o_g(t)$ and policy π_g of each GT g, the system generates a reward value $r(t) : O \times A \rightarrow R$. The system's reward can be calculated using the following formula:

$$r(t) = -(\omega_d \sum_{g \in G} \frac{\Phi_g^{sum}(t)}{N_g(t)} + \omega_e \sum_{g \in G} \frac{E_g^{sum}(t)}{N_g(t)}) \tag{21}$$

The system reward is based on the computation of the total task completion time and total energy consumption of all GTs at the current time t, divided by the number of completed tasks. Furthermore, the reward of GT g can be computed by subtracting the reward obtained by excluding GT g from the system's reward, as follows:

$$r_g(t) = r(t) - r_{-g}(t) \quad (22)$$

where $r_{-g}(t)$ represents the reward value obtained by excluding GT g. The set of different rewards for GTs is denoted as $r_{\mathrm{G}}(t) = \{r_1(t), r_2(t), ..., r_{|\mathrm{G}|(t)}\}$.

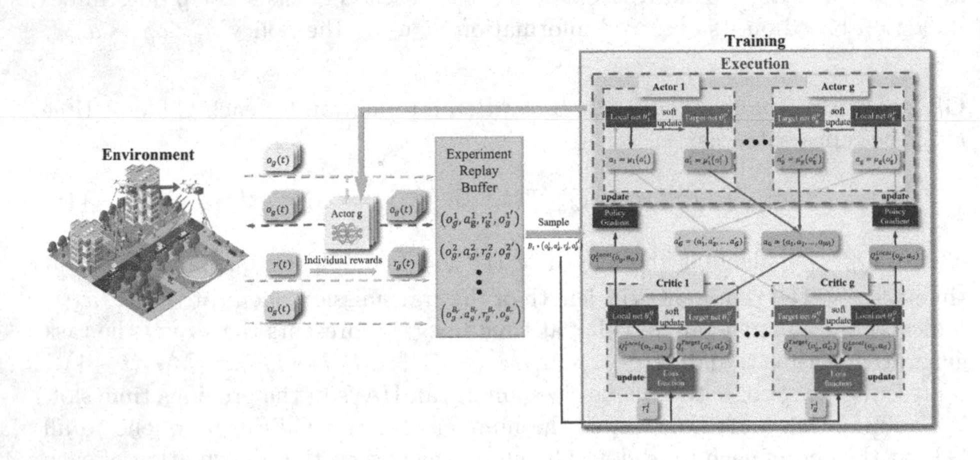

Fig. 2. framework of the proposed algorithm

4.2 MADDPG-Based Algorithm

Based on the framework diagram shown in Fig. 2, we employ the MADDPG-based algorithm to address the partially observable Markov game involving |G| GTs. Each GT independently executes its own DDPG algorithm, which consists of an Actor and a Critic component. The Actor of each agent g learns a policy function μ_g through a neural network, enabling it to select actions based on observed values o_g at each time slot. The Critic, on the other hand, learns a state-action function through another neural network, allowing it to evaluate the value of the current state-action pair. Since the system reward depends on the actions of all agents, the Critic for agent g takes as input the actions of other agents, the current agent's observation, and its own action, represented as $\{a_1, a_2, ..., a_{|G|}, o_g\}$. Each Actor and Critic has two instances of neural networks: a local network that is updated based on the current parameters, and a target network that is updated based on the local network parameters. Each agent g has a local network with parameters θ_g^Q and θ_g^μ for the Actor and Critic, and target network parameters $\theta_g^{Q'}$ and $\theta_g^{\mu'}$. Additionally, since DDPG is an offline algorithm, it employs an experience replay mechanism. We use a replay buffer B of size B_r to store the transition data. During the training phase, each agent g randomly samples mini-batches of size B_t from the replay buffer, denoted as $(o_g^{i:i+B_t}, a_g^{i:i+B_t}, r_g^{i:i+B_t}, o_g^{i':i'+B_t})$. Here, o_g^i represents the observation obtained by agent g from the i-th transition, and $o_g^{i'}$ represents the observation after agent

g takes action a_g^i. By utilizing these transition data, agent g can perform learning and optimization. The procedure of our proposed MADDPG-based solution is illustrated in Algorithm 1.

Algorithm 1: DMARL

```
1  Initialize actor networks (θ_g^μ, θ_g^μ') and critic networks (θ_g^Q, θ_g^Q') for each
   GT g;
2  Initialize replay buffer B for each GT g;
3  foreach episode do
4  |  Receive initial system state s(0);
5  |  foreach step t in T do
6  |  |  foreach GT g in G do
7  |  |  |  Receive local observation o_g(t);
8  |  |  |  Select action a_g(t) = μ_g(o_g(t));
9  |  |  end
10 |  |  Execute a(t) = {a_1(t), a_2(t), ..., a_|G|};
11 |  |  Receive system reward r(t);
12 |  |  Obtain new system state s'(t) and observation o'(t);
13 |  |  foreach GT g in G do
14 |  |  |  Compute individual rewards r_g(t) based on Eq. (22);
15 |  |  |  Store (o_g(t), a_g(t), r_g(t), o'_g(t)) into its replay buffer;
16 |  |  end
17 |  end
18 |  foreach GT g in G do
19 |  |  Sample B_r transitions randomly from its own replay buffer B;
20 |  |  Update (θ_g^μ, θ_g^Q);
21 |  end
22 |  Update (θ_g^μ', θ_g^Q');
23 end
```

The loss function for the local critic network of each agent g can be computed using the following equation:

$$L(\theta_g^Q) = \frac{1}{B_t}\sum_{i=0}^{B_t}(Q_g^{Target} - Q_g^{Local}(o_g^i, a_{\mathrm{G}}^i|\theta_g^Q))^2 \tag{23}$$

where Q_g^{Target} represents the state-action function computed by the target network of the critic, given by $Q_g^{Target} = r_g^i + \tau Q_g^{Target}(o_g^{i'}, a_{\mathrm{G}}^{i'}|\theta_g^{Target})$. Each agent g determines its policy based on its own observations, and the local actor's network parameters are updated by maximizing the state-action function. The objective function of the actor network can be expressed as:

$$J(\theta_g^\mu) = \frac{1}{B_t}\sum_{i=0}^{B_t} Q_g((o_g^i, a_g)|a_g = \mu_g(o_g^i)) \tag{24}$$

Since the decision actions of each agent are continuous, the policy function μ_g is also continuous, allowing the use of policy gradient methods to update the parameters θ_g^{μ} by computing the direction of $\nabla_{\theta_g^{\mu}} J(\theta_g^{\mu})$. In updating the local network parameters, we employ a soft update approach. Specifically, the actor and critic target networks $\theta_g^{Q'}$ and $\theta_g^{\mu'}$ are updated using the following equations:

$$\theta_g^{Q'} = \varepsilon\theta_g^{Q} + (1-\varepsilon)\theta_g^{Q'} \tag{25}$$

$$\theta_g^{\mu'} = \varepsilon\theta_g^{\mu} + (1-\varepsilon)\theta_g^{\mu'} \tag{26}$$

where ε is a tunable parameter with $\varepsilon \ll 1$.

5 Performance Evaluation

In this chapter, we validate our proposed method through simulation experiments. The experimental environment is built using Python 3.9 and PyTorch 1.11. We consider a scenario with 10 GTs and 2 UAVs. The simulation experiments are conducted on a computer running Win11, equipped with an i7-12700H processor, 32 GB of memory, and an NVIDIA RTX 3070 GPU. Both the Actor and Critic networks are constructed using fully connected layers. The Critic network consists of two fully connected hidden layers with neuron counts of [256, 128], while the Actor network has two fully connected hidden layers with neuron counts of [64, 32]. The main experimental environment parameters are provided in Table 2 for detailed reference.

Table 2. Experimental Environment Parameters

Parameters	Value	Parameters	Value
Length of area	200 m	Task of data size	[1,20] MB
Number of CPU cycles per bit	500	UAV bandwidth (W)	20 MHz
Effective switched capacitance (k)	10^{-27}	Task request rate (λ)	0.6
Computation capability of UAV (ω_u)	3 GHz	Height of UAV u	50 m
Computation capability of GT (ω_g)	10 GHz	Coverage of UAV u	100 m
Maximum transmit power of GT (p_g)	0.1	Length of time slot (δ)	1 s

To validate the performance of our proposed algorithm, we compared it with the following benchmark methods:

- Local execution(Local): Each GT performs all tasks locally without offloading, resulting in no upload power consumption and only considering local computing energy consumption.
- Random execution(Random): Each GT randomly selects the ratio of local processing and offloading, as well as the upload power.
- CCDDPG [19]: The central controller, assuming complete knowledge of all user information, utilizes the DDPG algorithm to determine the ratio of local processing, offloading, and upload power for each GT.

We have predefined three sets of weights (ω_d, ω_e) to represent the importance of delay and energy consumption in different scenarios. The first scenario, referred to as Scenario 1, focuses on energy-sensitive tasks with weights set as $(2, 8)$. The second scenario, referred to as Scenario 2, addresses general situations and aims to balance the importance of delay and energy consumption with weights set as $(5, 5)$. The third scenario, referred to as Scenario 3, emphasizes delay-sensitive tasks with weights set as $(8, 2)$. We compared the convergence of the DMARL method and baseline experiments across these three scenarios. Furthermore, we evaluated and analyzed the delay, energy consumption, and final reward values per episode by comparing the trained models with the baseline experiments in these three different scenarios.

According to the results shown in Fig. 3, we compared the convergence of our proposed algorithm with Random and CCDDPG algorithms for three different scenarios. Since the three scenarios have different weights, the reward functions also differ for each scenario. In the first scenario, it can be observed that our proposed DMARL algorithm reaches convergence after approximately 300 iterations, indicating that each GT can learn the optimal policy from the rewards in this scenario. In the second scenario, the rewards of our algorithm consistently increase and stabilize to a suboptimal solution between 1000 and 2000 rounds, and then gradually increase again, exploring towards the optimal solution after 2000 rounds. In the third scenario, the algorithm gradually converges after around 1300 rounds, experiences a decline at 2700 rounds, and then gradually improves. In response to the abrupt fluctuations observed in the third scenario, it is noteworthy that when the agent takes actions in certain states, the corresponding reward function fails to effectively capture the quality of those actions. This discrepancy can lead to the agent learning erroneous behaviors. Additionally, the potential cause of sudden fluctuations can be attributed to the relatively high learning rate employed during the experiments. However, when compared with the CCDDPG approach across the three scenarios, the utilization of centralized decision-making in CCDDPG results in a larger action space dimension. Consequently, the rewards achieved in these scenarios do not surpass the performance of our proposed method.

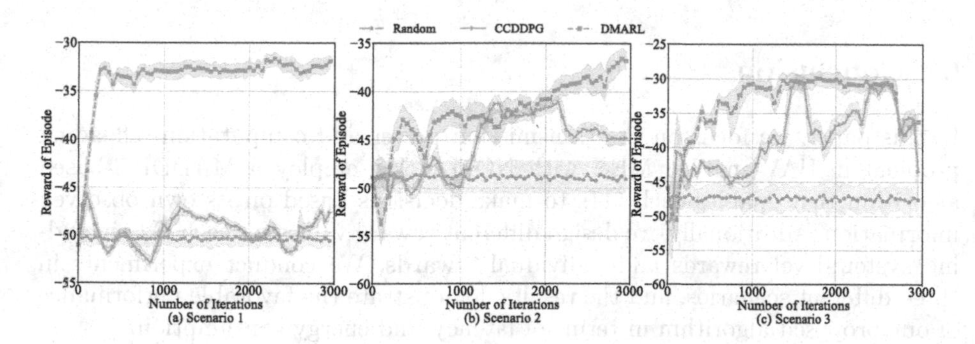

Fig. 3. convergence comparison

According to Fig. 4, we compared the performance of our trained model with the baseline experiment using three metrics: average latency, average energy consumption, and average reward per episode. In Scenario 1, which focuses on energy-sensitive tasks, our proposed algorithm effectively reduces latency and incurs only a 0.01 higher energy consumption compared to Local. Due to the higher weight on energy consumption in Scenario 1, the average reward per episode for Local is lower, but our algorithm still outperforms it. In Scenario 2, where the weight on latency and energy consumption is balanced (5:5), our algorithm significantly reduces latency but consumes more energy compared to Local. However, compared to CCDDPG, our algorithm improves both latency and energy consumption, and it achieves the lowest average reward per episode. In Scenario 3, our algorithm prioritizes lower latency at the expense of increased energy consumption, resulting in lower latency and lower average reward per episode compared to CCDDPG. Based on these scenarios, we observe that in Scenario 1, GTs tend to choose local execution to achieve lower energy consumption. In Scenario 2, GTs offload more tasks to UAVs to further reduce latency. In Scenario 3, GTs are willing to consume more energy in exchange for lower latency, given the lower weight on energy consumption.

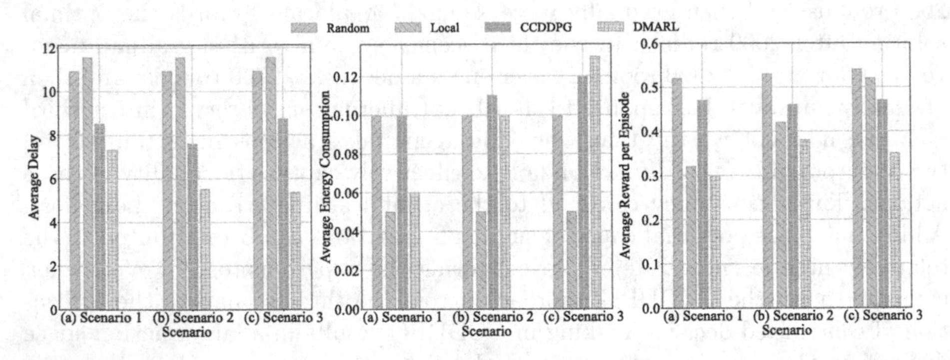

Fig. 4. performance comparison under difference metrics: (a) Average Delay;(b) Average Energy Consumption;(c) Average Reward per Episode

6 Conclusion

In this paper, we focus on the transmission power and computation offloading problem in UAV-enabled MEC with NOMA. We employ a MADDPG-based algorithm that enables each GTs to make decisions based on its own observed information. Additionally, we design different reward values for each GT, including system-level rewards and individual rewards. We conduct experiments in three different scenarios, and the results demonstrate the favorable performance of our proposed algorithm in terms of latency and energy consumption.

Acknowledgements. This work was supported in part by the Major Research plan of the National Natural Science Foundation of China under Grant (No. 92267104), National Natural Science Foundation of China under Grant (No. 62372242), Natural Science Foundation of Jiangsu Province of China under Grant (No. BK20211284), and Financial and Science Technology Plan Project of Xinjiang Production and Construction Corps under Grant (No. 2020DB005).

References

1. Xu, X., Tang, S., Qi, L., Zhou, X., Dai, F., Dou, W.: Cnn partitioning and offloading for vehicular edge networks in web3. IEEE Commun. Mag., 1–7 (2023)
2. Du, Z., Zheng, J., Yu, H., Kong, L., Chen, G.: A unified congestion control framework for diverse application preferences and network conditions. In: Proceedings of the 17th International Conference on emerging Networking Experiments and Technologies, pp. 282–296 (2021)
3. Yuben, Q., et al.: Service provisioning for uav-enabled mobile edge computing. IEEE J. Sel. Areas Commun. **39**(11), 3287–3305 (2021)
4. Mukherjee, M., Kumar, V., Lat, A., Guo, M., Matam, R., Lv, Y.: Distributed deep learning-based task offloading for uav-enabled mobile edge computing. In: IEEE INFOCOM 2020 - IEEE Conference on Computer Communications Workshops (INFOCOM WKSHPS). IEEE (2020)
5. Qi, X., Chong, J., Zhang, Q., Yang, Z.: Collaborative computation offloading in the multi-uav fleeted mobile edge computing network via connected dominating set. IEEE Trans. Veh. Technol. **71**(10), 10832–10848 (2022)
6. Yang, Z., et al.: Ai-driven uav-noma-mec in next generation wireless networks. IEEE Wireless Commun. **28**(5), 66–73 (2021)
7. Faraci, G., Grasso, C., Schembra, G.: Design of a 5g network slice extension with mec uavs managed with reinforcement learning. IEEE J. Sel. Areas Commun. **38**(10), 2356–2371 (2020)
8. Li, Z., et al.: A knowledge-driven anomaly detection framework for social production system. IEEE Trans. Comput. Soc. Syst., 1–14 (2022)
9. Waqar, N., Hassan, S.A., Mahmood, A., Gidlund, M., Jung, H.: Joint power and beamforming optimization of uav-assisted noma networks for b5g-enabled smart cities. In: Proceedings of the 1st Workshop on Artificial Intelligence and Blockchain Technologies for Smart Cities with 6G. ACM, New York (2021)
10. Zheng, J., Xu, H., Chen, G., Dai, H.: Minimizing transient congestion during network update in data centers. In: Proceedings of the 2014 CoNEXT on Student Workshop, pp. 4–6 (2014)
11. Zhang, X., Zhang, J., Xiong, J., Zhou, L., Wei, J.: Energy-efficient multi-uav-enabled multiaccess edge computing incorporating noma. IEEE Internet Things J. **7**(6), 5613–5627 (2020)
12. Xincao, X., et al.: Joint task offloading and resource optimization in noma-based vehicular edge computing: a game-theoretic drl approach. J. Syst. Architect. **134**, 102780 (2023)
13. Xia, X., et al.: Data, user and power allocations for caching in multi-access edge computing. IEEE Trans. Parallel Distrib. Syst. **33**(5), 1144–1155 (2022)
14. Tang, M., Wong, V.W.S.: Deep reinforcement learning for task offloading in mobile edge computing systems. IEEE Trans. Mobile Comput. (2020)

15. Zhao, N., Ye, Z., Pei, Y., Liang, Y.-C., Niyato, D.: Multi-agent deep reinforcement learning for task offloading in uav-assisted mobile edge computing. IEEE Trans. Wireless Commun. **21**(9), 6949–6960 (2022)
16. Guo, F., Zhang, H., Ji, H., Li, X., Leung, V.C.M.: Joint trajectory and computation offloading optimization for uav-assisted mec with noma. In: IEEE INFOCOM 2019 - IEEE Conference on Computer Communications Workshops (INFOCOM WKSHPS). IEEE (2019)
17. Xu, X., Liu, K., Dai, P., Xie, R., Luo, J.: Cooperative sensing and heterogeneous information fusion in vcps: A multi-agent deep reinforcement learning approach. ArXiv (2022)
18. Chen, X., et al.: Information freshness-aware task offloading in air-ground integrated edge computing systems. IEEE J. Sel. Areas Commun. **40**(1), 243–258 (2022)
19. Jiajie, X., Li, D., Wei, G., Chen, Y.: Uav-assisted task offloading for iot in smart buildings and environment via deep reinforcement learning. Build. Environ. **222**, 109218 (2022)

FedRKG: A Privacy-Preserving Federated Recommendation Framework via Knowledge Graph Enhancement

Dezhong Yao[1(✉)], Tongtong Liu[1], Qi Cao[2], and Hai Jin[1]

[1] National Engineering Research Center for Big Data Technology and System, Services Computing Technology and System Lab, Cluster and Grid Computing Lab, School of Computer Science and Technology, Huazhong University of Science and Technology, Wuhan 430074, China
{dyao,tliu,hjin}@hust.edu.cn

[2] School of Computing Science, University of Glasgow, Glasgow, UK
qi.cao@glasgow.ac.uk

Abstract. *Federated Learning* (FL) has emerged as a promising approach for preserving data privacy in recommendation systems by training models locally. Recently, *Graph Neural Networks* (GNN) have gained popularity in recommendation tasks due to their ability to capture high-order interactions between users and items. However, privacy concerns prevent the global sharing of the entire user-item graph. To address this limitation, some methods create pseudo-interacted items or users in the graph to compensate for missing information for each client. Unfortunately, these methods introduce random noise and raise privacy concerns. In this paper, we propose FedRKG, a novel federated recommendation system, where a global *knowledge graph* (KG) is constructed and maintained on the server using publicly available item information, enabling higher-order user-item interactions. On the client side, a relation-aware GNN model leverages diverse KG relationships. To protect local interaction items and obscure gradients, we employ pseudo-labeling and *Local Differential Privacy* (LDP). Extensive experiments conducted on three real-world datasets demonstrate the competitive performance of our approach compared to centralized algorithms while ensuring privacy preservation. Moreover, FedRKG achieves an average accuracy improvement of 4% compared to existing federated learning baselines.

Keywords: Federated learning · Recommendation systems · Knowledge graph · Graph neural network

1 Introduction

Recommendation systems are widely used in various domains, such as e-commerce and social recommendation, by alleviating users from the burden of sifting through vast amounts of data to discover suitable options [19]. These systems utilize user preferences and relevant information to provide personalized

H. Jin et al. (Eds.): GPC 2023, LNCS 14504, pp. 81–96, 2024.
https://doi.org/10.1007/978-981-99-9896-8_6

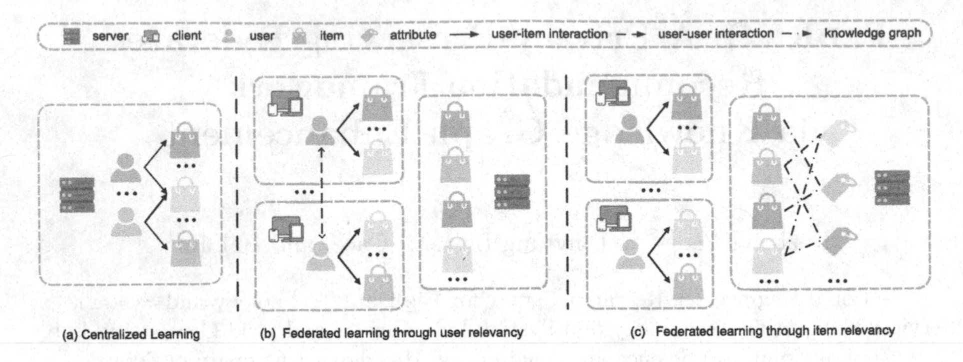

Fig. 1. Comparison of centralized learning, federated learning with enhanced user connections, and federated learning with enhanced project connections.

recommendations, making the process of finding relevant items more efficient and convenient [5]. However, the effectiveness of most recommendation methods heavily relies on centralized storage of user data [26]. User data generated from software usage has the potential to enhance user experiences, deliver personalized services, and provide insights into user behavior [22]. Nevertheless, user data inherently includes user preferences and involves personal privacy. With the increasing awareness of privacy and the implementation of relevant regulations such as the *General Data Protection Regulation* (GDPR) [14], service providers may face growing challenges in centrally storing and processing user data, as shown in Fig. 1(a).

The exclusive client access to local data leads to two challenges. Firstly, limited access to first-order interaction data hampers the effectiveness of the recommendation model. Secondly, privacy-preserving mechanisms are required to ensure secure communication between the client and server. To address these challenges, FL is introduced into the recommendation system. Existing works focus on the case of Fig. 1(b), where recommendations are achieved by directly finding correlations between users. For example, FedMF [4] and FedGNN [21] use only the local user-item interaction graph to find links between different users by *collaborative filtering* (CF). However, incorporating various types of information in the conventional graph recommendation task can significantly improve the recommendation accuracy while changing the graph structure [22]. Additionally, FeSoG [9] utilizes social networks as side information, adding direct connections between different users. Nevertheless, this method requires the server to possess the complete social network, which is a type of private data that is difficult to obtain for most recommendation systems. Furthermore, methods like FedGNN employ homomorphic encryption, which incurs substantial computational overhead and is not suitable as the primary encryption algorithm on edge devices with performance constraints.

To maximize the utilization of diverse data types while ensuring privacy protection on edge devices, we propose FedRKG[1], a GNN-based federated

[1] The source code is available at: https://github.com/ttliu98/FedRKG.

learning recommendation framework. Unlike CF or direct construction of connections between users using privacy-sensitive information, FEDRKG leverages publicly available item information (e.g., appearance, attributes) to establish higher-order connections between different items, as shown in Fig. 1(c).

The server firstly constructs and maintains *knowledge graphs* (KGs) by utilizing publicly available item information. Then, we employ on-demand sampling of KGs and distribute them to the client. Subsequently, we design a novel method to expand the local graph by merging KG subgraphs with the local user-item interaction graph, enabling the construction of high-order user-item interactions through KGs. Additionally, our framework introduces a request-based distribution mechanism. By obfuscating interaction items into request items, the server can efficiently distribute only the necessary request embeddings, significantly reducing communication overhead compared to previous methods while effectively protecting the privacy of raw interaction items. Simultaneously, we employ *local differential privacy* (LDP) to protect all uploaded gradients, further enhancing the privacy of the federated learning process. Our approach has been extensively evaluated on three real-world datasets, demonstrating its competitive performance compared to centralized algorithms while ensuring privacy preservation. Moreover, FEDRKG outperforms existing federated learning baselines, achieving an average accuracy improvement of approximately 4%. The major contributions of this work are summarized as follows:

- To the best of our knowledge, we are the first to introduce a knowledge graph to enhance the performance of the federated recommendation system while protecting privacy.
- We introduce an algorithm for user-item graph expansion using KG subgraphs to improve local training.
- We propose innovative privacy-preserving techniques for interaction items, while simultaneously reducing communication overhead through strategic distribution of embeddings.

2 Related Work

2.1 Knowledge Graph Based Recommendation

In recent years, significant research has focused on recommendation systems that utilize *Graph Neural Networks* (GNNs). GNNs have gained attention and popularity in recommendation systems due to their ability to learn representations of graph-structured data, which is well-suited for the inherent graph structures in recommendation systems. Knowledge graphs, as a typical graph structure, are often leveraged as side information in recommendation systems. By incorporating knowledge graphs, high-order connections can be established through the relationships between items and their attributes. This integration enhances the accuracy of item representations and provides interpretability to the recommendation results. One type of method is integrating user-item interactions into KG. Methods like KGAT [19], CKAN [20], and MKGAT [12] treat users as entities within KG, and relationships between users and items are incorporated as part of KG's relationships, too. This integration enables the merged graph to be processed using a generic GNN model designed for knowledge graphs. Another idea is employed by KGCN [18] and KGNN-LS [17], directly connecting KG to the user-item graph without any transformation. These methods utilize relation-aware aggregation and consider the user's preference for relationships when generating recommendations.

2.2 Federated Learning for Recommendation System

Federated learning is extensively utilized in privacy-preserving scenarios, as it ensures that the original data remains on local devices while allowing multiple clients to train a model together [7]. Considering the information required for recommendations, which includes users' preferences for items, the introduction of federated learning can help us prevent privacy breaches. FedSage [25] and FKGE [10] focus on cross-silo federation learning, they are not suitable for protecting the privacy of individual users on client devices. FCF [1] and FedMF [4] decompose the scoring matrix, retain user embeddings locally, and aggregate item embeddings on the server. FedGNN [21] utilizes homomorphic encryption for CF and protects the original gradients using pseudo-labeling and LDP. However, the computational requirements for homomorphic encryption pose challenges, particularly on performance-constrained devices. In contrast to methods that do not leverage any side information, FeSoG [9] introduces social networks to establish connections between users. Unfortunately, in many recommendation scenarios such as e-commerce, service providers do not offer social services, and social network information is considered private. Therefore, the lack of user connection on the server in Fig. 1(b), like a social network, restricts the method's ability to generalize [15]. Currently, there is a scarcity of federated learning algorithms that effectively utilize side information for cross-device scenarios.

3 Federated Recommendation with Knowledge Graph Enhancement

3.1 Problem Definition

User-item interactions can be represented by a typical bipartite graph $\mathcal{G} = (\mathcal{U}, \mathcal{T}, E)$, where $\mathcal{U} = \{u_1, u_2, \ldots, u_N\}$ and $\mathcal{T} = \{t_1, t_2, \ldots, t_M\}$ represent a set of users and items of size N and M, respectively. To describe the set of edges E, an interaction matrix $\mathbf{Y} \in \mathbb{R}^{M \times N}$ is employed. In particular, y_{ut} takes on the value 1 if an interaction exists in the user's history, and 0 otherwise.

For federated recommendation, each client c_i owned by corresponding user u_i can only access the interaction graph $\mathcal{G}_i$ stored locally, containing a set of items $\mathcal{T}_i$ that have been interacted with. Each $\mathcal{G}_i$ is a subgraph of the global interaction graph $\mathcal{G}$.

In addition to the client-side data, the server maintains a knowledge graph K, which is represented as a series of triples $\{(h, r, t) \mid h, t \in \mathcal{E}, r \in \mathcal{R}\}$. The entities h and t each refer to the head and tail, respectively, within the specific combination denoted by each triple, both belonging to the set of entities $\mathcal{E}$. The relationship r represents the connection between two distinct entities, belonging to a set of relations $\mathcal{R}$.

Our goal is to train a generalized GNN model using the local bipartite graphs $\mathcal{G}_i$ and the knowledge graph K while preserving user privacy. The model predicts the probability $\hat{y}_{ut}$ that a user u will be interested in an unexplored item t.

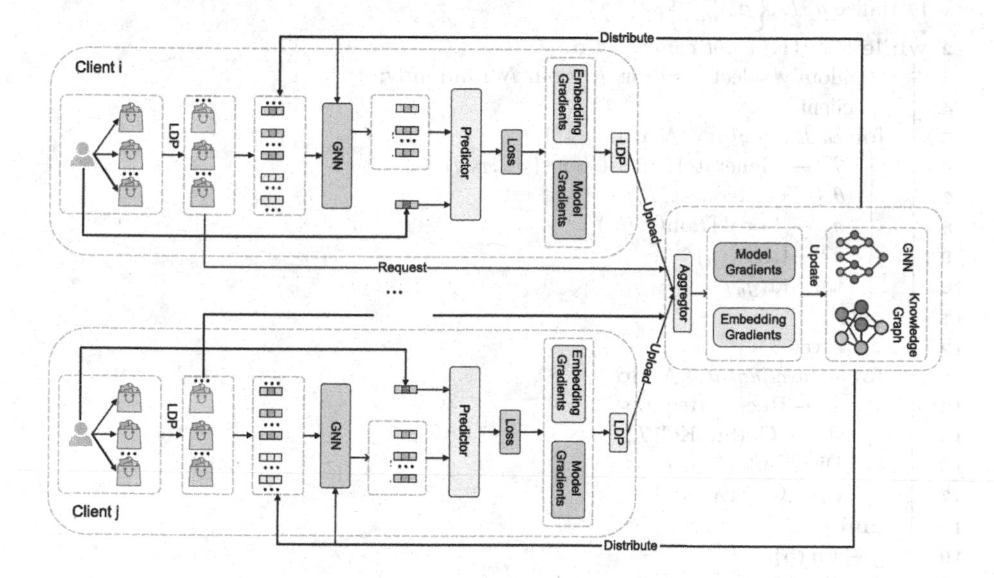

Fig. 2. The framework of FedRKG.

3.2 Framework Overview

To enable privacy-preserving recommendation tasks across diverse private devices, we introduce a federated learning framework, in Fig. 2, based on the knowledge graph named FedRKG. In the proposed framework, the client-server architecture is adopted. The client, which is the user's private device, is responsible for training a local graph neural network model. The server, on the other hand, is responsible for aggregating the models and embeddings, maintaining the knowledge graph, and constructing higher-order connections between clients.

The entire workflow is summarized in Algorithm 1, which concisely represents the complete workflow.

3.3 Client Design

In our framework, the client plays a crucial role in two tasks. First, it is responsible for ensuring the confidentiality of the user's private information during the communication process with the server, which is achieved through privacy-preserving algorithms. Second, the client utilizes the embeddings and models provided by the server to expand the local user-item graph and train the local model.

Algorithm 1: FedRKG

Input: Neighbor sampling size K; embedding size d; depth of receptive field H; learning rate η; client number N; item number M; pseudo items p; $(0, 1)$ flipping q;LDP parameter δ, λ; knowledge graph K; clients local graph $\left\{ \mathcal{G}_n|_{n=1}^{N} \right\}$

Output: GNN parameters and KG embeddings θ,user embeddings $\left\{ \mathbf{e}_u^*|_{n=1}^{N} \right\}$

1 Initialize θ, K,$\left\{ \mathbf{e}_u^*|_{n=1}^{N} \right\}$;
2 **while** FedRKG *not converged* **do**
3 | Randomly select a subset $\mathcal{N}$ from N randomly;
4 | // client
5 | **for** *each client* $n \in \mathcal{N}$ **do**
6 | | $\mathcal{T}_n' \leftarrow$ GenerateRequestItems($\mathcal{G}n, p, q$);
7 | | $\theta, \mathcal{G}_n \leftarrow$ Request($\mathcal{T}_n'$)
8 | | $g_n \leftarrow$ LocalTrain($\theta, \mathcal{G}_n$)
9 | | $\tilde{g}_n \leftarrow$LDP(g_n)
10 | | Upload($\tilde{g}_n$)
11 | **end**
12 | // server
13 | **for** *each client* $n \in \mathcal{N}$ **do**
14 | | $\mathcal{T}_n' \leftarrow$ReceiveRequest()
15 | | $\mathcal{G}_n \leftarrow$ GetSubKG($\mathcal{T}_n'$)
16 | | Distribute($\theta, \mathcal{G}_n$)
17 | | $\tilde{g_n} \leftarrow$ReceiveGrad()
18 | **end**
19 | $\overline{g} \leftarrow$Eq.(5)
20 | $\theta \leftarrow$Eq.(6)
21 **end**

Based on the knowledge graph shared by the server, we design a novel method to expand the local subgraph. During the request phase, the client applies a privacy protection mechanism to the interaction items $\mathcal{T}_n$, generating obfuscated request items $\mathcal{T}_n'$. These request items are then transmitted to the server. The client receives a GNN model and a knowledge subgraph that includes the request items and some of their neighboring entities in the complete KG. By merging this knowledge subgraph with local user-item interaction, the client generates a graph for local training. This approach guarantees the privacy of the user's interaction records by never disclosing them to the server, while also allowing the client to obtain more item-related information for training, thus indirectly enabling the construction of higher-order connections through knowledge subgraph.

Once the aggregated global model is received, the client proceeds to update its local model and initiates a training process. We use a relation-aware GNN as a recommendation model [3] that conforms to the message-passing paradigm [7],

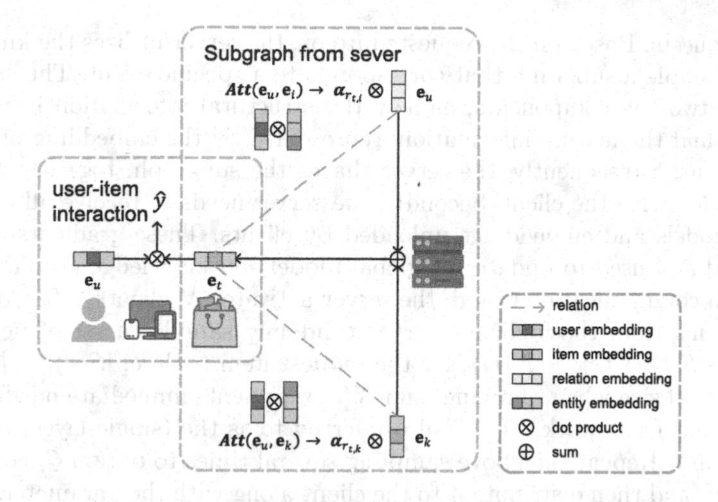

Fig. 3. Relation-aware aggregation in client.

as shown in Fig. 3. For a given user u, entity e_i, e_j, and $r_{i,j}$ as the relation between two entities, we follow node-wise computation at step t+1:

$$x_i^{(t+1)} = \phi\left(x_i^{(t)}, \rho\left(\left\{m_{r_{i,j}}^{(t+1)} : (u, e_j, r_{u,v}) \in \mathcal{E}\right\}\right)\right) \tag{1}$$

where $x_i^t \in \mathbb{R}^d$ is embedding of entity e_i in step t. We utilize a simple summation operation as the reduce function ρ and directly replace the original embeddings with the aggregated results as the reduce function, denoted as ϕ. e_j sends a relationship-aware message $m_{r_{i,j}}$ to its neighbor:

$$m_{r_{i,j}} = \alpha_{r_{i,j}}^u x_j \tag{2}$$

where the attention score $\alpha_{r_{i,j}}^u$ between user u and relation $r_{i,j}$ is derived from the following formula:

$$s_{r_{t,i}}^u = score(\mathbf{e}_u, \mathbf{e}_{r_{t,i}}) \tag{3}$$

$$Att(e_u, e_i) = \alpha_{r_{t,i}}^u = \frac{\exp\left(s_{r_{t,i}}^u\right)}{\sum_{i' \in \mathcal{N}(t)} \exp\left(s_{r_{t,i'}}^u\right)} \tag{4}$$

We calculate an attention score using a score function (e.g. inner product) and then normalize it. After obtaining the final embedding x_t of item t, we calculate the prediction $\hat{y}$ by a readout function and then train this GNN model using BCE as the loss function. Finally, client uploads encrypted gradient to server.

3.4 Server Design

Similar to clients, the server performs distinct tasks that are mainly distributed across two phases. Firstly, the server's primary responsibility is to respond to the client's requests. Based on the requested items, the server utilizes the knowledge graph to sample a subgraph that corresponds to a specific client. The subgraph comprises two key components, namely the structural information in the form of triples, and the feature information, represented by the embedding of entities and relations. Subsequently, the server shares the subgraph, together with the global model, with the client. Secondly, the server needs to receive all gradients of local models and embeddings uploaded by clients. These gradients are then aggregated and used to update the global model and knowledge graph.

In each communication round, the server activates $\mathcal{N}$ clients. After receiving request items from those clients, server randomly samples a set of neighbors, denoted as $S(t) \triangleq \{e|e \sim N(t)\}$, for the request item t. Here, $|S(v)| = K$ represents the fixed size when sampling, and $N(t)$ represents immediate neighbors for item t. In our framework, $S(v)$ is also referred to as the (single-layer) receptive field of item t. Repeat the above sampling several times to obtain $\mathcal{G}_i$ containing n iterations and then distribute it to the client along with the parameters θ, consisting of the model parameters θ^m and all embeddings of entities and relations in $\mathcal{G}_i$ denoted by θ^e. Finally, it receives the local gradients $\tilde{g}_i$ of these clients and aggregates them as follow:

$$\overline{\mathrm{g}} = \frac{\sum_{n \in \mathcal{N}} |\mathcal{T}'_n| \cdot \tilde{\mathrm{g}}^n}{\sum_{n \in \mathcal{N}} |\mathcal{T}'_n|} \tag{5}$$

After aggregation, the server updates all parameters $\boldsymbol{\theta}$ with gradient descent as:

$$\theta^* = \theta - \eta \cdot \bar{g} \tag{6}$$

where η is the learning rate.

3.5 Privacy-Preserving Communication

User Privacy. Within our proposed framework, user-related privacy pertains primarily to user embedding. Traditional embedding-based recommendation algorithms can derive both user and item embeddings and generate user-specific recommendations through a straightforward readout operation. However, user embeddings comprise the user's preference characteristics, which can lead to a compromise of their privacy. In the federated learning scenario where the server does not have access to the raw data, to avoid exposing user preferences directly to the server, it is obvious that we need to keep the user embeddings on the client side and isolate them from the server. Clients can simply protect user-related privacy by refraining from uploading user embeddings after the training phase.

Interaction Privacy. The interaction between users and items is considered highly sensitive information, susceptible to potential leaks during two stages. Firstly, due to the large size of the knowledge graph for items and limited transmission bandwidth, it is not practical to distribute all embeddings to client similar to FedGNN and FedSoG. Instead, we aim to complete the entire training process through the limited distribution of embeddings. However, this presents a challenge in determining which embeddings should be distributed by the server. Server can not explicitly obtain the required embeddings, as this would mean that it has access to the client's real interaction item. Therefore, we need to obfuscate the original interaction items to obtain encrypted request items, which can then be sent to the server to sample the corresponding subgraph required for training.

We have designed a *local differential privacy*(LDP) mechanism to generate request items from the interaction items. Specifically, user-item interaction for user u can be represented as a set $\{(t_i, y_{ui}) \mid y_{ui} \in \{0,1\}, i = 1, 2, \ldots, n\}$. This collection contains $|\mathcal{T}|$ elements, each of which is a binary, the first of which is an item and the second is either 0 or 1, indicating whether the user interacted with the item. Let the query for the t_i be y_{ui}, then the interaction can be privacy-preserving using an ϵ-LDP algorithm. The privacy budget ϵ indicates the maximum acceptable loss of privacy. Let the interaction for each item satisfy ϵ-LDP, and we have: for any item, keep the original interaction value with probability $\frac{e^\epsilon}{e^\epsilon+1}$ and invert it to another value with $\frac{1}{e^\epsilon+1}$ (0,1 flipping).

A potential privacy concern with the widely used pseudo-labeling method in previous work is that the interacted item will always generate gradients, even if pseudo-labeling is used. Additionally, the pseudo-labeling method applied during the training phase does not effectively reduce the communication overhead associated with distributing embeddings. To address this issue, we first sample several non-interactive samples, mix them with real interaction items, and further obfuscate them by the above LDP method to achieve privacy protection.

Gradients Privacy. Ensuring the privacy of users' sensitive information is a critical concern when maintaining a knowledge graph and updating the global model in a federated recommendation system. In each communication round, the server needs to aggregate gradients of entity embeddings, relational embeddings, and GNN models from different clients. However, it has been demonstrated, as exemplified by FedMF, that uploading users' gradients in consecutive steps can lead to the inadvertent exposure of sensitive data, such as users' ratings. Therefore, we need to obfuscate gradients to protect user privacy. However clients of recommendation systems, such as mobile devices, often have limited computational capabilities [13], and computationally intensive methods like homomorphic encryption may not be practical to implement on such devices. To tackle this, we employ LDP by injecting random noise into the local gradients before uploading them to the server. This approach effectively protects row gradients without compromising the accuracy of the model. Moreover, it helps ensure that the computational overhead remains manageable and within acceptable limits.

To be more specific, gives all gradients as $g_n = \{g_n^e, g_n^m\} = \frac{\partial \mathcal{L}_n}{\partial \theta}$, where $\mathcal{L}_n$ denotes loss of client n, the LDP is formulated as:

$$\tilde{\mathbf{g}}_n = \text{clip}\left(\mathbf{g}_n, \delta\right) + \text{Laplacian}\left(0, \lambda\right) \tag{7}$$

where $\tilde{g}_n$ is the encrypted gradient, $\text{clip}(x, \delta)$ denotes the gradient clipping operation with a threshold δ to limit x and prevent the gradient from being too large, after which we add to the gradient a mean value of 0 and an intensity of λ of Laplacian noise, denoted by Laplacian $(0, \lambda)$. This results in a ϵ-LDP, where the privacy budget ϵ is $\frac{2\delta}{\lambda}$.

4 Experiment

4.1 Datasets

In order to ensure the robustness of the algorithm, we aim to test the overall performance of the framework on a variety of datasets with different sizes, sparsity, and domains. Therefore, we have selected the following real-world datasets:

- **MovieLens-20M** [6] contains five-star ratings from MovieLens, a movie recommendation service, as of 2019. Each user in the dataset has provided a minimum of 20 ratings (ranging from 1 to 5) on the MovieLens website.
- **Book-Crossing** [27] contains user ratings (ranging from 0 to 10) of books extracted from the Book-Crossing community in 2004. In this dataset, a rating of 0 indicates an implicit interaction between the user and the book.
- **Last.FM** [2] contains musician listening recodes from the Last.FM music streaming service. We consider artists as items and the number of listens as ratings. In particular, we utilize the HetRec 2011 version in our study.

To adapt the dataset for the recommendation task in a federated learning environment, several steps are taken. Firstly, only the user-item interactions are retained from the original dataset, while other data are discarded. Then, the publicly available Microsoft Satori is utilized to create a knowledge graph by selecting triples with a confidence level greater than 0.9, where the tail corresponds to items in the dataset. Interactions, where the item is not present in the knowledge graph, are subsequently removed. Next, these three datasets are transformed into implicit feedback. We consider all artists listened to in Last.FM, all books with ratings present in book-cross, and all movies with ratings greater than or equal to 4 stars in MovieLens, as positive feedback. Conversely, items not meeting these criteria are treated as negative feedback. Lastly, since the original recommendation dataset already contains user information, each user's data is assigned to the corresponding client to generate a federated learning dataset. Details of the dataset are shown in Table 1

4.2 Baselines

We compare the proposed FEDRKG with the following baselines, in which the first two baselines are KG-free while the rest are all KG-aware methods. Hyperparameter settings for baselines are introduced in the next subsection.

- **SVD** [8] is a classical CF recommendation algorithm based on matrix decomposition. Here we use an unbiased version.

Table 1. Dataset basic information and hyperparameters, notation is consistent with Algorithm 1.

	MovieLens-20M	Book-Crossing	Last.FM
users	138,159	19,676	1,872
items	16,954	20,003	3,846
interactions	13,501,622	172,576	42,346
entities	102,569	25,787	9,366
relations	32	18	60
KG triples	499,474	60,787	15,518
K	4	8	8
d	32	64	16
H	2	1	1
λ	10^{-7}	2×10^{-5}	10^{-4}
η	2×10^{-2}	2×10^{-4}	5×10^{-4}
$\mathcal{N}$	32768	64	32

- **LibFM** [11] is a method based on Factorization Machines that captures the similarity between features
- **PER** [23] is an algorithm based on a personalized attention mechanism and constructs a Meta-path between users and items through a heterogeneous graph (KG).
- **CKE** [24] is a knowledge graph-based collaborative embedding recommendation algorithm that combines data from CF and other modalities.
- **RippleNet** [16] is a memory-network-like approach that simulates and exploits the ripple effect between users and items to propagate information on the knowledge graph
- **KGCN** [18] is a KG-based method, that achieves efficient recommendations by merging KG and CF data.
- **FedMF** [4] is a recommendation algorithm based on matrix decomposition while protecting privacy through an encryption mechanism.
- **FedGNN** [21] is a GNN-based recommendation algorithm that uses homomorphic encryption for aggregation and protects the original gradient by differential privacy and pseudo-labeling.

4.3 Experimental Settings

Table 2 shows the hyperparameter for the experiments. We split the datasets into training, validation, and testing sets in a 6:2:2 ratio. AUC and F1 scores are used as evaluation metrics for *click-through rate* (CTR) prediction.

For the Last.FM, Book-Crossing, and MovieLens-20M datasets, the SVD method is applied with imensions $(8, 8, 8)$ and learning rates $(0.1, 0.5, 0.5)$. For LibFM, the dimensions are $(8, 1, 1)$. PER utilizes the user-item-attribute-item

Table 2. Results for CRT prediction. KGCN achieves the best AUC among the first five centralized learning methods. Our method performs best in the next three federal learning methods, while the gap with KGCN is acceptable.

Model	MovieLens-20M		Book-Crossing		Last.FM	
	AUC	F1	AUC	F1	AUC	F1
SVD	0.952(±0.013)	0.909(±0.014)	0.665(±0.058)	0.628(±0.051)	0.760(±0.026)	0.688(±0.022)
LibFM	0.960(±0.018)	0.907(±0.024)	0.692(±0.046)	0.619(±0.063)	0.779(±0.019)	0.711(±0.011)
PER	0.824(±0.119)	0.780(±0.121)	0.611(±0.101)	0.557(±0.100)	0.627(±0.125)	0.593(±0.107)
CKE	0.918(±0.050)	0.866(±0.056)	0.673(±0.057)	0.607(±0.055)	0.739(±0.044)	0.669(±0.046)
RippleNet	0.964(±0.010)	0.909(±0.020)	0.712(±0.023)	0.648(±0.032)	0.777(±0.016)	0.699(±0.015)
KGCN	0.978(±0.002)	0.932(±0.001)	0.738(±0.003)	0.688(±0.006)	0.794(±0.002)	0.719(±0.003)
FedMF	0.865(±0.012)	0.852(±0.015)	0.657(±0.039)	0.605(±0.060)	0.720(±0.018)	0.660(±0.013)
FedGNN	0.939(±0.011)	0.891(±0.021)	0.671(±0.024)	0.620(±0.037)	0.753(±0.014)	0.681(±0.028)
FedRKG	**0.970**(±0.002)	**0.919**(±0.002)	**0.724**(±0.004)	**0.667**(±0.006)	**0.785**(±0.004)	**0.708**(±0.002)

meta-path, with dimensions $(64, 128, 64)$ and learning rates $(0.1, 0.5, 0.5)$. The learning rates for KG in CKE are $(0.1, 0.1, 0.1)$, while the dimensions are $(16, 4, 8)$ and the H values are $(3, 3, 2)$. RippleNet's dimensions are $(16, 4, 8)$, H values are $(3, 3, 2)$, learning rates are $(0.005, 0.001, 0.01)$, regularization parameters λ_1 are $(10^{-5}, 10^{-5}, 10^{-6})$, and λ_2 are $(0.02, 0.01, 0.01)$. Other hyperparameters remain the same as in the original papers, and the federated learning settings are consistent with this paper.

4.4 Overall Comparison

We conduct a comprehensive comparison of multiple models under various settings. Given the dataset's specific characteristics, only including knowledge graphs and user-item graphs, many federated learning algorithms simplify to FedGNN in this dataset. Therefore, we select FedGNN and FedMF as the baseline methods, representing GNN and matrix decomposition approaches in federated learning. The experimental results for CTR prediction are presented in Table 2, while Fig. 4 illustrates the outcomes of top-k recommendation. Based on those results, we draw the following conclusions:

- On the one hand, GNN-based algorithms, such as KGCN and FedRKG, outperform matrix decomposition-based algorithms like SVD and FedMF. This is due to the superior performance of GNNs in automatically capturing user preferences and enabling the spreading of user or item embeddings to neighboring nodes. On the other hand, algorithms that require manual design such as meta-paths for PER and *knowledge graph embedding* (KGE) method for CKE, often underperform due to the complexity of graph data.
- The experimental results consistently demonstrate that the appropriate utilization of additional side information can significantly improve the accuracy of recommendation systems. For example, KGCN and RippleNet outperform other centralized algorithms regarding both AUC and F1 metrics, while FedRKG, as a knowledge graph-based algorithm, performs best in

federated learning. However, it should be noted that not all methods that leverage side information deliver satisfactory outcomes. This holds true for methods like PER and CKE, which encounter difficulties in effectively harnessing side information.
- Knowledge graphs are well-suited for integration into recommendation systems as side information, especially using GNNs, given their inherent graph structure and the ability to combine multi-domain knowledge. Algorithms incorporating relation-aware aggregation, such as KGCN and FEDRKG, achieve the best performance in their respective settings, confirming the effectiveness of introducing relational attention mechanisms.

Overall, our framework outperforms existing federated learning algorithms and achieves competitive performance compared to centralized algorithms.

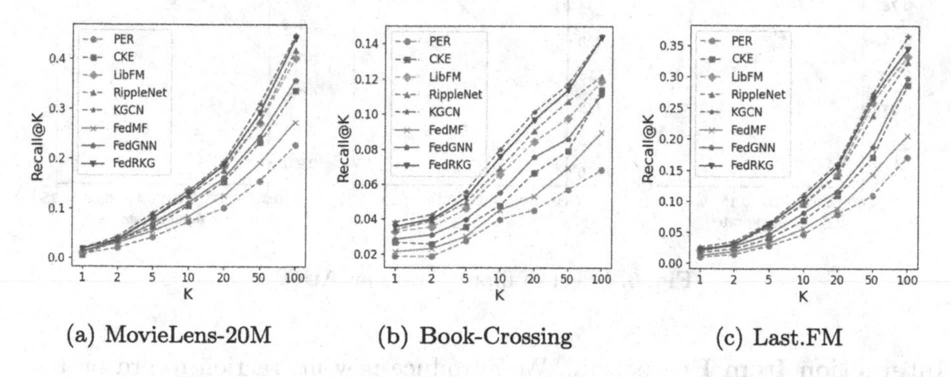

Fig. 4. Results for top-K recommendation. The dashed line represents centralized learning, while the solid line represents federated learning. Our method surpasses all federated baselines and, furthermore, achieves competitive results compared to centralized learning.

4.5 Sensitivity Analysis

Activated Client Number. In general, a smaller number of activated clients in each training round will speed up the model convergence and conversely better capture the global user data distribution. We test the algorithm on three datasets with three different numbers of activation clients, and the results are shown in the figure above. Probably due to the sparse data and a large number of clients, a small adjustment has a limited impact on the final results and the Last.FM and Book-Crossing datasets both show a small decrease in AUC when 64 clients are activated.

Receptive Field Depth. By testing different receptive field depths, we note that an excessive receptive field reduces model prediction accuracy. As data sparsity decreases, better performance needs a larger receptive field, while a one-layer perceptual region is sufficient to achieve better performance on those sparse data sets.

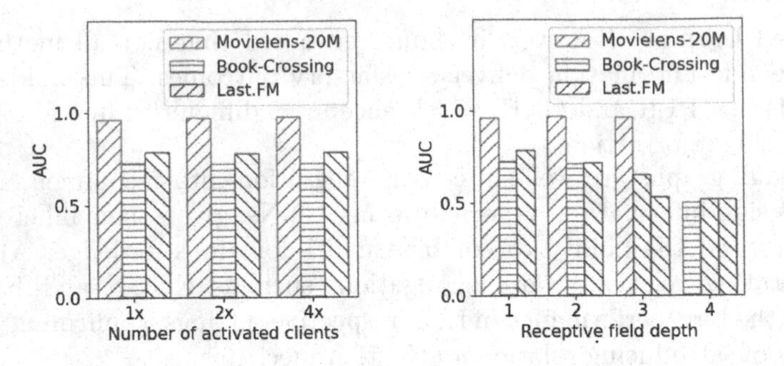

Fig. 5. Sensitivity analysis of activated clients and receptive field depth.

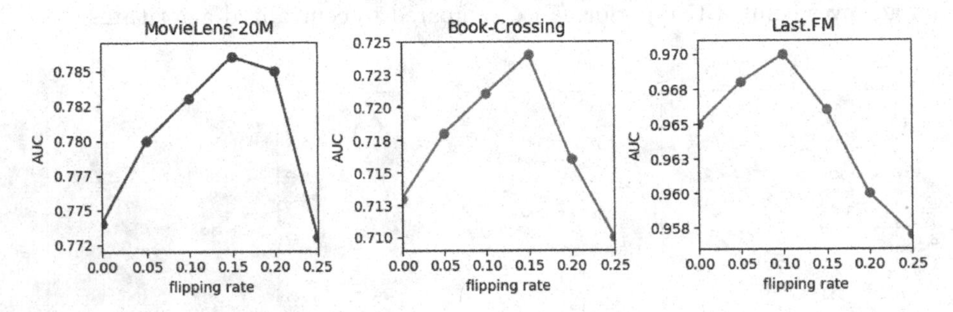

Fig. 6. Effect of flipping rate on AUC.

Interaction Item Protection. We introduce new interaction record protection and assess diverse flipping rates, with corresponding results depicted in Fig. 6. Generally, integrating privacy-preserving mechanisms often diminishes recommendation accuracy. However given limited client-side graph data, our scenario tends to induce model overfitting. Hence, proper regularization effectively enhances recommendation accuracy and privacy protection. Notably, excessive flip rates can compromise system performance despite heightened privacy. Our experiments indicate a balance between accuracy and privacy at a flipping rate of 0.1.

5 Conclusion

This paper introduces a novel federated learning framework, FedRKG, which employs GNN for recommendation tasks. Our approach integrates KG information while upholding user privacy. The limitation here is the absence of user connections, and our forthcoming focus is on improving the efficiency and interpretability of utilizing existing user connections without introducing new private data. Specifically, a server-side KG is created from public item data, maintaining relevant embeddings. The client conceals local interaction items and requests server training data. The server samples a KG subgraph and distributes it with the GNN model to the client. The client then expands its user-item graph with the KG subgraph for training, uploading the gradient for server aggregation. Our framework creates higher-order interactions without extra privacy data, relying solely on public information for KG. Sampled KG subgraphs enhance local training by capturing interactions between users and items without direct links. We employ LDP and pseudo-labeling to protect privacy and reduce overhead by requesting partial data. Gradients are encrypted using LDP for user preference protection and local user embedding storage. Experimental results on three datasets demonstrate our framework's superiority over SOTA federated learning recommendation methods. It also performs competitively against centralized algorithms while preserving privacy.

Acknowledgements. This work is supported by the National Key Research and Development Program of China under Grant No.2021YFB1714600 and the National Natural Science Foundation of China under Grant No.62072204 and No.62032008. The computation is completed in the HPC Platform of Huazhong University of Science and Technology and supported by the National Supercomputing Center in Zhengzhou.

References

1. Ammad-Ud-Din, M., et al.: Federated collaborative filtering for privacy-preserving personalized recommendation system. arXiv preprint arXiv:1901.09888 (2019)
2. Cantador, I., Brusilovsky, P., Kuflik, T.: Second workshop on information heterogeneity and fusion in recommender systems (HetRec2011). In: Proceedings of the 2011 ACM Conference on Recommender Systems, RecSys, pp. 387–388 (2011)
3. Cao, W., Zheng, C., Yan, Z., Xie, W.: Geometric deep learning: progress, applications and challenges. Sci, China Inf. Sci. **65**(2), 126101 (2022)
4. Chai, D., Wang, L., Chen, K., Yang, Q.: Secure federated matrix factorization. IEEE Intell. Syst. **36**(5), 11–20 (2020)
5. Guo, Q., et al.: A survey on knowledge graph-based recommender systems. IEEE Trans. Knowl. Data Eng. **34**(8), 3549–3568 (2020)
6. Harper, F.M., Konstan, J.A.: The movielens datasets: history and context. ACM Trans. Interactive Intell. Syst. **5**(4), 19:1–19:19 (2016)
7. Jin, H., et al.: Personalized edge intelligence via federated self-knowledge distillation. IEEE Trans. Parallel Distrib. Syst. **34**(2), 567–580 (2023)
8. Koren, Y.: Factorization meets the neighborhood: a multifaceted collaborative filtering model. In: Proceedings of the 14th ACM SIGKDD International Conference on Knowledge Discovery and Data Mining, KDD, pp. 426–434 (2008)
9. Liu, Z., Yang, L., Fan, Z., Peng, H., Yu, P.S.: Federated social recommendation with graph neural network. ACM Trans. Intell. Syst. Technol. **13**(4), 1–24 (2022)
10. Peng, H., Li, H., Song, Y., Zheng, V., Li, J.: Differentially private federated knowledge graphs embedding. In: Proceedings of the 30th ACM International Conference on Information & Knowledge Management, CIKM, pp. 1416–1425 (2021)
11. Rendle, S.: Factorization machines with libfm. ACM Trans. Intell. Syst. Technol. **3**(3), 1–22 (2012)
12. Sun, R., et al.: Multi-modal knowledge graphs for recommender systems. In: Proceedings of the 29th ACM International Conference on Information & Knowledge Managemen, CIKM, pp. 1405–1414 (2020)
13. Tian, Y., Wan, Y., Lyu, L., Yao, D., Jin, H., Sun, L.: FedBERT: when federated learning meets pre-training. ACM Trans. Intell. Syst. Technol. **13**(4), 1–26 (2022)

14. Voigt, P., von dem Bussche, A.: The EU General Data Protection Regulation (GDPR). Springer, Cham (2017). https://doi.org/10.1007/978-3-319-57959-7
15. Wang, G., et al.: Temporal graph cube. IEEE Transa. Knowl. Data Eng. 1–15 (2023)
16. Wang, H., et al.: Ripplenet: propagating user preferences on the knowledge graph for recommender systems. In: Proceedings of the 27th ACM International Conference on Information and Knowledge Management, CIKM, pp. 417–426 (2018)
17. Wang, H., et al.: Knowledge-aware graph neural networks with label smoothness regularization for recommender systems. In: Proceedings of the 25th ACM SIGKDD International Conference on Knowledge Discovery and Data Mining, KDD, pp. 968–977 (2019)
18. Wang, H., Zhao, M., Xie, X., Li, W., Guo, M.: Knowledge graph convolutional networks for recommender systems. In: Proceedings of the World Wide Web Conference, WWW, pp. 3307–3313 (2019)
19. Wang, X., He, X., Cao, Y., Liu, M., Chua, T.S.: Kgat: knowledge graph attention network for recommendation. In: Proceedings of the 25th ACM SIGKDD International Conference on Knowledge Discovery and Data Mining, KDD, pp. 950–958 (2019)
20. Wang, Z., Lin, G., Tan, H., Chen, Q., Liu, X.: Ckan: collaborative knowledge-aware attentive network for recommender systems. In: Proceedings of the 43rd International ACM SIGIR Conference on Research and Development in Information Retrieval, SIGIR, pp. 219–228 (2020)
21. Wu, C., Wu, F., Cao, Y., Huang, Y., Xie, X.: FedGNN: federated graph neural network for privacy-preserving recommendation. arXiv preprint arXiv:2102.04925 (2021)
22. Wu, S., Sun, F., Zhang, W., Xie, X., Cui, B.: Graph neural networks in recommender systems: a survey. ACM Comput. Surv. **55**(5), 1–37 (2022)
23. Yu, X., et al.: Personalized entity recommendation: a heterogeneous information network approach. In: Proceedings of the 7th ACM International Conference on Web Search and Data Mining, WSDM, pp. 283–292 (2014)
24. Zhang, F., Yuan, N.J., Lian, D., Xie, X., Ma, W.Y.: Collaborative knowledge base embedding for recommender systems. In: Proceedings of the 22nd ACM SIGKDD International Conference on Knowledge Discovery and Data Mining, KDD, pp. 353–362 (2016)
25. Zhang, K., Yang, C., Li, X., Sun, L., Yiu, S.M.: Subgraph federated learning with missing neighbor generation. In: Proceedings of the Annual Conference on Neural Information Processing Systems, NeurIPS. vol. 34, pp. 6671–6682 (2021)
26. Zhou, J., et al.: Graph neural networks: a review of methods and applications. AI Open **1**, 57–81 (2020)
27. Ziegler, C.N., McNee, S.M., Konstan, J.A., Lausen, G.: Improving recommendation lists through topic diversification. In: Proceedings of the World Wide Web Conference, WWW, pp. 22–32 (2005)

Efficient and Reliable Federated Recommendation System in Temporal Scenarios

Jingzhou Ye[1], Hui Lin[1(✉)], Xiaoding Wang[1(✉)], Chen Dong[2], and Jianmin Liu[1]

[1] The College of Computer and Cyber Security, Fujian Normal University, Fuzhou 350117, China
{linhui,wangdin1982}@fjnu.edu.cn

[2] College of Computer and Data Science, Fuzhou University, Fuzhou 350116, China

Abstract. Addressing privacy concerns and the evolving nature of user preferences, it is crucial to explore collaborative training methods for federated recommendation models that match the performance of centralized models while preserving user privacy. Existing federated recommendation models primarily rely on static relational data, overlooking the temporal patterns that dynamically evolve over time. In domains like travel recommendations, factors such as the availability of attractions, introduction of new activities, and media coverage constantly change, influencing user preferences. To tackle these challenges, we propose a novel approach called FedNTF. It leverages an LSTM encoder to capture multidimensional temporal interactions within relational data. By incorporating tensor factorization and multilayer perceptrons, we project users and items into a latent space with time encoding, enabling the learning of nonlinear relationships among diverse latent factors. This approach not only addresses the privacy concerns by preserving the confidentiality of user data but also enables the modeling of temporal dynamics to enhance the accuracy and relevance of recommendations over time.

Keywords: Federated Learning · Business Transaction · Tensor Factorization · Time Interaction Learning

1 Introduction

Recommendation systems have been widely used to create personalized prediction models, helping individuals identify content that interests them. This not only brings convenience to users but also creates economic benefits for businesses, achieving a win-win situation for both users and enterprises. Many businesses have applied recommendation systems to their respective business scenarios to collect various personal characteristics, such as demographic features, explicit feedback through ratings, or implicit feedback through user interactions with specific projects.

H. Jin et al. (Eds.): GPC 2023, LNCS 14504, pp. 97–107, 2024.
https://doi.org/10.1007/978-981-99-9896-8_7

The most direct technique for generating recommendations is collaborative filtering [1,2]. In the past, in the case of centralized learning, service providers commonly collected user profile information and user-project interaction data to be processed in a data center. However, this approach is becoming increasingly impractical in today's society. On the one hand, this is due to the introduction of privacy and data protection laws and regulations such as the General Data Protection Regulation (GDPR) [3], as well as an increasing awareness of privacy protection among users. Specifically, GDPR and other laws and regulations require that no organization or institution may collect personal data of data subjects without their consent. On the other hand, in traditional recommendation systems, a user's preferences and learning feature vectors may reveal sensitive information, and some advanced techniques may result in data de-anonymization. Generally speaking, traditional machine learning algorithms are often unable to train an effective model without obtaining enough user data. Therefore, a method is needed to construct a recommendation system that performs as close as possible to centralized training without leaking private user data.

In conventional recommendation methods, users are required to transmit their data to a central server, limiting their role to data generation and transmission. All data processing and model construction tasks are then performed by the server. However, this approach raises concerns regarding unauthorized access to user data during the transmission process, leading to potential risks such as data theft, fraud, or identity theft. To address these concerns and alleviate the burden on servers dealing with massive data storage and processing, Google introduced federated learning in 2017 [4]. This innovative technology takes advantage of the increased storage and computing power of mobile devices used in everyday life. By performing machine learning locally on the user's device without transmitting raw data, federated learning offers significant benefits in terms of privacy preservation and efficient distributed computing. The potential combination of federated learning with recommendation systems has attracted considerable interest from both industry and academia. The field of federated recommendation still holds untapped potential, particularly in exploring neural network and matrix factorization techniques. Traditional federated recommendation models, based on matrix factorization or neural collaborative filtering, have demonstrated great success in modeling relational data, such as user-item interactions. Further advancements and research can be made to enhance the capabilities of federated recommendation systems.

However, the previous federated recommendation models have limitations in assuming the relationship data to be static, as these models did not consider the changing user preferences over time and the underlying factors driving the changes in user-item relationships. Our idea is to extend the user-item interaction matrix represented by local dataset to a three-dimensional tensor that includes temporal information. Subsequently, tensor factorization techniques can be used to project users and items into a latent space with time encoding. In addition, traditional federated matrix factorization models use dot product for link prediction, ignoring the possibility of modeling non-linearity between latent factors.

To address these challenges and limitations, we use a neural network-based federated tensor factorization model (FedNTF) for time-aware interaction learning in the recommendation scenario. This model integrates long short-term memory (LSTM) networks and tensor factorization technology into the existing FedNCF model. The LSTM module captures the dependencies of multidimensional interactions based on the learned time intervals. It transforms the current user and item to be predicted, as well as the interaction time between the user and item, into their respective embeddings, concatenates them together, and inputs them into a multi-layer perceptron architecture. Thus, the learned representation encodes nonlinear interactions between different dimensions. The gradients of the relevant embeddings and predictors are sent to the server for secure aggregation. The aggregated global model is then distributed to the currently selected client at the beginning of the next round and updated through multiple rounds of training iterations until the global model converges.

The remainder of this paper is structured as follows. In Sect. 2, we delve into the existing research in the field of federated recommendation. Section 3 provides an overview of the background knowledge, encompassing topics such as matrix factorization in recommendation systems, the framework of neural collaborative filtering, and the training process of federated learning. Moving on, Sect. 4 presents a detailed explanation of our proposed algorithm, FedNTF. Subsequently, in Sect. 5, we evaluate the performance of FedNTF specifically in the context of the link prediction task. Finally, Sect. 6 serves as the concluding section, summarizing the key findings and contributions of this paper.

2 Related Work

Ammad et al. [5] proposed the first federated collaborative filtering recommendation algorithm (FCF). The algorithm addresses the issue of local user feature vector updates and item feature vector gradient computation in the ALS (alternating least square)-based collaborative filtering algorithm, where uploading item gradient information may lead to information leakage [6]. In addition, in their system, the coordinating server waits for updates from all available participants before initiating the update process. Therefore, this approach is an asynchronous FL framework [7], where the coordinating server waits for updates from one to multiple clients, which may result in delayed updates [8], i.e., received updates may be based on outdated model calculations.

Dolui et al. [9] proposed a federated matrix factorization algorithm (FederatedMF) to address the issue of traditional matrix factorization algorithms requiring collection of user and item feature matrices at the server-side. In FederatedMF, each user's (user, item) rating vector is stored locally on the client-side and updates for user feature vector U_u and item feature vector V_i are performed locally, followed by sending the item feature matrix to the server. The server performs weighted averaging of the received item feature matrices to obtain the latest item feature matrix.

Perifanis et al. [10] proposed a federated version of the Neural Collaborative Filtering approach (FedNCF) that enables training without users needing to expose or transmit their raw data. Experimental results showed that FedNCF achieved recommendation quality comparable to the original NCF system. By integrating enhanced privacy protection with a secure aggregation scheme, the approach satisfies security requirements for honest-but-curious (HBC) entities without compromising the quality of the original model.

3 Preliminaries

3.1 Matrix Factorization

Matrix factorization (MF) is a collaborative filtering technique that represents users and items using latent feature vectors inferred from their interaction patterns [11]. In general, a recommendation service provider can access a set of M users, $U = \{u_1, u_2, ..., u_M\}$, and a set of N items, $I = \{i_1, i_2, ..., i_N\}$. Each user u_i has interacted with a subset of items n, and this interaction can be explicit, such as a rating, or implicit, such as clicking or browsing an item. The interactions between user i and a subset of items can be expressed as $r_{i,j} \in \mathbb{R}$, where j represents the item index. All user-item interactions can be represented by a sparse matrix $R \in \mathbb{R}^{M \times N}$, since most users only interact with a small subset of items. The goal of a collaborative filtering system is to provide a ranked list of top-K items that a specific user has not interacted with yet but align with their preferences. The user interaction matrix R is factorized into the product of two low-dimensional matrices, $R \approx X^T Y$, where $X \in \mathbb{R}^{D \times M}$ and $Y \in \mathbb{R}^{D \times N}$ are the latent feature matrices for users and items, respectively. D represents the dimensionality of the latent space.

3.2 Neural Collaborative Filtering

Cheng et al. [12] introduced Neural Collaborative Filtering (NCF), a collaborative filtering framework. While traditional filtering techniques like matrix factorization rely on linear models, they may overlook more intricate relationships within user-item interaction data. In contrast, the NCF framework leverages deep neural networks to potentially capture deep associations between users and items, as deep neural network models have demonstrated their capability to approximate any continuous function. The proposed framework by Cheng et al. combines two approaches: one is a generalization of the matrix factorization concept known as Generalized Matrix Factorization (GMF), and the other involves utilizing multi-layer perceptrons (MLP) to learn the relationship between the latent features of users and items.

The authors demonstrated that the integration of the linear GMF model with the nonlinear MLP model in NeuMF yields improved recommendation quality and faster convergence rates. In NeuMF, the GMF generates a product vector based on latent vectors, while the MLP concatenates the latent vectors and feeds them into the deep neural network. These two outputs are then combined in the final hidden layer to generate prediction results for recommendations.

3.3 Federated Learning

In conventional machine learning techniques, the training data is typically centralized and stored prior to training. However, federated learning offers an alternative approach. With the increasing usage of mobile devices, these devices have become valuable data sources for data-driven companies aiming to develop powerful algorithms. Nonetheless, much of this data is private and unsuitable for storage on a central server. Federated learning addresses this challenge by distributing the training process to mobile devices, enabling training without the need to exchange sensitive data. The fundamental steps of the training process are outlined as follows:

In the collaborative training process of federated setting, the central server initially shares the current global model with participating devices. Each device then independently trains the model using its own data. After local training, the devices send their personalized model updates back to the central server. Finally, the central server aggregates these updates to create a new global model, which serves as the baseline for the next round of training. This iterative process ensures collaborative training while preserving the privacy of individual data.

4 The Implementation Details of the Proposed Federated Neural Tensor Factorization (FedNTF)

Assume we have a recommendation system with sets of m items and n users, denoted as I and U respectively. In this collaborative setting, the clients aim to jointly train a global model while ensuring the privacy of their personal data. The model parameters θ of the recommendation system consist of four components: an item model θ_{item} that maps item IDs to item embedding vectors, a user model θ_{user} that generates user interest embedding vectors based on user profiles such as user ID or previous interactions with items, a time model θ_{time} that utilizes embedding vectors from previous time slots to produce the current time embedding vector, and a prediction model θ_{pred} that takes the corresponding embedding vectors of items, users, and time as input to predict the probability of user-item interactions. During each training round, the server initially distributes the current global model parameters $[\theta_{item}; \theta_{time}; \theta_{pred}]$ to a randomly selected subset of n_c clients.

Our framework employs embedding matrices $E_u \in \mathbb{R}^{l \times n}$ and $E_i \in \mathbb{R}^{l \times m}$ to implement the user model θ_{user} and item model θ_{item}, respectively. Here, the embedding dimension is represented by l. To obtain the user latent vector U_k and item latent vector I_v, we feed the one-hot encodings of users and items into the embedding layer. The time model θ_{time} is composed of an LSTM encoder, a time embedding matrix $E_t \in \mathbb{R}^{l \times o}$, and a projection layer, where o denotes the number of time intervals. The LSTM encoder generates embeddings to capture the evolving temporal hidden factors. We formally introduce the hidden states c_t and h_t to describe the encoding process of the contextual sequence.

$$(c_1, h_1) = LSTM(T_{e-s}, c_0, h_0) \tag{1}$$

$$(c_2, h_2) = LSTM(T_{e-s+1}, c_1, h_1) \tag{2}$$

$$\ldots$$

$$(c_s, h_s) = LSTM(T_{e-1}, c_{s-1}, h_{s-1}) \tag{3}$$

Through the utilization of the last hidden state vector ht, we can generate embedding vectors $\hat{T}_e$ by employing a projection layer. This process is depicted below:

$$\hat{T}_e = \sigma(W_t h_s + b_t) \tag{4}$$

In this context, the projection weight matrix is denoted as W_t, and the projection bias vector is represented by b_t. The activation function used is the sigmoid function, denoted as σ.

In the traditional tensor factorization approach, the predicted value $\hat{x}_{k,v,e}$ is obtained by taking the dot product of U_k, I_v, and T_e. This represents the probability of the interaction between user k and item v at time slot e. However, this approach has significant limitations due to its linear nature, which cannot effectively handle the complex nonlinear interactions found in real-world relationship data. Furthermore, it fails to capture the dynamic nature of time since the predicted interaction value $\hat{x}_{k,v,e}$ solely depends on the current time slot T_e. To address these issues, we propose a federated neural tensor factorization approach. We concatenate the embedding vectors of U_k, I_v, and T_e and feed them as input to a multilayer perceptron to obtain the output $\hat{x}_{k,v,e}$. This approach leverages the neural network architecture to capture complex nonlinear interactions by combining latent factors from previous embedding layers.

$$z_1 = f_{a1}(W_1 \left[U_k, I_v, \hat{T}_e\right] + b_1) \tag{5}$$

$$z_2 = f_{a2}\left(W_2 z_1 + b_2\right) \tag{6}$$

$$\ldots$$

$$z_{n-1} = f_{an-1}(W_{n-1} z_{n-2} + b_{n-1}) \tag{7}$$

$$\hat{x}_{k,v,e} = \sigma(W_n z_{n-1}) \tag{8}$$

The i-th hidden layer in our model is characterized by the weight matrix w_i, bias vector b_i, and activation function f_{ai}, where we specifically employ the ReLU function in our experiments. Inspired by He et al.'s approach in designing a neural collaborative filtering model, we incorporate an unbiased fully connected layer as the final layer in our model. This layer maps the vectors to a one-dimensional space and utilizes a sigmoid activation function to obtain the predicted probability value of the interaction between users and items.

After client selection, each client calculates the update gradient g based on its local data. The training data for the local link prediction task consists of positive instances D^+ and negative instances D^-, where the negative instances can be either all uninteracted items or a random subset of them. Since the training process is treated as a binary classification task, binary cross-entropy is employed to train the local model. Subsequently, the client uploads $[g_{item}; g_{time}; g_{pred}]$ to

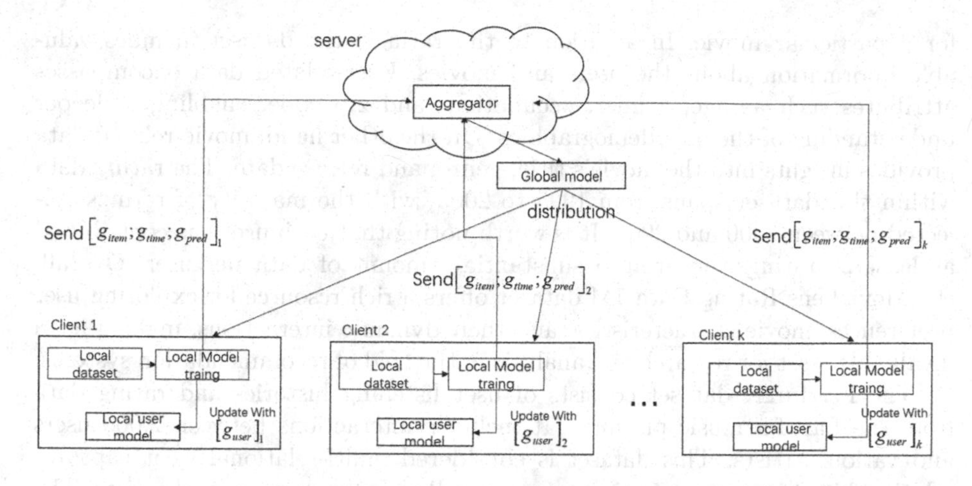

Fig. 1. The training process of the federated recommendation system

the server and updates its local user model θ_{user} using g_{user}. However, in order to preserve user privacy preferences and following the approach proposed by Wu et al. [13,14], g_{user} is not uploaded. Finally, the server aggregates all the received gradients using specific aggregation rules and updates the global model $[\theta_{item}; \theta_{time}; \theta_{pred}]$. This training process is repeated until convergence.

In summary, our FedNTF is a multi-layer representation learning model in a federated environment, using neural methods for tensor decomposition to explicitly model the temporal evolution interactions between different dimensions. Our federated recommendation system training process is depicted in Fig. 1.

5 Experiment

In this section, we describe our experimental setup and present our experimental results. We assess the effectiveness of FedNTF in the context of link prediction, a concrete real-world application that involves inferring the existence of links in dynamic scenarios. This evaluation focuses specifically on the task of predicting links in dynamic relational data. We also perform horizontal comparisons of our model with FedMF and FedNCF on two different datasets.

5.1 Experimental Setting

We use two real-world datasets commonly used to evaluate recommendation systems in collaborative filtering algorithms: MovieLens 1M and Lastfm2K to evaluate the performance of the FedNTF system.

The MovieLens Rating Data 1M dataset comprises a vast collection of 1 million ratings provided by more than 6,000 users for over 4,000 movies. Each rating is represented by an integer ranging from 1 to 5, reflecting the user's preference for a particular movie. In addition to the ratings, the dataset includes valuable information about the users and movies. User-related data encompasses attributes such as age, gender, occupation, and zip code, enabling a deeper understanding of the user demographics. On the other hand, movie-related data provides insights into the movie's title, genre, and release date. The rating data within this dataset spans from 1995 to 2003, with the majority of ratings collected between 2000 and 2003. It is worth noting that each user has contributed at least 20 ratings, ensuring a substantial amount of data per user. Overall, the MovieLens Rating Data 1M dataset offers a rich resource for exploring user preferences, movie characteristics, and their dynamic interactions, making it an invaluable asset for research and analysis in the field of recommendation systems.

The Lastfm2K dataset consists of user listening histories and rating data from the Last.fm music platform. It includes interactions between 2,000 users and various artists. This dataset is considered multi-relational as it captures relationships between users and songs, as well as between songs and artists. The dataset covers a time span of 3 months, from March to June 2005. However, we observed that in this dataset, the same user may have multiple interactions with the same artist simultaneously due to the presence of multiple music labels describing different attributes of the artist or music work, such as style, genre, and emotion. For our training task, which focuses on determining user interest in specific artists, we retain only one interaction between a user and an artist at a given time for training purposes. Furthermore, we filter out users with fewer than 5 interactions and artists who have never interacted with any users in this dataset to ensure data quality and relevance for our training.

During the initial data processing phase, a common technique employed in recommendation systems involves transforming the numerical ratings in the MovieLens dataset into implicit feedback. In the case of the Lastfm2K user interaction data, it is already in implicit format. Subsequent to this step, we group the user interactions within the dataset. To evaluate the performance of the recommendation model, we utilize the well-established leave-one-out evaluation method. This involves setting aside the last interaction of each user as the test data, while employing the remaining interaction data for training purposes. Additionally, as a rapid evaluation approach, we establish a baseline by randomly pairing each item in the test set with 100 uninteracted items.

By employing the leave-one-out evaluation strategy, the collaborative filtering task undergoes a transformation into a ranking task, wherein the reserved items of each user are to be ranked among 100 unobserved items. The resulting ranking list is evaluated using two metrics: hit rate (HR) and normalized discounted cumulative gain (NDCG). To elaborate, HR assesses whether the true item appears within the top K ranking positions, while NDCG considers the position of the hit. In our experimental setup, we compute HR and NDCG at K = 10. Throughout the experiments, we calculate these two metrics for each user and subsequently average the scores.

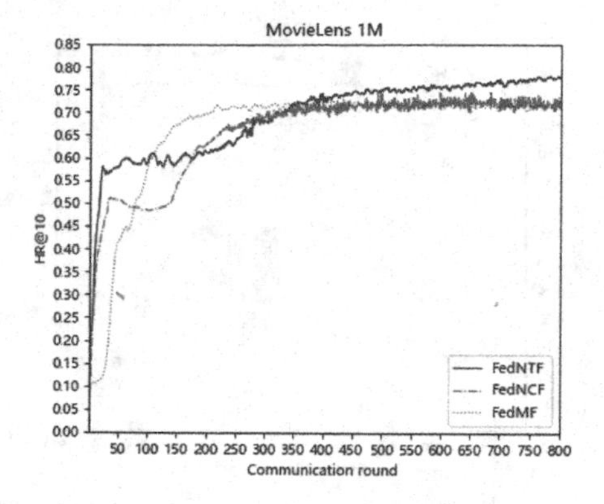

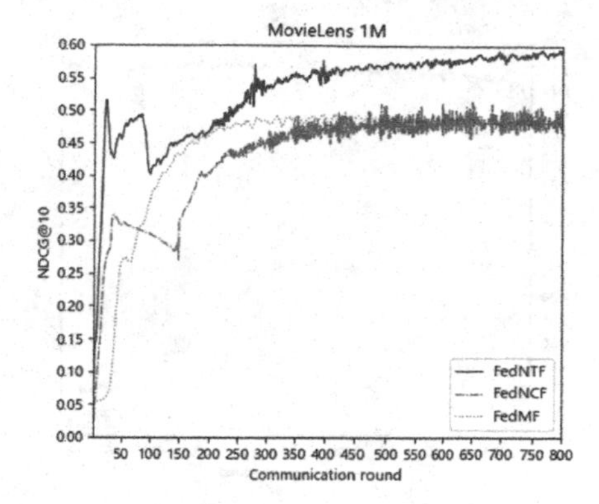

Fig. 2. Comparing FedNTF with other baseline methods on the MovieLens 1M dataset, analyzing their performance across multiple metrics.

5.2 Experimental Result

Figure 2 and Fig. 3 present the evaluation results of FedNTF using the FedAvg aggregation rule for link prediction tasks on the MovieLens 1M and Lastfm2K datasets, respectively. To conduct the evaluation, we partitioned the timestamps in both datasets into one-week time intervals. In each training round, the proportion of participating clients, denoted as C, was set to 0.4. For efficiency, we trained the model on a local dataset that included three negative instances for each positive interaction. These negative instances were generated through local inference on a random subset of unobserved items not present in the local client's data. To fill in the time information for interactions with negative items, we used the most recent interaction time between the user and the positive item. This assumption implies that the user remained disinterested in the negative item until their last interaction with the recommendation system.

In terms of recommendation performance, it is observed that FedNTF consistently outperforms FedNCF, which in turn outperforms FedMF. This provides evidence that FedNTF can be effectively utilized as an optimization method in collaborative filtering-based federated models. Evaluating the Lastfm2K dataset results, we find that FedNTF achieves nearly identical HR performance compared to the FedNCF model, while demonstrating improved NDCG performance. Although our model excels at capturing the temporal dynamics of user-item interactions, the advantage of incorporating LSTM into the federated recommendation model is not significant on the Lastfm2K dataset. This is because the dataset has a relatively limited time span, and users' preferences for items do not undergo significant changes. Hence, the HR performance remains similar to that of FedNCF. On the other hand, the NDCG metric focuses primarily on the ranking of items in the recommendation list. Even when user preferences do not exhibit substantial changes, the LSTM model with time information can effectively rank the relevant items at the top, resulting in improved NDCG performance.

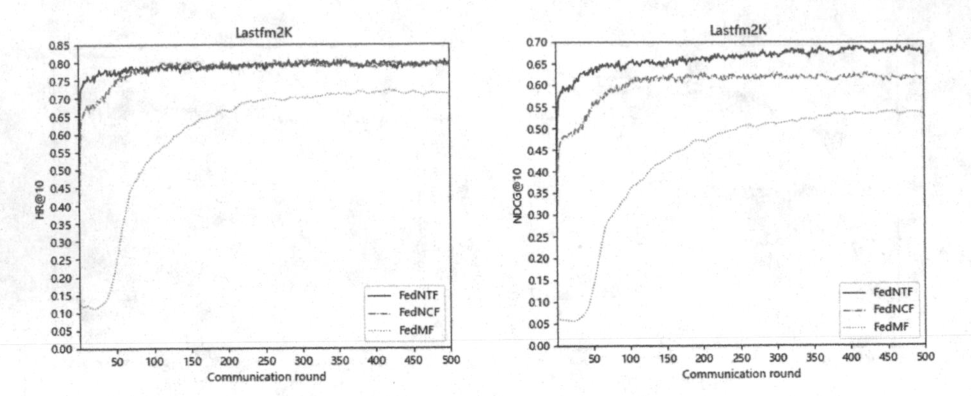

Fig. 3. Comparing FedNTF with other baseline methods on the Lastfm2K dataset, analyzing their performance across multiple metrics.

6 Conclusion

Current federated recommendation systems often ignore the crucial temporal information of user-item interactions, resulting in limited improvement of recommendation quality in the federated setting. In this paper, we propose a new method called Federated Neural network based Tensor Factorization (FedNTF) to model the evolving multidimensional interaction data of users and items over time, without requiring the sharing of private data among clients. FedNTF combines a multilayer perceptron architecture with an LSTM encoder to model time-varying factors, capturing users' preferences over time and the underlying factors driving the evolution of user-item relationships. To assess the effectiveness of our proposed method, FedNTF, we conduct extensive experiments on link prediction tasks utilizing the MovieLens1M and Lastfm2K datasets. Through these experiments, we demonstrate noteworthy advancements compared to existing baseline approaches, underscoring the superiority of FedNTF in the context of link prediction tasks.

References

1. Zhao, Z.-D., Shang, M.-S.: User-based collaborative-filtering recommendation algorithms on Hadoop. In: 2010 Third International Conference on Knowledge Discovery and Data Mining, pp. 478–481. IEEE (2010)
2. Sarwar, B., Karypis, G., Konstan, J., Riedl, J.: Item-based collaborative filtering recommendation algorithms. In: Proceedings of the 10th International Conference on World Wide Web, pp. 285–295 (2001)
3. Goddard, M.: The EU general data protection regulation (GDPR): European regulation that has a global impact. Int. J. Mark. Res. **59**(6), 703–705 (2017)
4. McMahan, B., Moore, E., Ramage, D., Hampson, S., y Arcas, B.A.: Communication-efficient learning of deep networks from decentralized data. In: Artificial Intelligence and Statistics, pp. 1273–1282. PMLR (2017)

5. Ammad-Ud-Din, M., et al.: Federated collaborative filtering for privacy-preserving personalized recommendation system (2019). arXiv preprint arXiv:1901.09888
6. Chai, D., Wang, L., Chen, K., Yang, Q.: Secure federated matrix factorization. IEEE Intell. Syst. **36**(5), 11–20 (2020)
7. Chen, Y., Ning, Y., Slawski, M., Rangwala, H.: Asynchronous online federated learning for edge devices with non-IID data. In: 2020 IEEE International Conference on Big Data (Big Data), pp. 15–24. IEEE (2020)
8. Damaskinos, G., Guerraoui, R., Kermarrec, A.-M., Nitu, V., Patra, R., Taiani, F.: FLeet: online federated learning via staleness awareness and performance prediction. ACM Trans. Intell. Syst. Technol. (TIST) **13**(5), 1–30 (2022)
9. Dolui, K., Gyllensten, I.C., Lowet, D., Michiels, S., Hallez, H., Hughes, D.: Towards privacy-preserving mobile applications with federated learning: the case of matrix factorization (poster). In: Proceedings of the 17th Annual International Conference on Mobile Systems, Applications, and Services, pp. 624–625 (2019)
10. Perifanis, V., Efraimidis, P.S.: Federated neural collaborative filtering. Knowl.-Based Syst. **242**, 108441 (2022)
11. Koren, Y., Bell, R., Volinsky, C.: Matrix factorization techniques for recommender systems. Computer **42**(8), 30–37 (2009)
12. He, X., Liao, L., Zhang, H., Nie, L., Hu, X., Chua, T.-S.: Neural collaborative filtering. In: Proceedings of the 26th International Conference on World Wide Web, pp. 173–182 (2017)
13. Chuhan, W., Fangzhao, W., Wang, X., Huang, Y., Xie, X.: Fairness-aware news recommendation with decomposed adversarial learning. In: Proceedings of the AAAI Conference on Artificial Intelligence, vol. 35, pp. 4462–4469 (2021)
14. Jinze, W., et al.: Hierarchical personalized federated learning for user modeling. In: Proceedings of the Web Conference, vol. 2021, pp. 957–968 (2021)

UAV-D2D Assisted Latency Minimization and Load Balancing in Mobile Edge Computing with Deep Reinforcement Learning

Qinglin Song and Long Qu(✉)

Faculty of Electrical Engineering and Computer Science, Ningbo University, Ningbo, Zhejiang, China
{2211100149,qulong}@nbu.edu.cn

Abstract. Now Unmanned Aerial Vehicle (UAV) with Mobile Edge Computing (MEC) severs and Device-to-Device (D2D) communications provide offload computing services for User Devices (UDs). However, the UAV has relatively high transmission latency. And D2D lacks the necessary flexibility. In this paper, we introduce a novel MEC system that utilizes the collaborative advantages of flexible movement of UAV and the low latency transmission of D2D communication to process tasks from UDs. We formulate an optimization problem focused on minimizing the tasks transmission and execution delay of UDs. The problem involves joint optimization of user scheduling, UAV trajectory, and resource allocation of Virtual Machines (VMs) on the MEC server. To tackle this nonconvex problem, we propose a Deep Reinforcement Learning (DRL) algorithm with Deep Deterministic Policy Gradient (DDPG). Through simulation results, we demonstrate that DDPG reduces the latency by 41% compared to Deep Q-Network (DQN) and Actor-Critic (AC) algorithm. Our collaborative UAV-D2D model has 16% and 32% lower latency than when only the UAV or D2D works alone.

Keywords: Mobile Edge Computing · Unmanned Aerial Vehicle · Device-to-Device · Virtual Machines · Deep Deterministic Policy Gradient

1 Introduction

In the era of 5G, mobile networks cater to a wide range of devices, including computers, mobile vehicles, and various types of sensors. With the rapid proliferation of Internet of Things (IoT) devices [1], there is a growing demand for applications with stringent requirements for low latency, such as Virtual Reality (VR), Augmented Reality (AR), and video streaming [2]. The traditional core networks are unable to meet the demands of latency-sensitive tasks. To address this challenge, Mobile Edge Computing (MEC) has emerged as a promising solution. MEC aims to reduce processing latency and enhance user experience by

H. Jin et al. (Eds.): GPC 2023, LNCS 14504, pp. 108–122, 2024.
https://doi.org/10.1007/978-981-99-9896-8_8

offloading compute-intensive tasks to edge servers close to the users [3]. Nevertheless, conventional MEC servers are typically deployed on static Base Stations (BSs) and lack the necessary flexibility. This limitation poses challenges in scenarios where infrastructure is scarce or compromised(such as post-disaster situations or remote mountainous areas) [4]. Therefore, there is a growing interest in leveraging Unmanned Aerial Vehicle (UAV) and Device-to-Device (D2D) assisted communications. These innovative approaches offer unique advantages in terms of mobility, adaptability, and coverage, which make them particularly suitable for addressing the limitations of traditional MEC deployments.

The research on UAVs primarily focuses on their trajectory and task scheduling [5]. UAV fly between fixed users to fulfill their computational offloading requirements. Meanwhile, there has been considerable research on the utilization of D2D communication as an emerging technology for MEC services [6]. The short-range and low-latency advantages of D2D communication assist MEC in task processing [7]. In the context of MEC servers, Virtual Machine (VM) reuse is a fundamental technique [8]. MEC servers utilize multiple VMs to perform parallel computing tasks, resulting in a significant reduction in computational latency. However, existing work has rarely explored the collaborative efforts of UAV-D2D communications and VM workloads on UAV to meet user task demands.

In this paper, we propose a system that leverages the collaborative advantages of UAV and D2D communication to assist MEC. UD splits and offloads tasks to UAV and D2D for joint calculation. We also optimize the workload on MEC servers to achieve optimal computational efficiency. The main contributions of this work are summarized as follows:

(1): We present the system of using UAV and D2D communication together to support MEC services. This novel approach harnesses the unique advantages of UAV mobility and D2D communication low latency. Compared to scenarios using only UAV or D2D, the task execution delay is reduced by 16% and 32%, respectively.

(2): We address the workload on MEC servers to maximize their computing capacity. By optimizing the allocation of VMs on the servers, we achieve a balanced workload and efficient resource utilization. Compared with no VMs allocation, task execution delay decrease 6%.

(3): We compare our proposed Deep Deterministic Policy Gradient (DDPG) with Deep Q-Network (DQN) and Actor-Critic (AC). Through simulations and experiments, we demonstrate that the DDPG reduces latency of tasks about 41%.

In the rest of this article is organized as follows. The Sect. 2 discusses related work. We introduced the system model in Sect. 3. In Sect. 4 we present the algorithm. The simulation and experimental results are presented in Sect. 5. The Sect. 6 summarizes.

2 Related Work

In recent years, there has been extensive research in the academic community focused on MEC assisted offloading. Arash *et al.* [9] aimed to minimize delay and energy consumption by finding the Pareto optimal frontier. Li *et al.* [10] proposed an online learning method that reduces task processing cost through multi-hop assisted collaboration. However, the static deployment of MEC servers mentioned above is not adaptable to various scenarios. Asim *et al.* [11] tackled the issue by minimizing system energy consumption through the optimization of the hovering position for each time slot of the Unmanned Aerial Vehicle (UAV). Wang *et al.* [12] employed Deep Reinforcement Learning (DRL) to plan multiple UAV trajectories while considering UAV load balancing to minimize energy consumption. Umber *et al.* [13] focused on D2D shared spectrum and aimed to minimize the sum of all device task execution delays under energy constraints, utilizing an offloading framework based on Orthogonal Frequency Division Multiple Access (OFDMA). Dai *et al.* [14] designed a framework that integrates migration and offloading willingness in D2D communication, aiming to minimize task latency and migration costs. However, there are limited studies on UAV and D2D co-assisted MEC computations. Pu *et al.* [15] found that opening multiple VMs in the same Physical Machine (PM) could impact overall performance due to I/O interference between VMs. Koushik *et al.* [16] employed the DQN algorithm, to design UAV trajectories and optimize network throughput. However, the DQN algorithm may face challenges in scenarios with continuous action spaces due to the curse of dimensionality, making convergence difficult. To address this, Ding *et al.* [17] proposed the DDPG algorithm to handle high-dimensional continuous motion of UAV and achieve improved performance.

In comparison to the reviewed related studies, we propose the DDPG algorithm to jointly leverage UAV and D2D communication for MEC offloading. Our approach aims to optimize the UAV trajectory and workload of MEC servers, leading to a reduction in task execution delay. By employing DDPG, we effectively address the challenge of high-dimensional continuous motion in UAV. In addition, offloading between close D2D is able to get low transmission latency. This jointly utilize advantages which enables efficient task distribution, ultimately resulting in a smaller task execution delay.

3 System Model

In this section, we will consider the issue of minimizing the latency of UDs. As shown in Fig. 1, we assume the D2D-assisted UAV-MEC system without BSs, which consists of a UAV and M UDs, denoted by the set $\mathcal{M} = \{1,2,\ldots,M\}$. Besides, We assume that the UDs are divided into two groups, one for D2D transmitter and one for D2D receivers, which set $\mathcal{I} = \{1, 2, \ldots, i, \ldots, I\}, \forall i \in \mathcal{M}$ and $\mathcal{J} = \{1, 2, \ldots, j, \ldots, J\}, \forall j \in \mathcal{M}$, respectively. The UAV is equipped with MEC servers and provides offloading services for the D2D transmitters. Simultaneously, the D2D receivers also assist in offloading tasks for the transmitters.

3.1 UAV Trajectory Model

In our model, we set a square region in Cartesian coordinates. Then, we assume that the UAV has sufficient power to maintain flying [2] at a fixed altitude H and serve the users dynamically during the flight cycle time of T. The flight period is divided into equal and sufficient small time slots N denoted by the set $\mathcal{N} = \{1,2,\ldots,n,\ldots,N\}$. Besides, We assume that the UAV remains hovering in each time slot n, so the position coordinate of the UAV is $\mathbf{q}(n) = [X(n),\ Y(n),\ H]$, $n \in \mathcal{N}$. The flight direction and speed is controlled by the angle of $\delta(n) \in (0, 2\pi]$ and $v(n) \in [0, v_{max}]$, respectively. Therefore, we get the coordinate of the UAV flies to the new hovering position at the nth time slot
$\mathbf{q}(n+1) = [X(n) + v(n)t\cos\delta(n), Y(n) + v(n)t\sin\delta(n), H]$
with a flight time t. Moreover, we have UAV flight constraints as the following

$$0 \le X(n+1) \le X(n) + v_{max}t\cos\delta(n), \forall n \in \mathcal{N} \tag{1}$$

$$0 \le Y(n+1) \le Y(n) + v_{max}t\sin\delta(n), \forall n \in \mathcal{N} \tag{2}$$

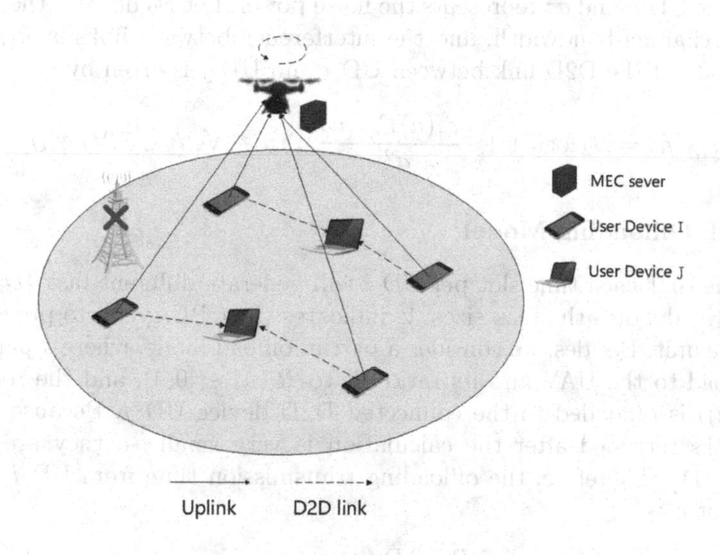

Fig. 1. UAV-D2D MEC System.

3.2 Communication Model

We assume that the UAV schedules one UD i per time slot to communicate, while this UD i generates a D2D link with the nearest UD j. Besides, we assume the positions of transmitter i and receiver j are fixed in our model, which are denoted as $\mathbf{q}_i = (x_i, y_i, 0)$ and $\mathbf{q}_j = (x_j, y_j, 0)$, respectively. The communication link between the UAV and the UD i is dominated by the line-of-site(LoS) channel, so their channel gain in time slot n be expressed as

$$g_i(n) = \beta_0 d_i^{-2}(n) = \frac{\beta_0}{\|\mathbf{q}(n) - \mathbf{q}_i\|^2}, \forall n \in \mathcal{N}, \forall i \in \mathcal{I} \tag{3}$$

where β_0 is the channel power gain at a reference distance of 1 m, and $d_i(n)$ denotes the Euclidean distance between UD i and UAV. Moreover, we obtain the D2D channel link between UD i and UD j be modeled as

$$\begin{aligned} g_{i,j}(n) &= h(n)\beta_0 d_i^{-2}(n) \\ &= \frac{h(n)\beta_0}{\|\mathbf{q}_i - \mathbf{q}_j\|^2}, \forall n \in \mathcal{N}, \forall i \in \mathcal{I}, \forall j \in \mathcal{J} \end{aligned} \tag{4}$$

where $h(n)$ represents the small-scale fading coefficient of obeying $\mathcal{CN} \sim (0,1)$. Then, the transmission rate between UAV and UD i is given as

$$r_i(n) = B_1 \log_2(1 + \frac{g_i(n)P_i(n)}{\sigma^2}), \forall n \in \mathcal{N}, \forall i \in \mathcal{I} \tag{5}$$

where B_1 is the ground-to-air channel bandwidth, $P_i(n)$ denotes the transmission power of the UD i, and σ^2 represents the noise power. Let B_2 denotes the ground-to-ground channel bandwidth, and the interference between links is ignored, so the data rate of the D2D link between UD i and UD j is given by

$$r_{i,j}(n) = B_2 \log_2(1 + \frac{g_{i,j}(n)P_{i,j}(n)}{\sigma^2}), \forall n \in \mathcal{N}, \forall i \in \mathcal{I}, \forall j \in \mathcal{J} \tag{6}$$

3.3 Task Offloading Model

We assume that each time slot per UD i will generate different task $[D_i(n), V]$, where $D_i(n)$ denotes the task sizes, V indicates the CPU cycles to process each byte of the unit. Besides, we consider a partial offload mode, where a part of the tasks offload to the UAV and its ratio set to $R_i(n) \in [0, 1]$, and the remaining $(1 - R_i(n))$ is offloaded to the connected D2D device UD j. Because the size of the tasks returned after the calculation is very small, so they are usually negligible [11]. Therefore, the offloading transmission time from UD i to UAV at time slot n is

$$t_i^{tran}(n) = \frac{R_i(n)D_i(n)}{r_i(n)}, \forall n \in \mathcal{N}, \forall i \in \mathcal{I}. \tag{7}$$

The transmission time from UD i to UD j is given as

$$t_{i,j}^{tran}(n) = \frac{(1 - R_i(n))D_i(n)}{r_{i,j}(n)}, \forall n \in \mathcal{N}, \forall i \in \mathcal{I}, \forall j \in \mathcal{J}. \tag{8}$$

Hence, the transmission time of the scheduled UD i to offload tasks in time slot n is

$$T_{tran}(n) = t_i^{tran}(n) + t_{i,j}^{tran}(n), \forall n \in \mathcal{N}, \forall i \in \mathcal{I}, \forall j \in \mathcal{J}. \tag{9}$$

3.4 Task Computing Model

In our model, we incorporate load balancing for the MEC server. As depicted in Fig. 2, upon receiving tasks, the MEC server creates multiple VMs on the same PM to process the tasks in parallel. However, turning on more VMs leads to increased load, which negatively impact the overall performance of the MEC server. We denote $Z > 0$ [18] as the attenuation factor, representing the percentage of overall computing capability degradation when multiple VMs are simultaneously active. Additionally, we assume that the tasks received by the MEC server can be randomly divided into multiple sub-tasks, with the number of sub-tasks denoted as $k(n)$ in time slot n. Consequently, the parallel computing time of $S(n)$ VMs on the MEC server in time slot n can be expressed as follows:

$$t_i^{comp}(n) = t_{\max}^{k(n)}(1+Z)^{S(n)-1}, \forall n \in \mathcal{N}, \forall i \in \mathcal{I} \tag{10}$$

where $t_{\max}^{k(n)}$ denotes the maximum computing time for a VM to process parallel sub-tasks, and it is expressed as

$$t_{\max}^{k(n)} = \frac{D_{\max VM}^{sub}(n)V}{f_{VM}}, \forall n \in \mathcal{N} \tag{11}$$

where $D_{\max VM}^{sub}(n)$ represents the maximum sub-task size for a VM computing, f_{VM} denotes computing capability of virtual machine. Moreover, the computing time of UD j is

$$t_{i,j}^{comp}(n) = \frac{(1-R_i(n))D_i(n)V}{f_j}, \forall n \in \mathcal{N}, \forall i \in \mathcal{I}, \forall j \in \mathcal{J}, \tag{12}$$

where f_j denoted the computing capability of UD j. Therefore, the total computing time of the task $D_i(n)$ at time slot n is

$$T_{comp}(n) = t_i^{comp}(n) + t_{i,j}^{comp}(n), \forall n \in \mathcal{N}, \forall i \in \mathcal{I}, \forall j \in \mathcal{J}. \tag{13}$$

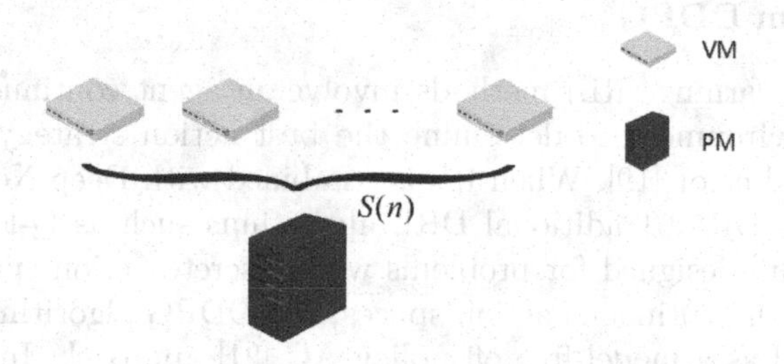

Fig. 2. VMs Parallel Computing.

3.5 Problem Formulation

In this paper, we jointly optimize user scheduling, UAV trajectory, UD i launch power, offload ratio, and number of VMs to achieve MEC server load balancing and minimize tasks execution latency. Specifically, we minimize the delay with transmission time and computation time which is formulated as

$$\min_{\substack{\alpha_i(n),\mathbf{q}(n+1),R_i(n),\\ P_i(n),P_{i,j}(n),S(n)}} \sum_{n=1}^{N}\sum_{i=1}^{I} \alpha_i(n)(T_{tran}(n) + T_{comp}(n)) \tag{14}$$

$$s.t. \quad \alpha_i(n) \in 0,1, \forall n \in \mathcal{N}, \forall i \in \mathcal{I}, \tag{14a}$$

$$\sum_{i=1}^{I} \alpha_i(n) = 1, \forall i \in \mathcal{I}, \tag{14b}$$

$$0 \le X(n+1) \le X(n) + v_{max} t \cos\delta(n), \forall n \in \mathcal{N}, \tag{14c}$$

$$0 \le Y(n+1) \le Y(n) + v_{max} t \sin\delta(n), \forall n \in \mathcal{N}, \tag{14d}$$

$$1 \le k(n) \le K_{max}, \forall n \in \mathcal{N}, \tag{14e}$$

$$1 \le S(n) \le k(n), \forall n \in \mathcal{N}. \tag{14f}$$

The constraint (14a) and (14b) ensure that only one user is scheduled for offloading in time slot n. Constraint (14c) and (14d) guarantee UAV flight trajectory is not exceeding its capacity limits. Constraint (14e) denotes the number of sub-tasks split on the MEC sever does not exceed the maximum. Constraint (14f) limit the number of VMs no more than the amount of sub-tasks.

4 Proposed Approach

In this section, so we propose a DRL algorithm DDPG to slove the above complex optimization problem with multiple non-convex constraints and multiple optimization objectives.

4.1 Algorithm DDPG

Reinforcement Learning (RL) methods involve an agent continuously interacting with the environment to determine the best action strategy for each step through trial and error [19]. When RL is combined with Deep Neural Networks (DNN), it forms DRL. Traditional DRL algorithms such as Q-learning, Sarsa, and DQN [20] are designed for problems with discrete action spaces. However, when dealing with continuous action spaces, the DDPG algorithm, as shown in Fig. 3, is utilized as a model-free off-policy AC [21] approach. In the DRL, the environment is typically modeled as a discrete-time Markov Decision Process (MDP). Following the Markov framework, the agent selects an action based on the current state of the environment and receives an immediate reward, which guides its subsequent actions. The primary objective of the agent is to maximize the accumulated reward by making optimal decisions based on the current environment state.

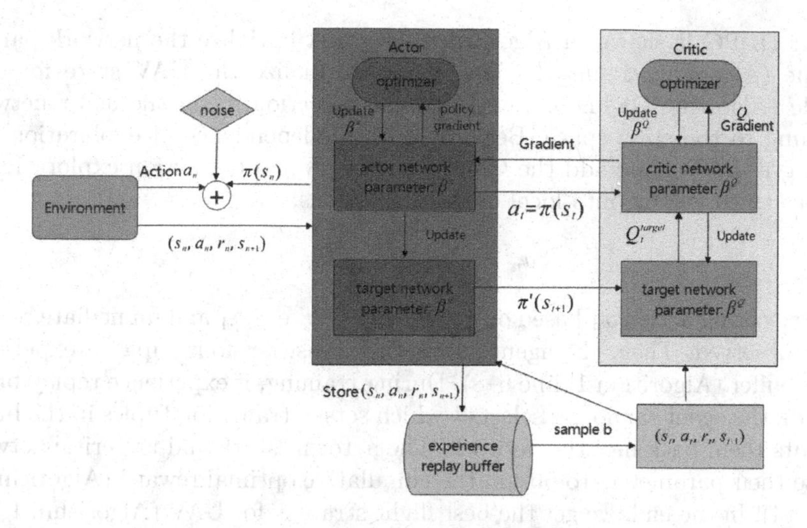

Fig. 3. DDPG Schematic Diagram.

The DDPG as a deterministic policy algorithm where the continuous action spaces output is a deterministic action. The actor network π is defined as a function

$$a_n = \pi(s_n|\theta^\pi) \tag{15}$$

where s_n is current state to get deterministic action a_n, and θ^π is actor network training parameters. The critic network is $\mathcal{Q}(s_n, a_n|\theta^\mathcal{Q})$ to approximate Q-value function. Besides, both the actor network and the critic network contain the same structural target network as they are, which updates the approximate $\pi'(s_n|\theta^{\pi'})$ and $\mathcal{Q}'(s_n, a_n|\theta^{\mathcal{Q}'})$, respectively. In addition, DDPG has an experience replay mechanism that randomly selects the mini bitch b input network in the experience buffer to accelerate convergence. The critic network minimize the loss function to update

$$L(\theta^\mathcal{Q}) = \frac{1}{b}\sum_{t=1}^{b}[\mathcal{Q}^{target} - \mathcal{Q}(s_t, a_t|\theta^\mathcal{Q})]^2 \tag{16}$$

where $\mathcal{Q}^{target} = r_t + \gamma\mathcal{Q}(s_{t+1}, \pi'(s_t|\theta^{\pi'})|\theta^{\mathcal{Q}'}))$. Moreover, the strategy gradient update formula is

$$\nabla_{\theta^\pi} J = \mathbb{E}_{\pi'}[\nabla_a \mathcal{Q}(s, a|\theta^\mathcal{Q})|_{s=s_t, a=\pi(s_t|\theta^\pi)} \nabla_{\theta^\pi}\pi(s|\theta^\pi)|_{s=s_t}]. \tag{17}$$

Hence, the parameters of the target network are updated as

$$\theta^{Q'} \leftarrow \tau\theta^\mathcal{Q} + (1-\tau)\theta^{\mathcal{Q}'} \tag{18}$$

$$\theta^{\pi'} \leftarrow \tau\theta^\pi + (1-\tau)\theta^{\pi'} \tag{19}$$

where $\tau \in (0, 1)$ is a constant to update target network softly.

Our DDPG is shown in Algorithm 1, we first initialize the network parametersthe (Algorithm 1: line 1–3). Then, we initialize the UAV state for each episode (Algorithm 1: line 4–5). UAV chooses action a_n in the actor network according to the state space. Because of the independence of exploration and learning in DDPG, we add the Gaussian noise N_n to the action exploration in order to avoid getting into local optimal solutions,

$$a_n = \pi(s_n|\theta^\pi) + N_n. \tag{20}$$

After executing action a_n based on s_n, the next state s_{n+1} and immediate reward r_n are observed. Then, the agent stores the transition four tuple in experience replay buffer (Algorithm 1: line 6–8). During training, if experience replay buffer B is full, the agent randomly selects b which sets of transition tuples in the buffer and puts them back into the network. The actor network and the critic network update their parameters to obtain the cumulative optimal reward (Algorithm 1: line 9–14). In the end, we get the best flight strategy for UAV (Algorithm 1: line 16–17).

Algorithm 1. DDPG-based Dynamic Computation Resource Allocation and Task Offloading algorithm

1: Initialize actor network with weights θ^π and critic network with weights θ^Q.
2: Initialize the weights of target network $\theta^{\pi'} = \theta^\pi$ and $\theta^{Q'} = \theta^Q$, respectively.
3: Set the experience replay buffer $B = 0$.
4: **for** each episode **do**
5:     Reset the UAV initial position and observe the initial state s_1
6:     **for** $n = 1, 2, \ldots, N$ **do**
7:         Perform exploration actions $a_n = \pi(s_n|\theta^\pi)+N_n$, get the reward r_n and observe next state s_{n+1}.
8:         Store tuple (s_n, a_n, s_{n+1}, r_n) in the experience replay buffer B.
9:         **if** B is full, **then**
10:             Randomly sample tuple with mini-batches of b from B.
11:             Update the θ^Q of critic network by minimizing the loss (16).Update the θ^π of actor network by policy gradient (17).
12:             Update target network of critic network and actor network by (18) and (19), respectively.
13:         **end if**
14:     **end for**
15: **end for**
16: **return** θ^π
17: Select the optimal action a_n^{op}.

4.2 MDP Model

In our system, UAV act as agent to creat MDP. We model the MDP as three tuples (S, A, R), where S indicates state space, A is a set of action, and R represents the reward function.

State Space. The state space of the environment in our model consists of UD i task size $D_i(n)$, the number of sub-tasks $k(n)$ and UAV position $\mathbf{q}(n)$ in time slot n. Therefore, the state space is given as

$$s_n = [D_i(n), k(n), \mathbf{q}(n)]. \tag{21}$$

Action Space. The action space consists of continuous flight actions and scheduling calculation of UAV, including flight speed $v(n)$, flight angle $\delta(n)$, offloading ratio $R_i(n)$, scheduling UD $i(n)$, UD launch power $P_i(n), P_{i,j}(n)$, and number of VMs $S(n)$. Thus, the action space is modeled as

$$a_n = [v(n), \delta(n), R_i(n), i(n), P_i(n), P_{i,j}(n), S(n)]. \tag{22}$$

Since the output actions of the actor network are continuous, the action variables $i(n)$, $S(n)$ need to be discretized, e.g. if $i(n) = 0$,then discretization $i' = 1$; if $i(n) \neq 0$, the $i' = \lceil i(n) \rceil$, where $\lceil \cdot \rceil$ is rounding up.

Reward Function. The reward function is a crucial component in evaluating the rationality of actions chosen by the agent. In our approach, we utilize the optimization objective as the basis for the reward function. Additionally, we incorporate a penalty factor, denoted as p_n, to account for the UAV flying out of the designated boundary. Consequently, the reward function can be expressed as follows:

$$r_n = -\sum_{n=1}^{N}\sum_{i=1}^{I} \alpha_i(n)(T_{tran}(n) + T_{comp}(n)) + p_n \tag{23}$$

Table 1. Simulation parameters

Parameter	Value	Parameter	Value
H	100 m	σ^2	–100 dBm
v_{max}	50 m/s	V	1000 cycles/bit
β_0	–50 dB	f_j	0.6 GHz
B_1	1 MHz	f_{VM}	1.2 GHz
B_2	0.8 MHz	Z	0.2
t	1 s	$D_i(n)$	[1.5, 2]Mbits

5 Simulation Results

In this section, we conduct simulations using the DDPG algorithm with specific parameter values, and compare its performance with other baseline algorithms. Our simulations use the CPU of AMD 5800H with 3.2 GHz. All algorithms are implemented in Python 3.6 and Tensorflow 1.5.0. The DDPG algorithm uses a 4-layer fully connected neural network with two hidden layers [300,10] neurons in both actor and critic networks. Our model considers a square area with dimensions of 100 m $\times$ 100 m. We have a total of $I = 4$ UDs positioned at [75, 19], [40, 88], [47, 17], and [93, 55] meters, respectively. Additionally, there are $J = 2$ destination points located at [0, 0] and [100, 100] meters. The initial position of the UAV is set to [50, 50] meters. For the simulation, we utilize various parameters which are specified as follows Table 1.

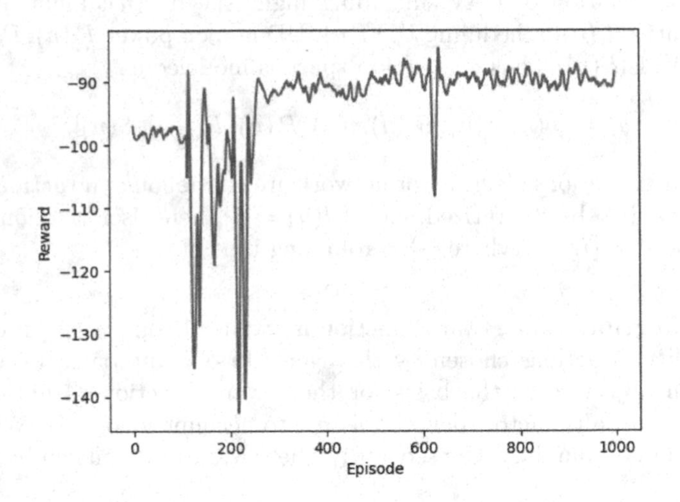

Fig. 4. Reward Convergence of DDPG.

First, we analyze the reward of DDPG as shown in Fig. 4. After conducting tests, we have observed that the best convergence performance when the learning rate of the critic network and the actor network is set to $\psi_{actor} = 0.001, \psi_{critic} = 0.002$, respectively. Meanwhile, our discount factor set as $\gamma = 0.001$, exploration parameter set as $\sigma_e = 0.01$. Initially, due to the lack of previous knowledge about the environment, the UAV explores actions in an almost random manner. As a result, the reward experiences significant fluctuations. However, as the UAV accumulates enough samples and gains more information about the environment, the reward gradually increases and eventually converges. This convergence indicates that the UAV has found the optimal flight strategy.

Furthermore, in Fig. 5, we compare the delay performance of various algorithms. As the training episodes increase, the AC algorithm fails to converge

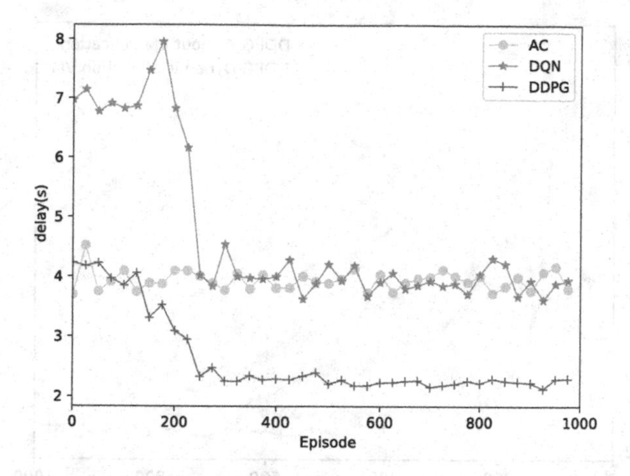

Fig. 5. Different Algorithm Delay Performance Comparison.

due to simultaneous updates of its actor and critic networks. The difficulty in converging the critic network prevents the accurate guidance of the optimal action through the value function. In contrast, both DQN and DDPG employ evaluation networks and target networks, which ensure relatively independent training data and enable convergence. Because the DQN is limited to scenarios with discrete actions, when dealing with problems involving a large number of action dimensions, we need to quantify continuous actions into finite discrete values which will result in lower cumulative rewards. The DDPG converges to smaller delay values eventually because of the extensive exploration of continuous actions. As a result, DDPG has 41% lower latency than DQN and AC.

Figure 6 (a) illustrates a comparison of the delay performance between DDPG with and without dynamic VM allocation. We observe that both systems converge after 300 episodes as the UAV finds the best strategy. However, the allocation of VM results in lower task processing latency. This is because optimizing the distribution of VM leads to a further reduction in computational latency, causing a decrease in latency of 6%.

In Fig. 6 (b), we simulate the delay performance of UAV or D2D working alone, and compare them with our system. When only D2D is working, due to the limited computing capability of D2D receiver devices, they are unable to quickly complete all task processing. Besides, only relying on UAV operation unable guarantee low-latency transmission for all tasks. Our model allows the user to partially offload to the UAV and partially offload to the D2D receiver device, which leverages the respective strengths of UAV and D2D. Therefore, the latency of our model is 16% or 32% smaller than when only UAV or D2D works alone, respectively.

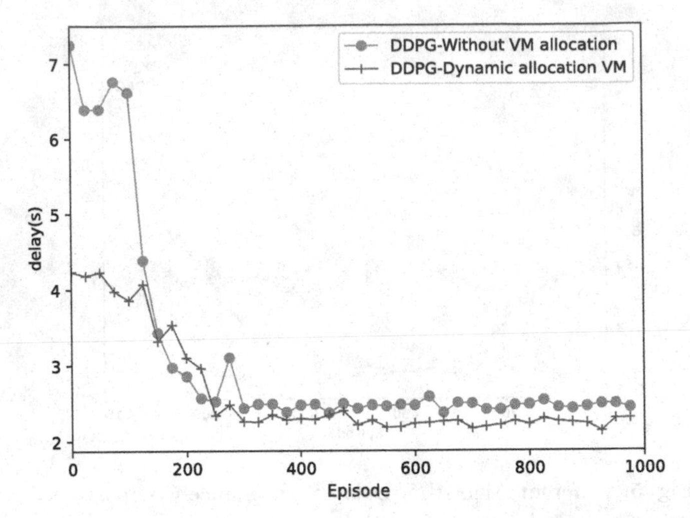

(a) Delay Performance of DDPG with and without VM Allocation.

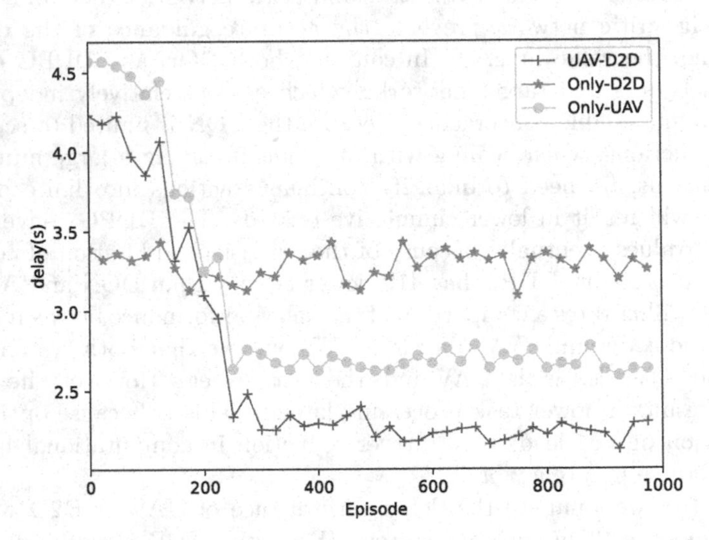

(b) Comparison of UAV-D2D combined and solo operation delay.

Fig. 6. Delay comparison of different system models.

6 Conclusion

In this paper, we consider a system that D2D and UAV-MEC collaborate to assist in user task offloading. We optimize the UAV trajectory and achieve load balancing on the MEC server to minimize the sum of the tasks' transmission delay and computation delay. Specifically, to solve the integer nonlinear problem, we propose the DDPG algorithm to obtain the optimal strategy.

Through extensive simulations, we evaluate the performance of DDPG in terms of processing task latency and compare it with DQN and AC. The results demonstrate that DDPG approach outperforms the DQN and AC about 41% in terms of task latency reduction. This indicates the effectiveness and superiority of our proposed solution in optimizing the UAV trajectory and achieving load balancing on the MEC server.

By leveraging the collaboration between D2D and UAV-MEC, our system demonstrates improved efficiency and reduced latency in task offloading. And compared with only UAV or D2D wokrs alone, the task processing delay reduces 16% and 32%, respectively. In addition, the latency for dynamic VM allocation is 6% lower than for fixed VM numbers.

Acknowledgment. This work was supported in part by the Ningbo Natural Science Foundation under Grant 2021J070, in part by the Zhejiang Natural Science Foundation under Grant LY20F010004, and National Natural Science Foundation of China under Grant 61801254.

References

1. Cui, G., He, Q., Chen, F., Zhang, Y., Jin, H., Yang, Y.: Interference-aware Game-Theoretic device allocation for mobile edge computing. IEEE Trans. Mob. Comput. **21**(11), 4001–4012 (2022)
2. Zhang, J., Wang, Z.-J., Wang, K., Guo, S., Wang, B., Guo, M.: Improving power efficiency for online video streaming service: a self-adaptive approach. IEEE Trans. Sustain. Comput. **4**(3), 308–313 (2019)
3. Tiankui Zhang, Y.X., Loo, J., Yang, D., Xiao, L.: Joint computation and communication design for UAV-assisted mobile edge computing in IoT. IEEE Trans. Ind. Inf. **16**(8), 5505–5516 (2020)
4. Li, X., Feng, W., Chen, Y., Wang, C., Ge, N.: UAV-Enabled accompanying coverage for hybrid satellite-UAV-terrestrial maritime communications. In: Proceedings of 28th Wireless and Optical Communications Conference, pp. 1–5 (2019)
5. Wang, Z., Duan, L., Zhang, R.: Adaptive deployment for UAV-aided communication networks. IEEE Trans. Wirel. Commun. **18**(9), 4531–4543 (2019)
6. Chatzopoulos, D., Bermejo, C., Haq, E.U., Li, Y., Hui, P.: D2D task offloading: A dataset-based Q and A. IEEE Commun. Mag. **57**(2), 102–107 (2019)
7. Cheng, Y., Liang, C., Chen, Q., Yu, R.: Energy-efficient D2D-assisted computation offloading in NOMA-Enabled cognitive networks. IEEE Trans. Veh. Technol. **70**(12), 13441–13446 (2021)
8. Liang, Z., Liu, Y., Lok, T., Huang, K.: Multiuser computation offloading and downloading for edge computing with virtualization. IEEE Trans. Wireless Commun. **18**(9), 4298–4311 (2019)
9. Bozorgchenani, A., Mashhadi, F., Tarchi, D., Salinas Monroy, S.A.: Multi-objective computation sharing in energy and delay constrained mobile edge computing environments. IEEE Trans. Mob. Comput. **20**(10), 2992–3005 (2021)
10. Li, Y., Wang, X., Gan, X., Jin, H., Fu, L., Wang, X.: Learning-Aided computation offloading for trusted collaborative mobile edge computing. IEEE Trans. Mob. Comput. **19**(12), 2833–2849 (2020)

11. Asim, M., Mashwani, W.K., Shah, H., Belhaouari, S.B.: A Load-Balanced and Energy-Efficient navigation scheme for UAV-Mounted mobile edge computing. Soft Computing, pp. 1–14. Springer, Berlin, Germany (2021)
12. Wang, Z., Rong, H., Jiang, H., Xiao, Z., Zeng, F.: An evolutionary trajectory planning algorithm for multi-UAV-assisted MEC system. IEEE Trans. Netw. Sci. Eng. **9**(5), 3659–3674 (2022)
13. Saleem, U., Liu, Y., Jangsher, S., Tao, X., Li, Y.: Latency minimization for D2D-Enabled partial computation offloading in mobile edge computing. IEEE Trans. Veh. Technol. **69**(4), 4472–4486 (2020)
14. Dai, X., et al.: Task Co-Offloading for D2D-Assisted mobile edge computing in industrial Internet of Things. IEEE Trans. Ind. Inf. **19**(1), 480–490 (2023)
15. Pu, X., Liu, L., Mei, Y., Sivathanu, S., Koh, Y., Pu, C.: Understanding performance interference of I/O workload in virtualized cloud environments. In: Proc. IEEE 3rd International Conference on Cloud Computing, pp. 51–58 (2010)
16. Koushik, A.M., Hu, F., Kumar, S.: Deep Q-learning-based node positioning for throughput-optimal communications in dynamic UAV swarm network. IEEE Trans. Cogn. Commun. Netw. **5**(3), 554–566 (2019)
17. Ding, R., Gao, F., Shen, X.S.: 3D UAV trajectory design and frequency band allocation for energy efficient and fair communication: a deep reinforcement learning approach. IEEE Trans. Wireless Commun. **19**(12), 7796–7809 (2020)
18. Liu, Y., Yan, J., Zhao, X.: Deep reinforcement learning based latency minimization for mobile edge computing with virtualization in maritime UAV communication network. IEEE Trans. Veh. Technol. **71**(4), 4225–4236 (2022)
19. Orhean, A.I., Pop, F., Raicu, I.: New scheduling approach using reinforcement learning for heterogeneous distributed systems. J Parallel Distrib Comput. **117**, 292–302 (2018)
20. Mnih, V., Kavukcuoglu, K., Silver, D.: Human-level control through deep reinforcement learning. Nature **518**(7540), 529–533 (2015)
21. Cheng, N., et al.: Space/Aerial-assisted computing offloading for IoT applications: a learning-based approach. IEEE J. Sel. Areas Commun. **37**(5), 1117–1129 (2019)

Mobile Sensing and Computing

A Study of WSN Localization Based on the Enhanced NGO Algorithm

Qiang Sun[2], Yiran Tian[1](✉), and Yuanjia Liu[1]

[1] School of Physics and Electronic Engineering, Mudanjiang Normal University, Mudanjiang 157011, China
tyr19981@163.com

[2] School of Computer and Information Technology, Mudanjiang Normal University, Mudanjiang 157011, China

Abstract. In this paper, an enhanced INGO optimization algorithm is proposed to solve the problem of large positioning error of the original DV-Hop algorithm in wireless sensor networks. By introducing cubic chaotic mapping and increasing the diversity of population initialization to expand the search scope, the sensor node location information can be collected more widely, so that the algorithm can search for the best solution as far as possible. In addition, a hybrid method of optimal - worst reverse learning and lens imaging reverse learning strategy is added to help the algorithm get rid of the local extreme value easily and improve the positioning accuracy. By comparing with the localization results of the classical DV-Hop localization algorithm, SSADV-Hop algorithm, and WOADV-Hop algorithm, the INGO algorithm reduces the average normalized localization error when the beacon node, communication radius, and total number of nodes are different.

Keywords: DV-Hop Positioning · North Goshawk Optimization Algorithm · Cubic chaotic mapping · Hybrid reverse learning strategy

1 Introduction

Wireless Sensor Networks (WSN) are network systems composed of a multitude of distributed wireless sensor nodes. They are utilized for sensing, collecting, and transmitting diverse information within the environment [1]. In WSN networks, node positioning technology serves as the foundation for network construction. This technology enables the determination of location information for other nodes in the network based on the data from a few known nodes. However, imprecise positioning can lead to decreased communication efficiency among sensor nodes and cannot ensure the reliability of transmitted node information. Therefore, the accurate localization of unknown node coordinates holds significant importance [2].

Traditional positioning technology typically relies on global positioning systems to provide positioning and navigation information through satellites. However, satellite-based positioning systems suffer from drawbacks such as high cost

H. Jin et al. (Eds.): GPC 2023, LNCS 14504, pp. 125–138, 2024.
https://doi.org/10.1007/978-981-99-9896-8_9

and energy consumption. Additionally, achieving accurate positioning in large-scale practical scenarios, such as forest rescue and personnel search, poses significant challenges [3]. To address these limitations, the DV-Hop(distance vector-hop) algorithm was developed as a range-free positioning algorithm. It offers high accuracy and robustness without requiring additional hardware equipment. Its simplicity and ease of operation make it well-suited for large-scale wireless sensor network positioning. The DV-Hop algorithm determines node coordinates by measuring the number of hops and distances between nodes. However, the traditional DV-Hop algorithm suffers from significant positioning errors. To overcome this, researchers both domestically and internationally have proposed numerous improved algorithms aimed at achieving the most accurate calculation of node location information [4–6].

In the literature [7], an improvement to the traditional DV-Hop algorithm is proposed by utilizing the weighted least square method to enhance node positioning accuracy. However, this improvement solely focuses on the third step, without addressing the positioning error caused by the second step. Another improved algorithm, described in the literature [8], modifies the estimated distance between unknown nodes and different anchor nodes based on fractional hop information and relatively accurate anchor node coordinate information. This adjustment aims to reduce localization errors by improving the minimum hop count and average distance in the second step. With the emergence of swarm intelligence optimization algorithms, many scholars have employed swarm intelligence algorithms to optimize the DV-Hop algorithm. In the literature [9], a method utilizing double communication radius is proposed to mitigate distance estimation errors. Additionally, the SSA (Social Spider Algorithm) is employed for node position estimation. Furthermore, in the literature [10], the parallel whale algorithm is employed to address the distance error between anchor nodes and unknown nodes in the second step of the DV-Hop algorithm.

In this study, we introduce a novel optimization algorithm called the Northern Goshawk Optimization (NGO) algorithm, aiming to enhance positioning accuracy. The key innovations of this research are as follows:

Firstly, we propose a positioning algorithm named INGODV-Hop, which combines the enhanced Northern Goshawk Optimization algorithm with the traditional DV-Hop algorithm.

Secondly, the INGODV-Hop method incorporates a chaos mechanism to expand the population's distribution area, enhance the quality of initial solutions, and reduce the error rate.

Thirdly, we introduce a hybrid reverse learning strategy to improve global search capabilities and prevent the original Northern Goshawk algorithm from converging to local optima.

To evaluate the superiority of the improved algorithm, we conduct experiments to test the positioning accuracy under different parameters, comparing it with algorithms such as Whale Optimization Algorithm(WOA), Social Spider Algorithm(SSA), and other improved algorithms. The experimental results demonstrate that INGO effectively enhances the localization accuracy of the original DV-Hop algorithm, providing more accurate node location information.

2 Construction of WSN Node Location Model

2.1 Classic DV-Hop Algorithm

DV-Hop is a wireless sensor network location algorithm based on hop number. It estimates the position of nodes by hop number measurement and the least square method. To better comprehend the methodology for calculating unknown node coordinates, The DV Hop algorithm's basic idea is explained in depth in Fig. 1. i, j, k are set beacon node coordinates respectively, u_1, u_2... represents unknown node distribution, d_1, d_2 and d_3 represent the distance between two pairs of beacon nodes. Three stages are taken to ascertain the node's position:

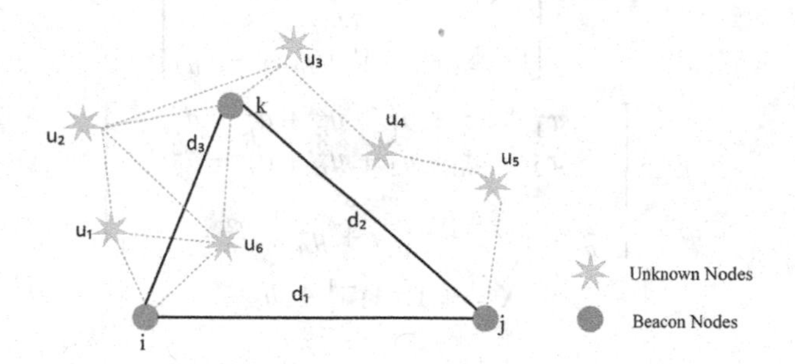

Fig. 1. Schematic diagram of DV-Hop principle

I. The Each sensor measures its distance to all of its neighbors and sends this distance information to its connected anchor node.
II. The anchor node then calculates the average number of hops between each sensor and the anchor node based on the distance information it receives. Among them(x_i, y_i), (x_j, y_j) represent the coordinates of beacon nodes i and j; h_{ij} is the number of hops between two beacon nodes. $HopSize_i$ represents the average jump distance of the beacon node i, n is the number of beacons. H_i is the number of hops between the unknown node and the nearest beacon node; The distance L between the unknown node and the beacon node is calculated by the hop information according to formula (2).

$$HopSize_i = \sum_{i \neq j}^{n} \sqrt{(x_i - x_j)^2 + (y_i - y_j)^2} / \sum_{i \neq j}^{n} h_{ij} \quad (1)$$

$$L_{ij} = HopSize_i * H_i \quad (2)$$

III. The position of each node is determined by polygon triangulation. The known beacon node's coordinates are (x_n,y_n) and the unknown node's coordinates are (x,y). d_1,d_2, $d_3$estimate distance between unknown node and beacon node.

Establish matrix model $AX = b$. The unknown node location information is estimated by the least square method. Where A, X, and $b$are as follows:

$$\begin{cases} \sqrt{(x-x_1)^2+(y-y_1)^2} = d_1 \\ \sqrt{(x-x_2)^2+(y-y_2)^2} = d_2 \\ \sqrt{(x-x_3)^2+(y-y_3)^2} = d_3 \\ \vdots \\ \sqrt{(x-x_n)^2+(y-y_n)^2} = d_n \end{cases} \tag{3}$$

$$A = 2\begin{bmatrix} (x_1-x_n)\&(y_1-y_n) \\ (x_2-x_n)\&(y_2-y_n) \\ \vdots\& \\ (x_{n-1}-x_n)\&(y_{n-1}-y_n) \end{bmatrix} \tag{4}$$

$$b = \begin{bmatrix} x_1^2+x_n^2+y_1^2+y_n^2+d_n^2-d_1^2 \\ x_2^2+x_n^2+y_2^2+y_n^2+d_n^2-d_2^2 \\ \cdots\& \\ x_{n-1}^2+x_n^2+y_{n-1}^2+y_n^2+d_n^2-d_{n-1}^2 \end{bmatrix} \tag{5}$$

$$X = (A^T A)^{-1} A^T b \tag{6}$$

2.2 Northern Goshawk (NGO) Algorithm

NGO algorithm is a population intelligence algorithm proposed by Mohammad Dehghani and others in 2022 [11]. It simulates the hunting behavior of the northern goshawk. The main core principles include two stages prey recognition and pursuit. The program searches for prey in a D-dimensional search space using a population size of N goshawks, which is the overall best solution to the relevant problem.

Prey recognition stage: in the first stage of hunting, the Northern Goshawk picks a target at random and attacks it quickly in the early stage of its hunt. This selection and attack of prey can be explained by the following formulas: (7)–(9):

$$P_i = X_k, i = 1, 2, ...N, k = 1, 2, 3...i-1, i+1, ...N \tag{7}$$

$$x_{i,j}^{new,P_1} = \begin{cases} x_{i,j} + r(p_{i,j} - Ix_{i,j}, Fp_i < F_i) \\ x_{i,j} + r(x_{i,j} - p_{i,j}, Fp_i \geq F_i) \end{cases} \tag{8}$$

$$X_i = \begin{cases} X_i^{new,P_1}, Fp_i^{new,P_1} < F_i) \\ X_i, Fp_i^{new,P_1} \geq F_i) \end{cases} \tag{9}$$

P_i is the location of the ith northern goshawk prey, F_i is the value of the ith northern goshawk's objective function. F_{P_i} is its objective function value, k is a random self-number in the range [1, N], X_i^{new,P_1} is the ith Northern Goshawk's new location, $x_{i,j}^{new,p_1}$ is the new position of the $j-dimension$ of the ith Northern Goshawk, F_i^{new,P_1} is the ith Northern Goshawk's revised objective function value during the initial hunting stage, i is a random number in the range [0,1], I is either 1 or 2.

Chase and escape phase: after selecting its prey, the goshawk will attack it, and pursuit of it at high speed, the prey cannot escape. Pretend that the hunt's second phase is at the attack position with radius R, this stage is described by the formula (10) to (12)

$$x_{i,j}^{new,P_2} = x_{i,j} + R(2 * r - 1)x_{i,j} \tag{10}$$

$$R = 0.02 * (1 - \frac{t}{T}) \tag{11}$$

$$X_i = \begin{cases} X_i^{new,P_2}, Fp_i^{new,P_2} < F_i) \\ X_i, Fp_i^{new,P_2} \geq F_i) \end{cases} \tag{12}$$

X_i^{new,P_2} is the new position of the ith Northern Goshawk, t is the current number of iterations, T represents the most iterations possible, $x_{i,j}^{new,P_2}$ is the new position of the ith Northern Goshawk in the jth dimension, F_i^{new,P_2} is the second stage's updated objective function value for the hth northern Goshawk.

3 Improve the NGO Algorithm

First, the NGO optimization algorithm uses a random distribution approach to initialize the population, which may cause some people to concentrate in one location while information gathering is lacking in other places and the population is insufficiently gathered to conduct further research. As the number of iterations rises and the population diversity falls in the later stage, the algorithm may enter the local optimal since all states in the solution space cannot be explored [12]. This study proposes corresponding improvement ideas in light of the deficiencies of the previous two points.

3.1 Population Initialization

Chaotic mapping refers to the dynamic behavior exhibited by a complex nonlinear system. The utilization of chaotic sequences for initialization addresses the limitations of the NGO algorithm [13]. In this study, Cubic maps (also known as cubic chaotic maps) with a relatively uniform distribution are selected. During the initialization process, a randomly generated weight factor is assigned to each individual. The expression for the mapping operator is as follows:

$$y(n+1) = 4y(n)^3 - 3y(n), y(n) \neq 0, n = 1, 2, 3... \tag{13}$$

$$W_i = Random(min_w, max_w) \tag{14}$$

The specific steps for initializing the cubic chaotic mapping function with the added weight factor are as follows:

Define the formula of the cubic chaotic mapping formula (13), which generates random and distributed numerical sequences.

Initialize the weight factor for each individual. Set the initial value of the cubic chaotic map and calculate it using formula (13). Then, map the result to the value range of the weight factor [min-w, max-w] to obtain the initial weight factor for the individual.

By assigning a random initial weight factor to each individual based on the cubic chaotic map, the diversity and randomness of individuals are increased. This ensures that the algorithm maintains diversity during the search process, facilitating global search.

3.2 Mixed Reverse Learning Strategies

This work presents a hybrid reverse learning approach by introducing lens image reverse learning and the optimal-worst reverse learning strategy to address the issue that the algorithm is prone to slip into the local extreme value with the increase of iterations in the later stage. Reverse learning of the specific positions of each iteration update in NGO can help the algorithm's overall search capability to some extent, preventing it from entering the extreme state too soon.

i. Lens imaging reverse learning strategy
Assume that in a two-dimensional space as shown in Fig. 2, the individual P with height h is projected onto the X-axis through a convex lens to obtain the global optimal position X_{best}, and P with height h^* is obtained by imaging; Range of axes $[a_j, b_j]$, the optimal position X^*_{best}(the point opposite x_{best}) is obtained on the coordinate axis. The formula is described as follows:

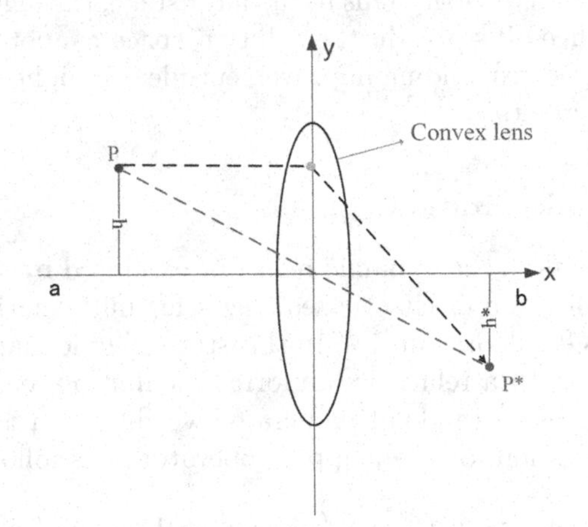

Fig. 2. Schematic diagram of lens imaging

$$\frac{(a)_j + b_j)/2 - x}{X^*_{best} - (a_j + b_j)/2} = \frac{h}{h^*} \tag{15}$$

$$n = \frac{h}{h^*} \tag{16}$$

$$X_{best}^{*} = \frac{a_j + b_j}{2} + \frac{a_j + b_j}{2 * n} - \frac{x^*}{n} \tag{17}$$

In formula (16), n is the adjustment parameter. When $n = 1$, lens imaging is a general reverse learning strategy; In this paper, n is set to $1.2 * 10^4$. Several studies have demonstrated that elite people have a broader search scope than average people. As a result, using the lens imaging reverse learning method can help NGOs escape local optimization when they reach local extremes.

ii. Optimal worst reverse learning strategy
For the specific goshawk in the worst position globally, the random reverse learning approach is used to address the local optimization and enhance the algorithm's search efficiency. In each iteration of the algorithm, position selection is carried out through formula (17) and formula (18). By comparing fitness values before and after reverse learning, the optimal position information and fitness values are updated. The expression is as follows:

$$X_{worst}^{*} = a_j + rand(b_j - x_{worst}) \tag{18}$$

where x_{worst} represents the worst position vector at present, and the rand is a random number in the range of [0,1]. In this paper, two individuals of the current optimal position and the worst position in the population are selected for location search. A dynamic boundary is used to make the search range of the algorithm more extensive and effective and improve the optimization accuracy of the algorithm; In elite reverse learning, the first few individuals in the population are generally used for processing. Individuals with little difference have little influence on the algorithm jumping out of the local extreme value, but increase the complexity of the algorithm.

3.3 Algorithm Steps in This Paper

I. The corresponding relationship between INGO algorithm and node location in wireless sensor networks is established. The INGO algorithm treats each solution to the node location problem as an individual goshawk. The set objective function evaluates the individual goshawk location at each iteration of the algorithm, and the minimal value of $fitness(i)$ is solved to determine the ideal node positioning. The following is how the fitness value function is expressed:

$$fitness = \sum_{i=1}^{n} \sqrt{(x - x_i)^2 + (y - y_i)^2 - d_i^2} \tag{19}$$

II. According to formula (2), the average number of hops between each sensor and the anchor node is calculated.

III. A cubic chaotic mapping strategy is adopted to initialize N individuals with relatively uniform location distribution in the search space.

IV. Prey recognition stage: formula (7), formula (8), formula (9) search and select prey; update the position information and objective function value of the first stage.

V. Select the best and worst individuals according to Eqs. (17) and (18), improve the search ability of the algorithm, increase the diversity of the population, and jump out of the local extreme value.
VI. The most recent position data for the northern goshawk is updated by the formula by comparing the target function values of prey and that of the goshawk formula (12).
VII. Judge whether the algorithm meets the stop condition, if so, end the cycle, and save the current optimal solution and objective function value, otherwise continue to repeat steps (4) to (6). The overall procedure flow chart is shown in Fig. 3.

4 Simulation Experiment and Result Analysis

4.1 Simulation Experiment

In this paper, simulation experiments were conducted using MatlabB2021. The proposed approach was compared with the classic original DV-Hop algorithm, as well as the SSADV-Hop algorithm [9], which is optimized by introducing a double communication radius combined with the sparrow optimization algorithm, and the WOADV-Hop algorithm [10], which is improved by combining two communication strategies: group communication and population average location value. The experimental area was a 100×100 m square region, and the environmental parameters were set as specified in Table 1.

Table 1. Experimental environment parameter Settings

Name	Parameter
population	N = 100
Maximum iterations	300
Number of nodes	100
Beacon node	30
Communication radius	30
Node distribution	Random

The performance of the four algorithms is compared in order to increase the results' accuracy, and the average value is taken 50 times. The localization error was normalized to evaluate the positioning accuracy. The expression is as follows:

$$Error = \frac{\sum_{i=1}^{n} \sqrt{(X_i - x_i)^2 + (Y_i - y_i)^2}}{N * R} \tag{20}$$

where, (X_i, Y_i) is the coordinate calculated by positioning algorithm, and (x_i, y_i) represents the real coordinate. N is the number of unknown nodes and R is the communication radius.

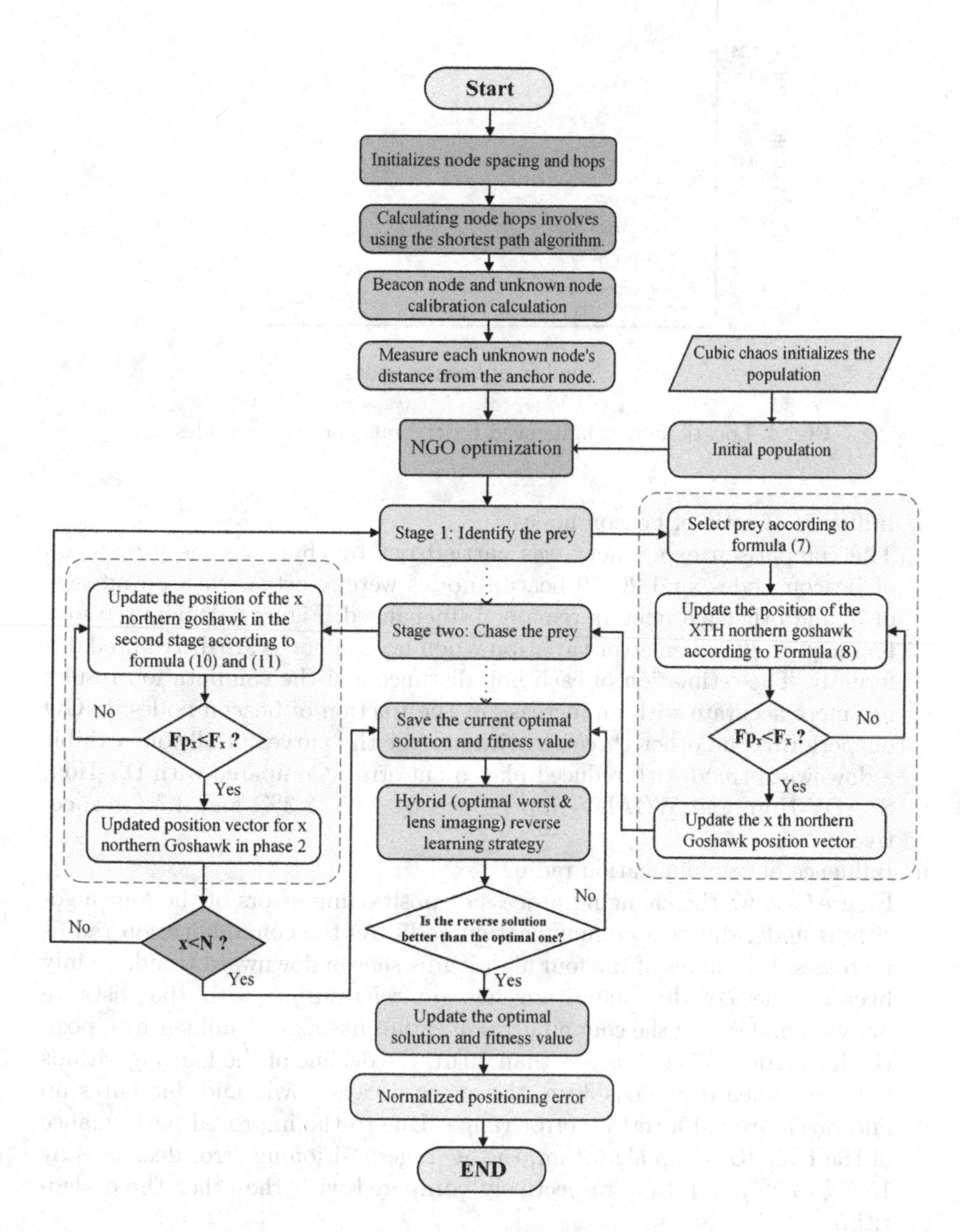

Fig. 3. Overall system flow chart

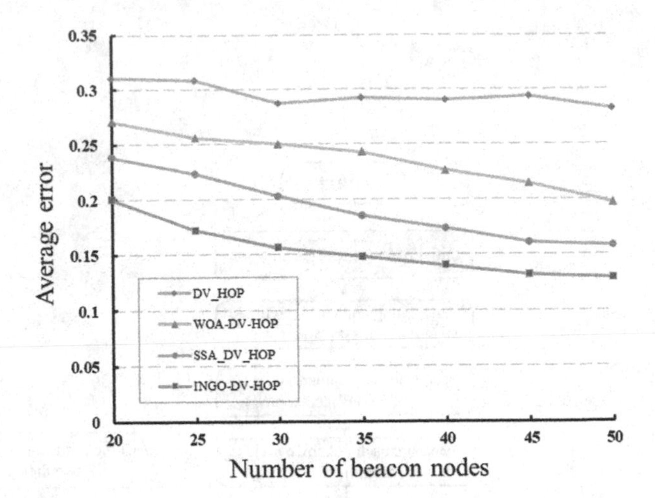

Fig. 4. Location error influenced by the ratio of beacon nodes

i. Influence of ratio of beacon nodes
The comparison experiment was carried out by changing the proportion of beacon nodes, and 20–50 beacon nodes were selected, with an interval of 5, and other parameters remained unchanged. Figure 4 depicts the four techniques' effects on error variation when beacon nodes are distributed differently. The estimation of each hop distance and the computation results are more accurate with an increase in the fraction of beacon nodes. INGO outperforms the other three algorithms, yet the curves of all four exhibit a downward trend with reduced placement error. Compared with DV-Hop, SSADV-Hop, and WOADV-Hop reduce by 14%, 8.3%, and 3.7% respectively.

ii. Influence of communication radius
Figure 5 shows the changing process of positioning errors of the four algorithms under different communication radii. As the communication radius increases, the curves of the four algorithms show a downward trend, mainly because the DV-Hop algorithm has no relationship with the distance between nodes, and the communication radius has a great influence on positioning errors. When R is less than 30 m, the decline of the four algorithms is large. When it exceeds 30 m, the curve decays slowly and fluctuates up and down around a certain error range. Due to the improved performance of the INGODV-Hop algorithm, the average positioning error decreases by 15.5%, 5.5%, and 3.1% respectively compared with the other three algorithms.

iii. Different number of nodes
Modify the overall number of nodes used in the wireless sensor network while maintaining the status quo for everything else. For the simulation, a range of 100 to 280 nodes was used, spaced 30 nodes apart. To check the precision of each algorithm's location, alter the total number of nodes. Figure 6

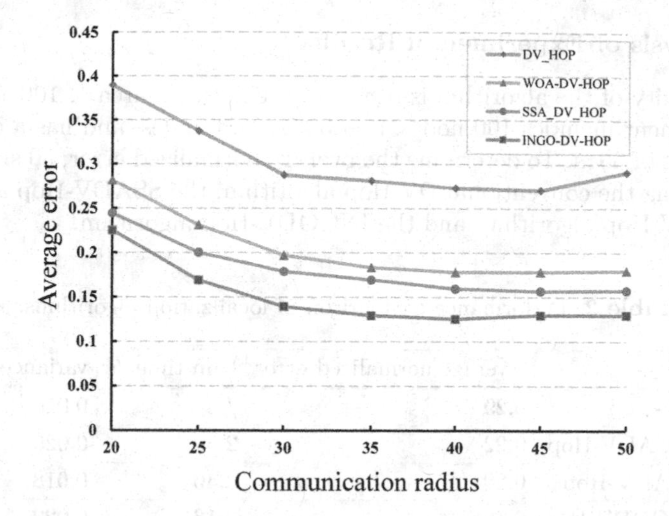

Fig. 5. Location error affected by communication radius

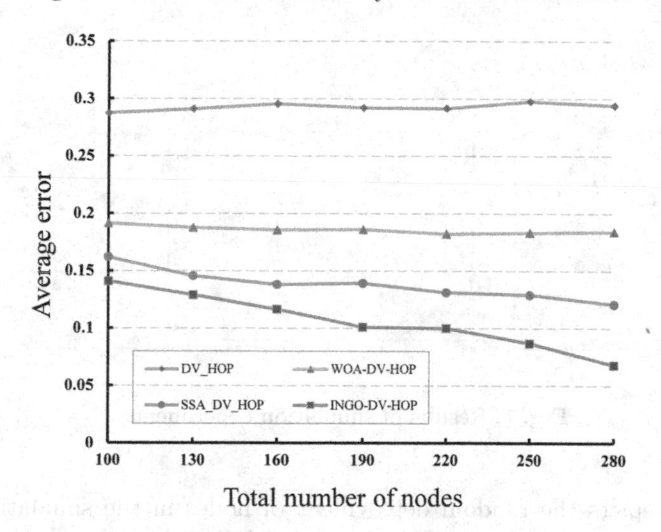

Fig. 6. The effect of the total number of nodes on the result

below displays the experimental data. The positioning inaccuracy showed a decreasing tendency as the total number of nodes increased, although the reducing range was not very wide. The placement inaccuracy of the revised technique in this research clearly lowers as the total number of nodes increases. The positioning accuracy of the enhanced NGODV-Hop algorithm is reduced by 18.6%, 7.9%, and 3.2%, respectively, compared with the other three algorithms.

4.2 Analysis of Experimental Results

The complexity of the algorithm is tested in the network area of 100 m × 100 m. The deployment includes 100 nodes, and 30 beacon nodes, and has a communication radius of 30 m. To determine the average normalized error, 30 simulations were run using the conventional DV-Hop algorithm, the SSADV-Hop algorithm, the WOADV-Hop algorithm, and the INGODV-Hop algorithm.

Table 2. Performance comparison of localization algorithms

	Average normalized error	Run time /s	variance
DV-Hop	0.29	/	0.026
WOADV-Hop	0.22	2.52	0.025
SSADV-Hop	0.20	3.40	0.018
INGODV-Hop	0.16	9.13	0.022

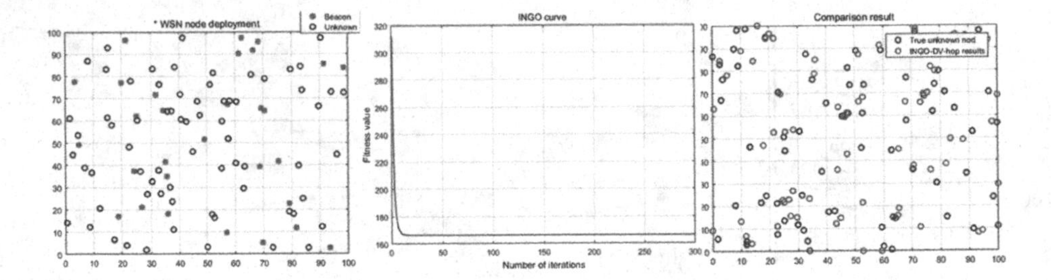

Fig. 7. Results of simulation experiments

Figure 7depicts the random deployment of nodes in the simulation experiments, the convergence curve of the improved NGO algorithm for localization, and the error comparison results. Table 2 shows the performance comparison of the four algorithms, the table shows that, when compared to the other three methods, the INGODV-Hop algorithm has a decreased average normalization error. Under the same conditions, the WOADV-Hop algorithm and SSADV-Hop algorithm added with a single improvement method take less time on average than the INGODV-Hop algorithm added with multiple improvement methods in this paper. The reason is that adding improvement strategies such as algorithm distribution and search mode may lead to better optimal solutions, but it will correspondingly increase the time of searching for the entire optimization algorithm. According to the variance comparison, the stability of the improved algorithm in this paper is slightly inferior to that of the SSADV-Hop algorithm.

5 Concluding

In this study, an enhanced NGO method is incorporated into the original DV Hop algorithm to improve the accuracy of node placement. MATLAB is employed for simulating the deployment of a wireless sensor network. To enhance the global search capability, the population initialization stage utilizes cubic chaotic mapping to introduce greater diversity in population locations. Furthermore, a mixed reverse learning strategy is employed for prey pursuit behavior parameters to prevent the algorithm from getting trapped in local extreme values. Experimental data suggests that the proposed approach offers several advantages over other algorithms in terms of improving positioning accuracy. Considering the algorithm's low operating efficiency, future research will primarily concentrate on achieving improved positioning accuracy while enhancing operational efficiency. Additionally, efforts will be made to expand the algorithm's applicability from two-dimensional space to three-dimensional space, aiming to further enhance the precision of positioning.

Acknowledgements. This research is jointly funded by the project Heilongjiang Province higher education teaching reform key entrusted project (SJGZ20200175; SJGZ20190063); key research topic of economic and social development of Heilongjiang Province (Special Base) (22356); basic operating expenses of Heilongjiang Provincial Education Department (1355JG011).

References

1. Xue, D.: Research of localization algorithm for wireless sensor network based on DV-Hop. EURASIP J. Wirel. Commun. Netw. **2019**, 1–8 (2019)
2. Ali, A., et al.: A comprehensive survey on real-time applications of WSN. Future Internet **9**(4), 77.T (2017)
3. Ghribi, H., Khelifa, F., Jemai, A., Salah, M.B.B.: A review of DV-hop localization algorithm. In: 2021 31st International Telecommunication Networks and Applications Conference (ITNAC), Sydney, Australia, pp. 121–126 (2021)
4. Khalaf, O.I., Sabbar, B.M.: An overview on wireless sensor networks and finding optimal location of nodes. Periodicals Eng. Nat. Sci. **7**(3), 1096–1101 (2019)
5. El Ghafour, A., Mohamed G., Kamel, S.H., Abouelseoud, Y.: Improved DV-hop based on squirrel search algorithm for localization in wireless sensor networks. Wirel. Netw. **27**, 2743–2759 (2021)
6. Omic, S., Mezei, I: Improvements of DV-Hop localization algorithm for wireless sensor networks. Telecommun. Syst. **61**, 93–106 (2016)
7. Kumar, S., Lobiyal, D.K.: An advanced DV-Hop localization algorithm for wireless sensor networks. Wirel. Personal Commun. **71**, 1365–1385 (2013)
8. Liu, G., Qian, Z., Wang, X.: An improved DV-Hop localization algorithm based on hop distances correction. China Commun. **16**(6), 200–214 (2019)
9. Lei, Y., De, G., Fe, L.: Improved sparrow search algorithm based DV-Hop localization in WSN. In: 2020 Chinese Automation Congress (CAC). IEEE (2020)
10. Chai, Q.-W., et al.: A parallel WOA with two communication strategies applied in DV-Hop localization method. EURASIP J. Wirel. Commun. Network. **2020**, 1–10 (2020)

11. Kumar, S., Bharathi, S.H.: Improving the performance of speech recognition feature selection using northern goshawk optimization. In: 2022 International Interdisciplinary Humanitarian Conference for Sustainability (IIHC). IEEE (2022)
12. Li, Y., Han, M., Guo, Q.: Modified whale optimization algorithm based on tent chaotic mapping and its application in structural optimization. KSCE J. Civil Eng. **24**(12), 3703–3713 (2020)
13. Feng, J., et al.: A novel chaos optimization algorithm. Multimedia Tools Appl. **76**, 17405–17436 (2017)

A Novel Framework for Adaptive Quadruped Robot Locomotion Learning in Uncertain Environments

Mengyuan Li[1], Bin Guo[1(✉)], Kaixing Zhao[1], Ruonan Xu[1], Sicong Liu[1], Sitong Mao[2], Shunbo Zhou[2], Qiaobo Xu[2], and Zhiwen Yu[1]

[1] Northwestern Polytechnical University, Xi'an 710072, China
guob@nwpu.edu.cn
[2] Huawei Cloud Computing Technologies Co. Ltd, Shenzhen 518000, China

Abstract. Learning diverse and flexible locomotion strategies in uncertain environments has been a longstanding challenge for quadruped robots. Although recent progress in domain randomization has partially tackled this difficulty by training policies on a wide range of potential factors, there is still a great need for improving efficiency. In this paper, we propose a novel framework for adaptive quadruped robot locomotion learning in uncertain environments. Our method is based on data-efficient reinforcement learning and learns simulation parameters iteratively. We also propose a novel Sampling-Interval-Adaptive Identification (SIAI) strategy that uses historical parameters to optimize sampling distribution and then improve identification accuracy. Final evaluations based on multiple robotic locomotion tasks showed superiority of our method over baselines.

Keywords: Reinforcement Learning · Robot Intelligence · Identification Strategy · Transfer Learning

1 Introduction

Quadruped robots have stronger adaptability and fewer terrain destruction in unstructured environments and are widely used in real-world missions, such as rescue, delivery, and industrial inspection [1]. However, the design of quadruped robots is not always easy: conventional approaches for legged locomotion require not only precise modeling of robot kinematics and dynamics, but also considerable expertise and manual tuning [2–4]. At the same time, benefiting from considerable progress in artificial intelligence technologies, deep reinforcement learning has gradually demonstrated superiority in perception and decision-making tasks [5]. In fact, deep reinforcement learning provides us with an alternative solution for helping quadruped robots learn flexible and stable control strategies autonomously [6,7].

In the domain of robotics learning, some of the earliest attempts to learn legged locomotion directly from data collected via the physical systems [8,9].

H. Jin et al. (Eds.): GPC 2023, LNCS 14504, pp. 139–154, 2024.
https://doi.org/10.1007/978-981-99-9896-8_10

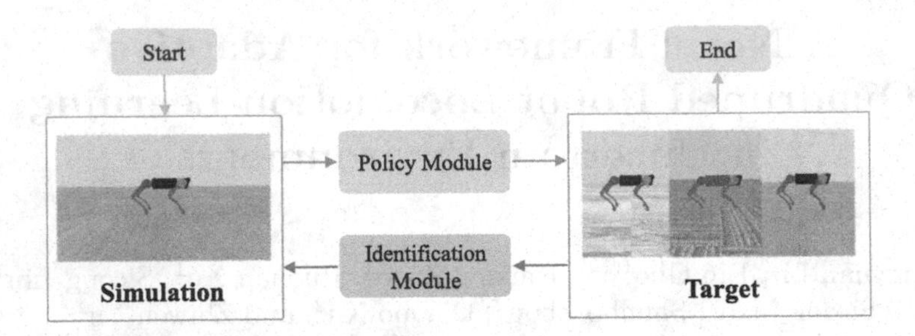

Fig. 1. The Simulation-to-Target-to-Simulation pipeline for narrowing policy-transfer gap. The policy module trains locomotion policies in simulation and the identification module learns correct simulation parameters using a few offline target data samples.

However, such policies are only limited to simple scenarios due to their complexity and safety, and nearly infeasible to scale to complex tasks. In contrast, learning in simulation for robots can collect data faster, safer, and at a lower cost, which has become a promising approach [10]. But meanwhile, robotic controllers learned in simulation often perform poorly in the real world due to the gap between reality and simulation [11,12]. Generally, this type of gap is caused by differences between two environments (i.e. reality and simulation), such as physical events that are not modeled. Otherwise, such differences would be further scaled and impair the learning process. To our knowledge, there are mainly two types of approaches that can be used to narrow the gap: the first one focuses on learning robust controllers using domain randomization [13] and dynamics randomization [14], which on one hand may lead to overly conservative strategies and on the other hand can be impossible to simulate every potential factor when the target environments are vastly different. The second method focuses on improving simulator fidelity, which requires to conduct system identification to find the correct simulation parameters. Different from the first type of method, the second one requires a large number of real-world samples [15,16] and may introduce safety risks.

More recently, several studies have begun to focus on forming a closed loop for simulation-to-reality transfer [17,18]. For example, Du et al. [17] used raw observation images from the real world to adjust the simulation parameters. Chebotar et al. [18] tried to update the distribution of parameters by comparing trajectories from simulation and the real world. However, these solutions are still very limited. On the one hand, it is infeasible to use fixed image recording for rapid locomotion tasks in their systems. On the other hand, general trajectory-matching task requires obtaining data simultaneously, which is often time-consuming and impractical. Moreover, there may be a large bias between the paired trajectories due to motion phase shifts.

To tackle the above-mentioned issues, in this paper, we propose to interleave parameter identification with policy improvement to further narrow the gap between reality and simulation. Our solution tried to only use a small amount of data samples in target domain to: 1) improve the fidelity of the simulator and 2) maximize data efficiency. More precisely, as shown in Fig. 1, we first train the locomotion policy based on reinforcement learning in simulation. Then the identification network is trained over an auto-generated simulated dataset, which directly encodes a mapping from partial observations to parameter values. However, the locomotion policy may be ineffective in new environments and the identification network would not generalize well outside the training set when the target domains differ greatly from simulation. An alternating training process is necessary to reduce distribution shifts and maximize data efficiency. We further propose to flexibly adapt the sampling interval during training to improve identification accuracy and ensure fault tolerance. In doing so, we are able to improve the policy transfer by fully training in simulation closer to the target environment.

The main contributions of this paper are summarized as follows:

- For quadruped robot locomotion tasks, we introduce a novel framework called Simulation-to-Target-to-Simulation for bridging the gap between simulation and the target environment.
- We propose a Sampling-Interval-Adaptive Identification (SIAI) strategy that adapts the simulation parameters using several observations in the target environment.
- We demonstrate that our proposed method outperforms baselines on a range of robotic locomotion tasks for policy transfer.

2 Related Work

Various locomotion planning and control methods have been widely used for quadruped robots for a long time, such as trajectory optimization [2], model predictive control [3,4], and whole-body control [19]. Due to complexity and variability of the environment, manual design is insufficient to account for all possible situations. Reinforcement learning can learn general strategies and adapt to challenging and uncertain terrains. These methods include curriculum learning [6,7,20], learning by cheating [21,22], hierarchical controller methods [23]. To narrow the reality gap and achieve zero-shot sim-to-real transfer, domain randomization methods have been used to train policies by randomizing system parameters to simulate diverse conditions a robot may encounter in the real world [7,10,14]. However, when the target domain differs significantly from the training environment, it also shows limited generalization.

Alternatively, various domain adaptation techniques can also reduce the sim-to-real gap. Prior works have shown that using small amounts of target data samples can significantly enhance the adaptability of policies, including fine-tuning [24,25], meta-learning [26], and general policy learning [27]. These methods directly optimize the learning policies to perform few-shot adaptation. But it is often difficult to obtain sufficient and diverse data, leading to data bias and limited generalization. While we also use a few target data samples in this work, we focus on improving the policy transfer by adapting the simulator closer to the target environment.

Classical system identification methods [16,28,29] use various gradient-free optimization techniques, which usually require a large amount of sample data and high computational costs. Recent work [30,31] has explored data-driven methods for high-fidelity simulators with only one batch of data collection, resulting in limited performance improvement. Several studies have begun to focus on forming a closed loop for simulation-to-reality. The closest to our approach are the methods from [17,18] that propose to learn simulation parameters and train policies iteratively. Du et al. [17] learn to adjust the simulation parameters by using raw observation images from the real world. It is infeasible to use fixed image recording for rapid locomotion tasks of quadruped robots. SimOpt [18] uses continuous object tracking in the real world to adapt simulation randomization. It takes manual effort to handle possibly large mismatches between paired trajectories because of the phase shift of quadruped locomotion. The current iteration optimization work is based on manipulator operation tasks. While our work focuses on the locomotion characteristics of quadruped robots, learning the locomotion policy and the identification network to achieve adaptive locomotion in uncertain environments.

3 Method

We propose a Simulation-to-Target-to-Simulation framework for quadruped robot locomotion learning in uncertain environments, as shown in Fig. 2. The architecture consists of two parts: the Simulation-to-Target part learns a reinforcement learning policy in simulation (Sect. 3.1), and the Target-to-Simulation part optimizes the identification network using few-shot methods (Sect. 3.2). The joint training process is described in Sect. 3.3. An iterative training procedure becomes essential for mitigating distribution shifts and optimizing data utilization. Additionally, we suggest dynamically adjusting the sampling intervals during training to enhance identification accuracy and ensure robustness against faults. This approach allows us to enhance policy transfer by effectively training in simulation, bringing it in line with the characteristics of the target environment. Our key insights are: 1) improve uncertain environments adaptation of quadruped robots by iteratively training locomotion policies and 2) maximize data efficiency and avoid distribution shifts by tuning simulation parameters.

3.1 Base Policy

In this section, we first formulate the locomotion task as a Partially-observable Markov Decision Process (POMDP). Formally, a POMDP can be described as a tuple: $(S, O, A, R, P, p_0, \gamma)$, where S represents the state space, O represents the observation space, A represents the action space, R is the reward function, P is the transition probability function, p_0 is the initial state distribution and $\gamma \in (0,1)$ is the discount factor. The agent learns a policy $\pi_\theta(a \mid s)$ with parameters θ that maps the current state to an action distribution. The objective of reinforcement learning methods is to learn the optimal policy π^* that maximizes

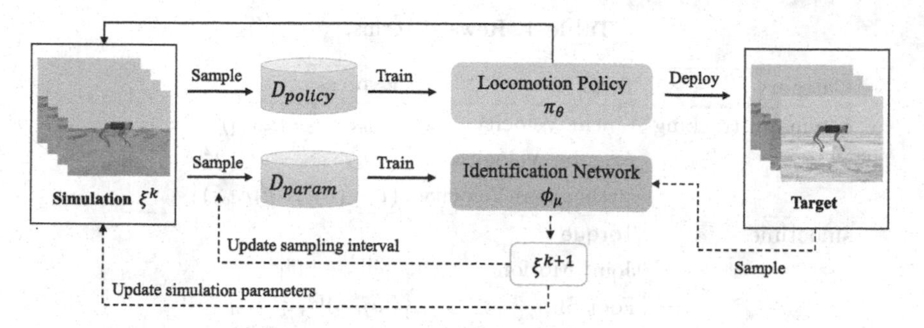

Fig. 2. Overview of our system: We train locomotion policy π_θ and identification network ϕ_μ in simulation. A few offline observations collected in target are used to correct simulation parameters ξ. While one identification is often not sufficient for two significantly different environments, we use new parameters ξ to tune the simulator so that it more closely approximates the target, and updates the sampling interval for higher identification accuracy. Our work interleaves parameter identification with policy improvement to maximize data efficiency and avoid distribution shifts.

the expected discounted reward $J(\theta) = E_{\pi_\theta}[\sum_{t=0}^{T-1} \gamma^t r_t]$ where $r_t \sim R(s_t, a_t)$, $s_{t+1} \sim P(s_{t+1} \mid s_t, a_t)$ and $a_t \sim \pi_\theta(a_t, s_t)$.

Observation. The observation space is similar to [6], which consists only proprioceptive measurements of the robot ($o_t \in R^{93}$) that can be obtained through sensors and state estimators on the physical robot system. The observations are composed of: velocity commands (3D), base linear velocity (3D), angular velocity (3D), joint positions (12D), velocities (12D), as well as previous foot target positions (12D). In addition, joint position errors (24D) and velocities (24D) measured at -0.01 and -0.02 s are also included in the observations.

Action. The design of the action space will affect the learning speed and policy quality. To accelerate model convergence, we use the Policies Modulating Trajectory Generator (PMTG) [32] framework as prior knowledge to generate cyclic and smooth actuation patterns. More precisely, the action space indicates foot position residuals ($a_t \in R^{12}$), which can be added to prior foot positions to obtain target foot positions. The target foot positions are converted to desired joint angles through inverse kinematics (IK) and tracked using proportional-derivative (PD) control.

Reward. The reward function is designed to follow the command velocities and to learn robust as well as efficient locomotion skills for quadruped robots. We design reward functions from three aspects: command tracking, action smoothness, and safety. We denote linear velocity as v, horizontal target linear velocity as v_{des}, angular velocity as ω, and target angular velocity as ω_{des}, all in robot's base frame. Besides, we define joint torques as τ, joint angles as q, joint velocities as $\dot{q}$, and joint accelerated velocities as $\ddot{q}$. We additionally express the velocity of the feet in the world frame as v_{foot}, ground reaction forces at the feet as F_{foot}, body contact state set as $I_{c,body}$, foot contact state set as $I_{c,foot}$, h_{foot} as the

Table 1. Reward terms.

Category	Term	Expression
command tracking	Linear Velocity	$(v_{des} \cdot v - \|v_{des}\|)^2$
	Angular Velocity	$(\omega_{des} \cdot \omega_z - \omega_{des})^2$
	Orthogonal Velocity	$(v - (v_{des} \cdot v)v_{des})^2$
smoothness	Torque	$\|\|\tau\|\|^2$
	Joint Motion	$\|\|\dot{q}\|\|^2 + \|\|\ddot{q}\|\|^2$
	Foot Slip	$\|\|v_{foot}\|\|^2, \forall F_{foot} \geq 0$
	Action Smoothness	$\|\|\tau_t - \tau_{t-1}\|\|^2$
safety	Body Collision	$\|I_{c,body}\ I_{c,foot}\|$
	Foot Clearance	$\sum_{foot}(h_{foot} - h_{des})^2\|\|v_{foot}\|\|$

foot height and h_{des} as the desired foot clearance. The total reward is a weighted sum of the nine quantities shown in Table 1.

3.2 Sampling-Interval-Adaptive Identification Strategy

The goal of system identification is to optimize the simulation parameters to match the policy behavior in simulation and the target world. Classical system identification methods often require a large number of samples and substantial expert knowledge. In this paper, we propose Sampling-Interval-Adaptive Identification (SIAI) strategy which is faster and more effective compared with existing methods. When given a short history of the states, SIAI can quickly identify parameters of corresponding physical system $\phi : (x_{t-L:t}) \to \xi$. More precisely, the training process of SIAI can be formulated as a supervised learning problem with the input being the partial observation X of length L and the output being the model parameters ξ:

$$\mu^* = \underset{\mu}{argmin} \sum_{X \in D_{param}} ||\phi_\mu(X) - \xi|| + \lambda ||\mu||^2 \tag{1}$$

where μ represents parameters of the neural network ϕ and λ is weight regularization constant.

In detail, during the training phase, datasets with known physical model parameters are used, which are auto-generated in simulation from the initial parameter distribution. There is no need to include hand-labeling or target domain data, which makes the training faster and more efficient. During the test phase, we collect one or several observations from the target domain to calculate new model parameters ξ. Given uniformly distributed training data over a certain interval $\Delta = [\xi - a, \xi + a]$, the probability densities of parameters ξ are equal to the constant $c = \frac{1}{2a}$, where a represents half of the interval length. In addition, it should be noted that in our design, the wider the parameter distribution, the smaller the probability density of ξ and the lower the identification accuracy. If the sampling interval is too narrow, the true parameters would be outside of the training distribution and the identification accuracy would be very low. Therefore, since reasonable design of the sampling interval is a key problem, we propose an iterative method to adapt the sampling interval δ and to gradually approach the true parameters.

More precisely, we design the sampling interval δ referring to the changes in historical parameter values. In order to avoid great differences between adjacent intervals, a^{k+1} is set to be not less than half of a^k, which can improve identification accuracy and ensure fault tolerance. Our method gradually shifts sampling distributions to contain the true values, which is effective to tune parameters that lie outside the ranges seen in training data. In this way, we improve the policy transfer by fully training in simulation closer to the target environment. The core formulas for this task are as follows:

$$\begin{aligned} a^{k+1} &= max[0.5a^k, 0.5|\xi^k - \xi^{k-1}|] \\ \Delta^{k+1} &= [\xi^k - a^{k+1}, \xi^k + a^{k+1}] \end{aligned} \tag{2}$$

3.3 Simulation-to-Target-to-Simulation Joint Learning

As mentioned before, our proposed solution interleaves simulation optimization with policy improvement to narrow the reality gap. If the target domain and simulation differ greatly, the original policy may be difficult to collect effective data and the identification approach would not generalize well outside the initial training set. Therefore, an alternating training process is necessary to reduce distribution shifts. In our solution, we first collect training data D_{policy} for reinforcement learning in simulation and generate original locomotion policy π_θ. We then set the parameter sampling interval $\Delta = [\xi - a, \xi + a]$, and use policy π_θ to collect training dataset D_{param} from uniform distribution in Δ. Actually, the dataset contains the robot's partial observations of length L and physical parameters, which can be used for training the identification network ϕ_μ via supervised learning. One or several offline data samples are collected in the target domain, which also contains the robot's observations of length L. Therefore, after new physical parameters ξ' are identified by the identification network ϕ_μ, we will update the simulator and the identification network sampling interval Δ using new system parameters. We repeat the above steps until the robot walks steadily in the target domain.

4 Experiment Setup

4.1 Robot and Terrain

We use PyBullet simulator [33] for rigid-body and contact dynamics simulation and conduct final evaluations on Unitree A1 quadruped robot [34]. The robot is equipped with 12 actuated joints and weighs about 12 kg. During tests, the control policy operates at 50 Hz and the predicted desired joint positions are

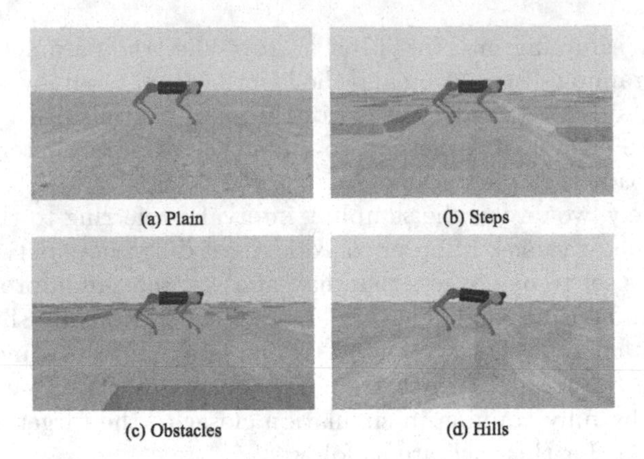

Fig. 3. Terrain types used in training and testing.

converted to torques by a low-level PD controller running at 500 Hz. Besides, to simulate different types of terrain, we introduced a parametric terrain generator to generate diverse uneven terrains, including Plain, Steps, Obstacles and Hills, as shown in Fig. 3. The Steps terrain consists of square steps of random height, with a maximum height of 10 cm. The Obstacles terrain consists of irregular shapes and their occurrence is random. The Hills terrain is based on Perlin noise, simulating slight undulating slopes in the environment.

4.2 Target Environments Setup

To improve the locomotion performance of quadruped robots in uncertain environments, we designed one source training environment and three target environments in our study. The training environment contains simply a quadruped robot with corresponding rugged rigid ground while the target environments are set by modifying robot and ground parameters (details are in Appendix). We designed these settings by taking into account possible external and ontological uncertainties and their details are summarized as follows:

Slide. In this case, we made the robot walks on slippery ground, like ice or oily patch. Our hypothesis is that it is difficult for robot to maintain its stability and move forward due to skid.

Heavy. Inspired by [30], in this case, we increased the mass of Front-Right (FR) leg upper link of the robot by 3 kg and moved the Center of Mass (CoM) of the leg down by 5 cm. Our hypothesis is that robot will be difficult to maintain balance and easy to overturn.

Soft. In this case, we made the robot walks on soft ground with a 3 kg payload on top of its body. Our hypothesis is that robot is prone to collisions and falls in this environment.

4.3 Implementation Details

We deployed an on-policy actor-critic reinforcement learning algorithm and leveraged the Proximal Policy Optimization (PPO) [35] to train the policy network π_θ. At the beginning of the training, we first randomly selected sample terrains from the same fixed difficulty and then generated a general policy that adapts to different complex terrains. During training, each episode lasted up to 500 steps, with early termination if the robot fell. We train the original policy π_θ 8,000 episodes to get better performance.

The identification network ϕ_μ consists of three fully-connected layers with ReLU activation functions. The inputs and outputs of the network are recorded partial observations and predicted physical system parameters respectively. To generate training data, we recorded the quadruped robot's linear velocity with the length of 200 timesteps as the input observed signal. We collected 900 simulations for the training phase and 100 for the testing phase to learn a mapping from states to parameter values. Then we leveraged the policy π_θ to sample 10 linear velocity observations from the target domain and updated simulation parameters as well as sampling interval with the mean value of the outputs of the identification network ϕ_μ to ensure result reliability. During subsequent fine-tuning iterations, we trained the locomotion policy in the optimized simulator for another 500 episodes. The whole training process was completed until the robot walked steadily in the target environment.

5 Experiments

To verify the proposed framework, we carried out further experiments and focused on two aspects: (1) Can the proposed method improve the locomotion performance in different target environments? (2) Can the proposed method correctly update the simulator using generated system parameters?

5.1 Locomotion Performance in the Target Environments

First, to evaluate the locomotion performance improvement of our method in the target environments, we conducted experiments by comparing with following baselines:

1) Original Policy (OP): It means to train the OP only in simulation, which represents the lower bound of the policy-transfer performance.
2) Domain Randomization (DR): It means to train the DR on a wide-range parameter distribution.
3) Fine-Tuning (FT): It means to fine-tune the original policy in each target environment for 500 episodes.
4) DR+FT: It means to fine-tune the domain randomization in each target environment for 500 episodes.

Table 2. Dynamics parameters of training and target environments.

Parameter	OP Range	DR Range	Slide	Heavy	Soft
Base Mass	[0.8, 1.2]	[0.5, 1.5]	1	1	**+3 kg**
Base Inertia	[0.8, 1.2]	[0.5, 1.5]	1	1	1
Leg Masses	[0.8, 1.2]	[0.5, 1.5]	1	**FR +3 kg**	1
Leg Inertia	[0.8, 1.2]	[0.5, 1.5]	1	1	1
Lateral Friction	[0.4, 1.0]	[0.2, 1.5]	0.6	0.6	0.6
Spinning Friction	[0, 0.2]	[0, 0.4]	0.2	0.2	0.2
Ground Friction	[0.8, 1.2]	[0.5, 1.5]	**0.3**	1	1
Contact Damping	\	\	\	\	**100**
Contact Stiffness	\	\	\	\	**100**

All locomotion policies were trained with the same architecture and reward function. But the dynamics parameters of the training and target environments were different. The selected parameters with their ranges (or values) are shown in the Table 2.

The evaluations were based on following domain-specific metrics: (1) average task reward; (2) moving distance (in meters) along the target direction; (3) speed reward generated by adding the linear velocity reward and the angular velocity reward; (4) joint motion reward; (5) body collision reward and (6) cost of transport (COT) defined by $\sum [\tau \dot{q}]^+/(mgv)$, which represents the positive mechanical power exerted by the actuator per unit weight and unit locomotion speed. In this paper, to better present the results, we conducted data preprocessing and made sure that metric values about speed reward, joint motion reward and body collision reward always be positive, and a larger value represents better performance.

We evaluated all methods (i.e. OP, DR, FT, DR+FT and our proposed method) by five trials having different random seeds. Final results were based on the mean value and standard deviation value of five trials. Table 3 shows the average task reward in the target environment, which clearly shows that our method outperforms the baselines in eleven of the twelve policy transfer experiments. In addition, results also confirmed that: 1) both domain randomization and fine-tuning methods can improve locomotion performance in target environments; 2) methods that collect target domain data (such as fine-tuning) have higher rewards than DR. However, as we mentioned before, training the policy directly in target environments is always not easy. Therefore, in this paper, we focused on improving the efficiency of policy learning with abundant simulation training data by making the simulator more representative of the target domain. On the other hand, our method requires a longer training time.

In terms of locomotion metrics, Table 4 shows distance covered, speed reward, joint motion reward and body collision reward in the Soft environment. More precisely, results revealed that domain randomization methods tended to trade

Table 3. Average task reward in target environments.

Environment	OP	DR	FT	DR+FT	Ours
Slide/Plain	626 ± 21	664 ± 14	677 ± 4	675 ± 6	$\mathbf{693 \pm 9}$
Slide/Steps	567 ± 24	619 ± 4	600 ± 43	678 ± 13	$\mathbf{682 \pm 17}$
Slide/Obstacles	574 ± 17	600 ± 8	593 ± 13	601 ± 22	$\mathbf{641 \pm 6}$
Slide/Hills	515 ± 11	547 ± 8	554 ± 7	566 ± 8	$\mathbf{596 \pm 20}$
Heavy/Plain	634 ± 9	649 ± 13	689 ± 4	686 ± 5	$\mathbf{720 \pm 3}$
Heavy/Steps	543 ± 13	607 ± 3	617 ± 5	629 ± 10	$\mathbf{634 \pm 17}$
Heavy/Obstacles	423 ± 3	598 ± 8	610 ± 8	606 ± 21	$\mathbf{624 \pm 16}$
Heavy/Hills	495 ± 21	499 ± 23	603 ± 17	593 ± 9	$\mathbf{628 \pm 7}$
Soft/Plain	460 ± 4	436 ± 2	537 ± 5	528 ± 9	$\mathbf{574 \pm 5}$
Soft/Steps	401 ± 5	419 ± 3	$\mathbf{486 \pm 9}$	438 ± 29	480 ± 25
Soft/Obstacles	405 ± 7	417 ± 5	484 ± 12	428 ± 34	$\mathbf{503 \pm 8}$
Soft/Hills	363 ± 6	379 ± 3	444 ± 4	436 ± 7	$\mathbf{507 \pm 4}$

Table 4. Locomotion performance in the Soft environment.

Method	Soft/Steps				Soft/Obstacles				Soft/Hills			
	distance (m)	speed reward	joint motion	body collision	distance (m)	speed reward	joint motion	body collision	distance (m)	speed reward	joint motion	body collision
OP	15.67	3.39	1.15	4.51	15.67	3.39	1.26	4.65	17.22	3.37	1.05	2.89
DR	9.51	3.24	**1.86**	3.06	9.51	3.24	**1.83**	3.19	14.56	2.91	**1.76**	1.67
FT	19.74	3.80	0.74	5.55	19.74	3.80	0.75	5.56	19.16	3.74	0.61	4.05
DR+FT	12.61	3.77	1.22	4.77	12.61	3.77	1.13	4.75	15.47	3.71	1.08	3.21
Ours	**23.35**	**3.99**	0.67	**6.57**	**23.35**	**3.99**	0.52	**6.65**	**22.96**	**3.90**	0.46	**5.02**

performance for robustness, which led to worse distance and speed reward than OP. Besides, the lower joint motion reward explained that our method adjusted the leg position more frequently to track speed commands and reduce body collisions.

In addition, we also evaluated our method in the Heavy environment by leveraging the cost of transport (COT) metric. As shown in Fig. 4, results clearly revealed that our method improves energy efficiency, especially in the Steps and Obstacles terrains.

We further analyzed locomotion skills learned by our method in the target environment. The captured walking gaits are shown in Fig. 5. As we can see, in the Slide environment, the robot learns to be more careful on uneven ground and when it comes to steps or obstacles, the robot will first conduct tentative action and stay still to avoid falling if current object is hard to cross. In the Heavy environment, we can see that the robot's hip joints increase their swing amplitude to maintain balance. Finally, in the Soft environment, the robot tries to increase its foot height to avoid collision as much as possible.

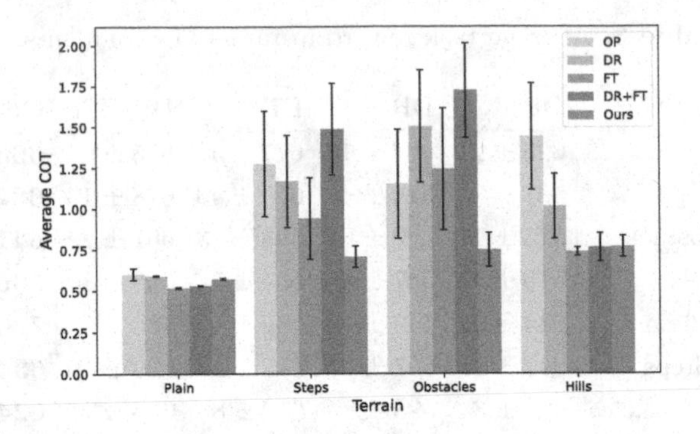

Fig. 4. Average COT in the Heavy environment.

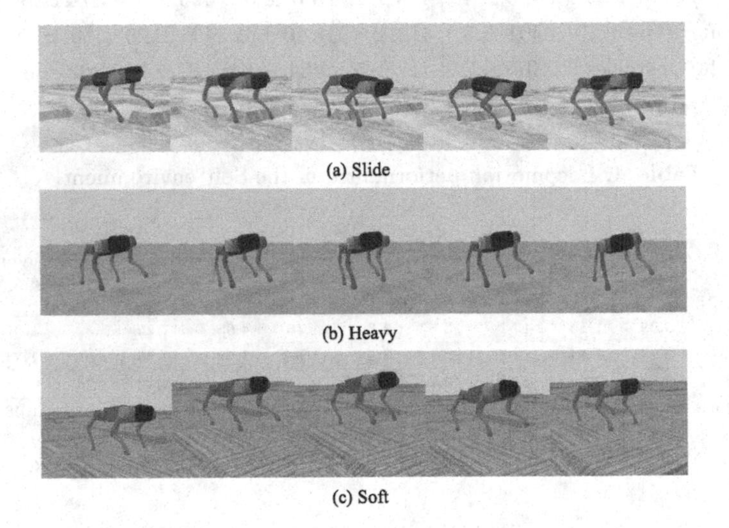

Fig. 5. Walking snapshots in three target environments.

5.2 Accuracy of Parameter Identification

In terms of parameter identification, results based on average value of five random seeds revealed that our method using interactive optimization is able to identify correct parameter.

In addition, our method performed also well in parameter-tuning task. In this paper, we focus on two metrics: mean absolute error (MAE) and mean square error (MSE) of the estimated friction coefficient in the Slide environment and three baselines: greedy entropy search [28] (GES), CMA-ES optimization [29], as well as neural network based on paired trajectories [31] (TuneNet). As shown in Table 5, prediction accuracy was both improved significantly for all four methods by narrowing the sampling interval iteratively. However, classical optimization methods such as GES and CMA-ES showed their ineffectiveness in early training stage while TuneNet encountered difficulty in converging to a good

Table 5. MAE and MSE in friction coefficient prediction.

Method	1 iteration		2 iterations		3 iterations	
	MAE	MSE	MAE	MSE	MAE	MSE
GES	0.4532	0.3091	0.1716	0.0502	0.0852	0.0113
CMA_ES	0.3903	0.1998	0.1346	0.0259	0.06089	0.0058
TuneNet	0.1417	0.0365	0.0601	0.0162	0.1058	0.0159
Ours	**0.1179**	**0.0241**	**0.0483**	**0.0364**	**0.0295**	**0.0014**

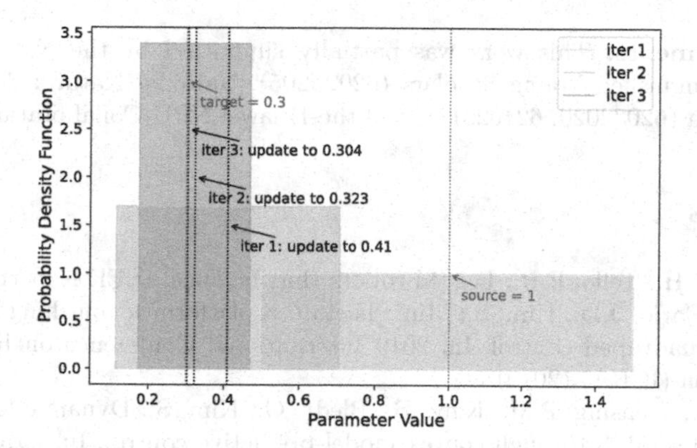

Fig. 6. Initial parameter value slowly changes to the target parameter value after 3 iterations.

value at the third iteration due to a large bias in paired trajectories. Instead, our method showed higher accuracy in the whole parameter prediction process. In the experimental setup, we chose to optimize only a small number of parameters. However, when multiple interrelated parameters are present for optimization, additional design considerations might be necessary. Figure 6 illustrated how the source friction parameter adapted to the true value in the Slide environment. As we can see, the first parameter update was relatively large (i.e. from 1 to 0.41), and successive iterations fine-tuned the estimate. Our method corrected the friction parameter with a small number of iterative tuning updates.

6 Conclusions

In this work, we propose a Simulation-to-Target-to-Simulation framework for learning adaptive locomotion in uncertain target environments. A novel sampling interval-adaptive identification method is implemented to correct simulation parameters even if the true values are outside the initial distribution range. The results demonstrate that our approach provides for highly robust and adaptive policies on a range of dynamical locomotion tasks. We further validate the accuracy of our identification method, outperforming strong baseline methods. There are some limitations of our current implementation. First, our approach uses longer training time in the optimized simulator. But we use fewer target observations to maximize data efficiency. Second, parameters may not be fully identifiable based on observations because the identifiable set of parameters may be coupled. Third, our framework is better suited for scenarios where there is a significant disparity between the target environment and the training environment. In cases where the target environment closely resembles the training environment, domain randomization methods tend to be more effective. In future work, we will address these limitations and apply this method on the real A1 robot.

Acknowledgements. This work was partially supported by the National Science Fund for Distinguished Young Scholars (62025205), National Natural Science Foundation of China (62032020, 62102317), and the Huawei-NPU Collaboration Project.

References

1. Raibert, M.H., Tello, E.R.: Legged robots that balance. IEEE Expert (1986)
2. Katz, B., Carlo, J.D., Kim, S.: Mini cheetah: a platform for pushing the limits of dynamic quadruped control. In: 2019 International Conference on Robotics and Automation (ICRA) (2019)
3. Carlo, J.D., Wensing, P.M., Katz, B., Bledt, G., Kim, S.: Dynamic locomotion in the MIT cheetah 3 through convex model-predictive control. In: 2018 IEEE/RSJ International Conference on Intelligent Robots and Systems (IROS) (2018)
4. Ding, Y., Pandala, A., Li, C., Shin, Y.H., Park, H.W.: Representation-free model predictive control for dynamic motions in quadrupeds. IEEE Trans. Robot. (2020)
5. Matas, J., James, S., Davison, A.J.: Sim-to-real reinforcement learning for deformable object manipulation. In: Conference on Robot Learning (2018)
6. Lee, J., Hwangbo, J., Wellhausen, L., Koltun, V., Hutter, M.: Learning quadrupedal locomotion over challenging terrain. Sci. Robot. (2020)
7. Miki, T., Lee, J., Hwangbo, J., Wellhausen, L., Koltun, V., Hutter, M.: Learning robust perceptive locomotion for quadrupedal robots in the wild. Sci. Robot. (2022)
8. Yang, Y., Caluwaerts, K., Iscen, A., Zhang, T., Tan, J., Sindhwani, V.: Data efficient reinforcement learning for legged robots. In: Conference on Robot Learning (2020)
9. Haarnoja, T., Ha, S., Zhou, A., Tan, J., Tucker, G., Levine, S.: Learning to walk via deep reinforcement learning. Robot. Sci. Syst. (2019)
10. Tan, J., Zhang, T., Coumans, E., et al.: Sim-to-real: Learning agile locomotion for quadruped robots. Robot. Sci. Syst. (2018)
11. Jakobi, N., Husbands, P., Harvey, I.: Noise and the reality gap: the use of simulation in evolutionary robotics. In: Advances in Artificial Life: Third European Conference on Artificial Life Granada, Spain, 4–6 June 1995, Proceedings, vol. 3 (1995)
12. Koos, S., Mouret, J.-B., Doncieux, S.: Crossing the reality gap in evolutionary robotics by promoting transferable controllers. in: Proceedings of the 12th Annual Conference on Genetic and Evolutionary Computation (2010)
13. Tobin, J., Fong, R., Ray, A., Schneider, J., Zaremba, W., Abbeel, P.: Domain randomization for transferring deep neural networks from simulation to the real world. In: 2017 IEEE/RSJ International Conference on Intelligent Robots and Systems (IROS) (2017)

14. Peng, X.B., Andrychowicz, M., Zaremba, W., Abbeel, P.: Sim-to-real transfer of robotic control with dynamics randomization. In: 2018 IEEE International Conference on Robotics and Automation (ICRA) (2018)
15. Farchy, A., Barrett, S., MacAlpine, P., Stone, P.: Humanoid robots learning to walk faster: from the real world to simulation and back. In: Proceedings of the 2013 International Conference on Autonomous Agents and Multi-agent Systems (2013)
16. Tan, J., Xie, Z., Boots, B., Liu, C.K.: Simulation-based design of dynamic controllers for humanoid balancing. In: 2016 IEEE/RSJ International Conference on Intelligent Robots and Systems (IROS) (2016)
17. Du, Y., Watkins, O., Darrell, T., Abbeel, P., Pathak, D.: Auto-tuned sim-to-real transfer. In: 2021 IEEE International Conference on Robotics and Automation (ICRA) (2021)
18. Chebotar, Y., Handa, A., Makoviychuk, V., et al.: Closing the sim-to-real loop: adapting simulation randomization with real-world experience. In: 2019 International Conference on Robotics and Automation (ICRA) (2019)
19. Mastalli, C., Havoutis, I., Focchi, M., Caldwell, D.G., Semini, C.: Motion planning for quadrupedal locomotion: coupled planning, terrain mapping, and whole-body control. IEEE Trans. Robot. (2020)
20. Rudin, N., Hoeller, D., Reist, P., Hutter, M.: Learning to walk in minutes using massively parallel deep reinforcement learning. In: Conference on Robot Learning (2022)
21. Sorokin, M., Tan, J., Liu, C.K., Ha, S.: Learning to navigate sidewalks in outdoor environments. IEEE Robot. Autom. Lett. (2022)
22. Agarwal, A., Kumar, A., Malik, J., Pathak, D.: Legged locomotion in challenging terrains using egocentric vision. In: 6th Annual Conference on Robot Learning (2022)
23. Tsounis, V., Alge, M., Lee, J., Farshidian, F., Hutter, M.: Deepgait: planning and control of quadrupedal gaits using deep reinforcement learning. IEEE Robot. Autom. Lett. (2020)
24. Smith, L., Kew, J.C., Peng, X.B., Ha, S., Tan, J., Levine, S.: Legged robots that keep on learning: fine-tuning locomotion policies in the real world. In: 2022 International Conference on Robotics and Automation (ICRA) (2022)
25. Peng, X.B., Coumans, E., Zhang, T., Lee, T.-W., Tan, J., Levine, S.: Learning agile robotic locomotion skills by imitating animals. arXiv preprint arXiv:2004.00784 (2020)
26. Nagabandi, A., Clavera, I., Liu, S., et al.: Learning to adapt in dynamic, real-world environments through meta-reinforcement learning. In: International Conference on Learning Representations (2018)
27. Yu, W., Tan, J., Liu, C.K., Turk, G.: Preparing for the unknown: learning a universal policy with online system identification. Robot. Sci. Syst. (2017)
28. Zhu, S., Kimmel, A., Bekris, K., Boularias, A.: Fast model identification via physics engines for data-efficient policy search. In: International Joint Conference on Artificial Intelligence (IJCAI) (2018)
29. Hansen, N.: The CMA evolution strategy: a tutorial. arXiv preprint arXiv:1604.00772 (2016)
30. Jiang, Y., Zhang, T., Ho, D., et al.: SimGAN: hybrid simulator identification for domain adaptation via adversarial reinforcement learning. In: 2021 IEEE International Conference on Robotics and Automation (ICRA) (2021)

31. Allevato, A., Short, E.S., Pryor, M., Thomaz, A.: Tunenet: one-shot residual tuning for system identification and sim-to-real robot task transfer. In: Conference on Robot Learning (2020)
32. Iscen, A., Caluwaerts, K., Tan, J., et al.: Policies modulating trajectory generators. In: Conference on Robot Learning (2018)
33. Coumans, E., Bai, Y.: Pybullet, a Python module for physics simulation for games, robotics and machine learning (2016). http://pybullet.org
34. Wang, X.: Unitree robotics. https://www.unitree.com/
35. Schulman, J., Wolski, F., Dhariwal, P., Radford, A., Klimov, O.: Proximal policy optimization algorithms. arXiv preprint arXiv:1707.06347 (2017)

NLP-Based Test Co-evolution Prediction for IoT Application Maintenance

Yuyong Liu(✉) and Zhifei Chen

Nanjing University of Science and Technology, Nanjing 210094, Jiangsu, China
lyy20010414@163.com

Abstract. The increasing deployment of the Internet of Things (IoT) leads to the diversified development of IoT-based applications. However, due to the fast updates and the growing scale of IoT applications, IoT developers mainly focus on the production code but overlook the co-evolution of the corresponding test code. To facilitate the maintenance of IoT applications, this paper proposes an NLP-based approach to predict whether the test code needs to be co-changed when its production code is updated. We collected data from the most popular projects on GitHub (top 1,000 with the highest stars). Three neural encoders were employed to capture semantic features of commit messages, production code changes, and related test code. We then generated our training samples, in which the features of each sample consist of < Commit Message, Production Code Change, Test Unit Code >. Finally, a neural network model was built by learning the correlations among these features to determine the possibility of test co-evolution. We evaluated the effectiveness of our NLP-based approach on 15 widely used Python projects in the IoT domain. The evaluation result shows that the prediction accuracy of our model achieves 93%, highlighting the practical significance of our approach in the maintenance of IoT applications.

Keywords: IoT Development · Test Co-evolution · Maintenance · NLP · Prediction Model

1 Introduction

In recent years, the Internet of Things (IoT) has been deployed in various fields [1–3], leading to the diversified development of IoT-based applications. However, as the development speed of IoT applications accelerates and the code size increases, the quality of IoT applications becomes a severe issue, which increases the security risks of IoT systems [4, 5]. In particular, when IoT developers often pay more attention to the updates of production code when they are contributing to IoT projects, but ignore the co-changes of the corresponding tests, which will lead to inaccurate test results and difficulties in later maintenance. To promote the wide application and reap the potential of IoT systems, a reliable solution should be proposed to determine whether the test code needs to be changed when the production code under test changes to facilitate the maintenance of IoT applications [6].

H. Jin et al. (Eds.): GPC 2023, LNCS 14504, pp. 155–171, 2024.
https://doi.org/10.1007/978-981-99-9896-8_11

As the best choice for IoT backend development and device software development, Python language is widely used by IoT developers [7]. In Python programs, several popular testing frameworks, such as Pytest, and Unittest, can help developers test the function and possible risks of important production code. Based on these testing frameworks, the existing studies mainly focus on how to generate test cases [8, 9] or eliminate redundant test code, to improve test efficiency [10]. However, to the best of our knowledge, there are currently no effective approaches for predicting the occurrence of test co-changes after the updates of production code.

To fill this gap, this paper proposes an NLP-based approach [11] to predict whether test code needs to be changed in concert with production code. In particular, we built a neural network model [12] that predicts co-changes in test codes of IoT applications. First of all, we collected data from the top 1,000 Python projects with the most stars on GitHub and extracted positive and negative training samples from them. The features of the samples consider the texts of commit messages and the semantic information of production code and test code. To capture these NLP features, we used three encoders, including the commit message encoder, the production code change encoder, and the test code encoder, to embed texts and code into context vectors respectively. By using these training samples, the neural network model was trained for a binary classification task according to the final probability scores. Therefore, it can quickly determine whether the test code needs to be updated after the production code under the test is modified.

We also evaluated the effectiveness of this NLP-based test co-change prediction approach on 15 Python projects. These projects are widely used in IoT fields, involving 150,000 commits, and covering the most popular testing frameworks including Pytest, Nose, and Unittest. The evaluation result shows that the prediction accuracy achieves 93%.

In summary, the main contributions of this paper are as follows:

(i) We propose an NLP-based approach to automatically predict whether test code needs to undergo co-changes when its corresponding production code is modified. The model incorporates the NLP features of commit messages, production code changes, and test code.
(ii) We build a large dataset by collecting positive and negative samples from the top 1,000 Python projects on GitHub, providing a usable dataset for future work in this field.
(iii) We evaluate our prediction approach on 15 popular projects in IoT fields, which verifies the practical significance of our approach in IoT development.

2 Approach

2.1 Task Definition

The motivation of this research is to investigate whether the test code needs to undergo co-change when the production code is modified. To achieve this goal, the task can be defined as a binary classification learning problem. For IoT projects with modular design, for a given commit, we define the commit message corresponding to this commit as CMG (M), the production code change part extracted from the production file in the diffs file as PCC (C), and the test code content related to the production code in the test file as TUC (T). The purpose of this paper is to automatically determine whether the test code needs covariance when the production code changes, which is recorded as state S. When covariation occurs in the test code, the state S is recorded as 1; otherwise, the state S is recorded as 0.

Therefore, the goal of this study is to train a model θ using the triplet $< M, C, T >$ to maximize the probability $P_\theta(S| < M, C, T >)$ on a given training dataset. Mathematically, this task can be defined as finding $\overline{y}$, as shown in the following formula:

$$\overline{y} = argmax_s P_\theta(S| < M, C, T >) \quad (1)$$

$P_\theta(S| < M, C, T >)$ can be seen as the conditional likelihood of predicting the final state S based on the given input triplet $< M, C, T >$. The overall process of the task is shown in Fig. 1.

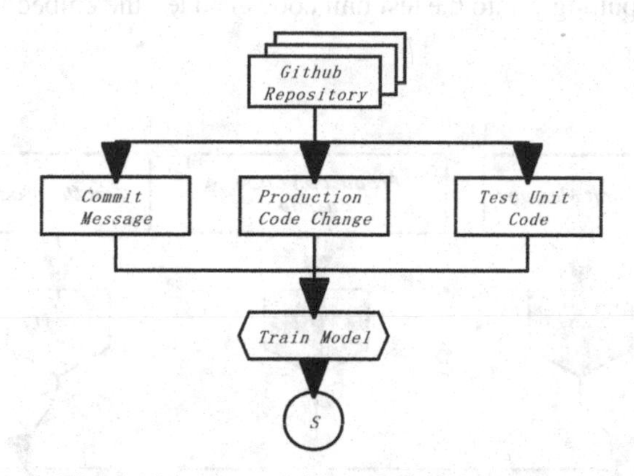

Fig. 1. Overview of the model training process

2.2 Model Overview

Encoders. In this paper, we use the $< M, C, T >$ triplet as the input of the model. Since the three types of data mentioned above may exist in different language types, such as source code and natural language, they cannot be easily mapped through simple lexical token matching. To bridge this gap, this model uses three encoders, namely the commit message encoder, the production code change content encoder, and the test unit code encoder, to embed the commit message, production code change, and test unit code into their respective word vector representations, allowing semantically similar concepts across the three modalities to be related in a high-dimensional vector space. Through embedding technology, heterogeneous data can be easily connected through their vectors.

In this experiment, BERT was used as the encoder template for the task. This design model consists of three encoders, namely the commit message encoder, the production code change encoder, and the test unit code encoder. These three encoders are almost identical in structure and are responsible for mapping the three types of inputs (i.e.,

CMG, PCC, and TUC) into their corresponding embeddings. The process of semantic embedding using the pre-trained BERT model can be seen in Fig. 2.

- Commit Message Encoder: The commit message M consists of N_M token values, denoted as $M = M_1, M_2, \ldots, M_{N_M}$. It can be embedded into a contextual vector h_m using BERT, which is pre-trained to capture the semantic meaning of the commit message.
- Production Code Change Encoder: The production code change encoder embeds code variations into vectors. A production code change is defined as $C = C_1, C_2, \ldots, C_{N_C}$, where N_C represents the number of token values, and BERT is used to embed it into vector h_C.
- Test Unit Code Encoder: The encoder embeds test unit code into vectors. A test unit code is defined as $T = T_1, T_2, \ldots, T_{N_T}$, where N_T represents the number of token values. After inputting T into the test unit code encoder, the embedded vector h_t can be obtained.

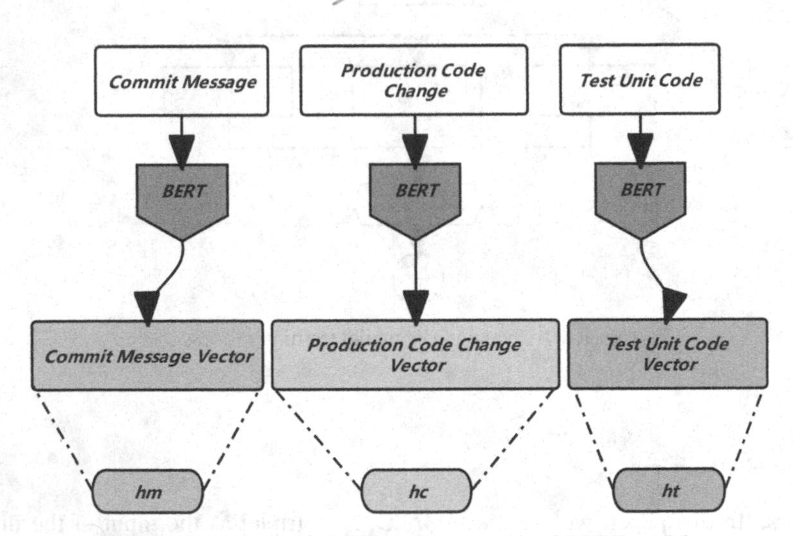

Fig. 2. Semantic embeddings using pre-trained BERT models

Multilayer Perceptron. Based on the theoretical and practical foundation described in Sect. 2.2.1, the current project has represented the triple <Commit Message, Production Code Change, Test Unit Code> as independent context vectors $<h_m, h_c, h_t>$. To capture the relationships between them, it is necessary to link and merge their information. In addition, a multi-layer perceptron (MLP) has been added to meet this requirement and is shared among the three encoders. This design first concatenates the context vector $<h_m, h_c, h_t>$ of the three encoders to merge semantic features.

To further capture the correlation and reference between the latent feature vectors, a standard MLP is added to the concatenated vector. This gives the model high flexibility and nonlinearity to learn the interaction between the three encoders. The MLP takes the contextual representations (i.e., h_m, h_c, h_t) as input and outputs the likelihood of the final state S = {0,1}. The definition of MLP is as follows:

$$z_1 = \phi_1(h_m, h_c, h_t) = \begin{bmatrix} h_m \\ h_c \\ h_t \end{bmatrix} \tag{2}$$

$$z_2 = \phi_2(z_1) = a_2(w_2^T z_1 + b_2) \tag{3}$$

...

$$z_L = \phi_L(z_{L-1}) = a_L(w_L^T z_{L-1} + b_L) \tag{4}$$

$$p(s = j| < C, T, M >) = \sigma(z_L) \tag{5}$$

where w_x, b_x and a_x represent the weight matrix, bias vector, and activation function of the x-th layer perceptron, respectively. σ is the sigmoid function $\sigma(x) = 1/(1 + e^{-x})$, which outputs the probability of the final state S between 0 and 1. The experiment uses binary cross-entropy loss function, which is a commonly used loss function for binary classification problems. This loss function measures the difference between the predicted probability and the actual label (resolved or unresolved). The formula for binary cross-entropy loss is:

$$L(y, \hat{y}) = -ylog(\hat{y}) - (1 - y)log(1 - \hat{y}) \tag{6}$$

where y represents the true label and $\hat{y}$ represents the predicted label of the model. The entire loss function is the sum of the two terms, so the model needs to optimize its ability to predict both positive and negative examples. The network structure of MLP is shown in Fig. 3.

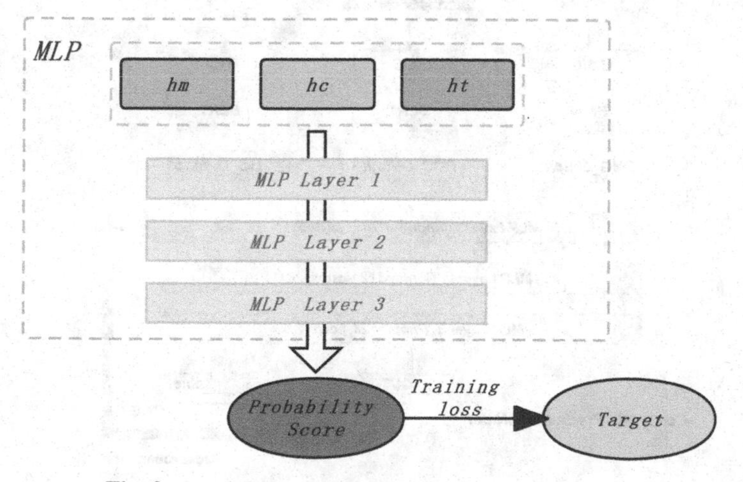

Fig. 3. Use MLP to build a binary classification model

2.3 Model Building

Section 2.1 outlines the general training process employed in this study. Specifically, the <Commit Message, Production Code Change, Test Unit Code> triplet serves as the model input. This triplet is fed into a pre-existing MLP classification model for training. The model's training score is computed, followed by the calculation of loss using the correct training label. Backpropagation is then performed to ensure the convergence of model parameters. As the source data in this paper is in natural language format, it cannot be directly inputted into the training model as a vector. Therefore, we divide the model into two modules: encoder and multi-layer perceptron, as explained in Sect. 2.2. The encoder module utilizes the pre-trained BERT model to convert the source data into vectors, establishing semantic connections between different data components to enhance semantic relevance. These converted vectors are then fed into the covariate prediction model for training. Figure 4 illustrates the overall model structure presented in this paper.

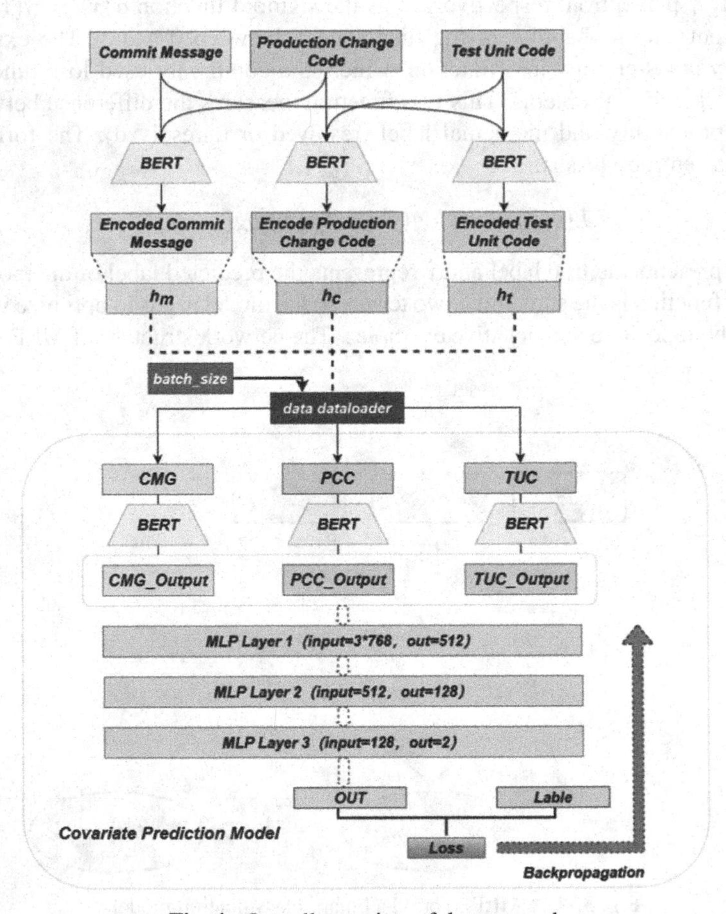

Fig. 4. Overall overview of the approach

3 Dataset Collection

We utilize crawler technology to remotely clone the top 1,000 Python repositories from GitHub based on their popularity. Following this, we employ heuristic rules to automatically collect both positive and negative samples.

3.1 Samples Collection

Positive Sample Description. This design is based on the triple <CMG, PCC, TUC> to predict whether the corresponding test code needs to be changed synchronously when the production code changes, such as functional parameter changes, content changes, etc. If the content of the test code corresponding to the production code also appears in the diffs file during a certain commit, it means that the test code has undergone a collaborative change, and the label corresponding to the triple extracted by the commit is recorded as 1, recorded as <CMG, PCC, TUC, 1>.

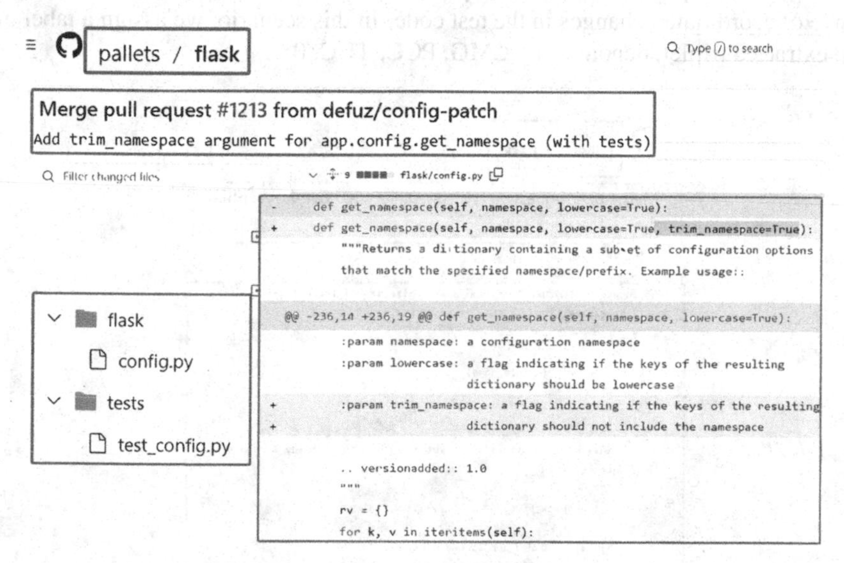

Fig. 5. Positive example display (partial)

As shown in Fig. 5 and Fig. 6, it is one of the positive examples. It can be seen from Fig. 5 that in a certain submission of the repository named "*pallets/flask*", the content of the submission information is "*Merge pull request # 1213 from defuz/config-patch Add trim_namespace argument for app.config.get_namespace (with tests)*", recorded as Commit Message. From the commit message, we can understand that the purpose of this commit is to modify the function named "*get_namespace*" in the code file "*config.py*", and its operation is to add function parameters. Then, it modifies the content of the corresponding test function in the corresponding code file named "*test_config.py*". From the figure, we can also intuitively see that the corresponding test changes have been made in the function "*test_get_namespace()*".

```
tests/test_config.py
@@ -182,3 +182,11 @@ def test_get_namespace():
182 182     assert 2 == len(bar_options)
183 183     assert 'bar stuff 1' == bar_options['STUFF_1']
184 184     assert 'bar stuff 2' == bar_options['STUFF_2']
    185 +   foo_options = app.config.get_namespace('FOO_', trim_namespace=False)
    186 +   assert 2 == len(foo_options)
    187 +   assert 'foo option 1' == foo_options['foo_option_1']
    188 +   assert 'foo option 2' == foo_options['foo_option_2']
    189 +   bar_options = app.config.get_namespace('BAR_', lowercase=False, trim_namespace=False)
    190 +   assert 2 == len(bar_options)
    191 +   assert 'bar stuff 1' == bar_options['BAR_STUFF_1']
    192 +   assert 'bar stuff 2' == bar_options['BAR_STUFF_2']
```

Fig. 6. Modified content of the test code in the positive sample

Negative Sample Description. Similarly, for a specific commit, if there is a production code change in the diffs file but the corresponding test code does not appear, it indicates that the IoT developer has not made any modifications to the test code, implying the absence of coordinated changes in the test code. In this scenario, we assign a label of 0 to the extracted triplet, denoted as <CMG, PCC, TUC, 0>.

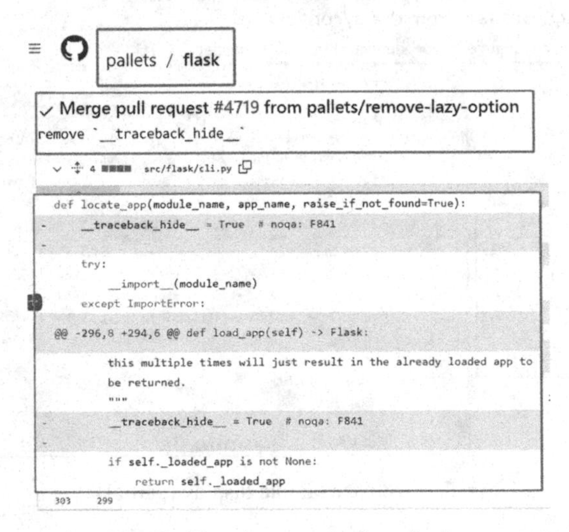

Fig. 7. Negative sample data display

As shown in Fig. 7, it is one of the negative examples. It can be seen from the figure that in a certain submission of the repository named "*pallets/flask*", the content of the submission information is "*Merge pull request #4719 from pallets/ remove-lazy-option remove '__traceback_hide__'*", recorded as Commit Message. From the submission information, we can roughly understand that the purpose of modifying the code file this time is to remove an item in the production code file, and since the corresponding test file has not undergone co-change, when it is necessary to restore the local repository file to the submitted version, to extract the corresponding test code.

3.2 Data Processing

Remove Abnormal Data. Due to the extensive dataset collection, it is necessary to filter out data that does not meet the criteria. For instance, certain samples may have incomplete production code changes (PCC) where only function definitions are present. This can occur when developers modify function names, making it challenging to extract the complete modified content from the diff file.

Merge Multiple Rows of Data. Since the data was not preliminarily processed during data interception, and most of the data was code content, there were a large number of spaces and line breaks. Therefore, for the robustness of the training, this design merged multiple lines of data into one line, and the redundant space deletion.

Canonical Commit Message. For the commit message, there is often some redundant information generated by the commit, which may interfere with the semantics of the experiment. Therefore, the issue IDs and commit IDs of GitHub are deleted to ensure semantic integrity.

Canonicalized Diff Information. This design separates the required production codes from the diffs file, but these data often have marked data, meaning different labels, that is, "+" and "−". If the standardized difference operation is not performed, the design model can be directly obtained from the difference labels, rather than learning the semantic content of the extracted code.

Data Partitioning. In this project, due to the difficulty of collecting positive and negative examples and the different code writing habits of developers, the data of positive and negative examples are quite different. After screening, the data of positive and negative examples will be integrated. In this experiment, the constructed data samples are divided into two chunks: 80% of the samples are used for training, and 20% are used for verification. The detailed data distribution is shown in Table 1.

Table 1. Data Statistics

Positive Examples	165602
Negative Examples	13571
Train Set	10857
Validation Set	2714

Manual Inspection. The above data are used to automatically construct positive and negative training samples for this design through heuristic rules, so this design cannot ensure that there are no abnormalities in the process of label establishment. Therefore, we performed a manual checking step to check that the labels for the training samples were correct. In this paper, 100 samples are randomly selected from the data set (including 50 positive samples and 50 negative samples). Then, the developer will check each sample manually. Finally, the data are cleaned as required. Therefore, we are confident in the dataset and labels collected in this experiment.

4 Evaluation

In order to assess our NLP-based test co-evolution prediction approach, we measure the effectiveness of the prediction model on IoT projects.

4.1 Subjects

We collect our test set on widely used IoT project repositories. For example, "Home Assistant" is an open-source home automation platform for controlling smart home devices and integrating various third-party services. It has already gained 60.9k stars on GitHub. Also, "Pycom MicroPython" is a MicroPython firmware provided by Pycom, a hardware supplier designed for IoT devices, that makes it easier to develop on Pycom hardware.

4.2 Evaluation Metrics

In this project, four statistical measures related to the task are defined: TP, TN, FP, and FN. Based on the above four statistical measures, the experiment uses widely accepted metrics, namely Accuracy, Precision, Recall, and F1-score, to evaluate the performance of the model.

The definitions of evaluation metrics are as follows:

- **Accuracy**: Accuracy represents the proportion of correctly predicted (true positive and true negative) cases among the total number of cases examined. The definition of the Accuracy metric is as follows:

$$Accuracy = \frac{TP+TN}{TP+TN+FP+FN} \tag{7}$$

- **Precision**: Precision represents the proportion of data correctly classified as needing covariation among all the data that have actually occurred with covariation. The definition of the Precision metric is as follows:

$$Precision = \frac{TP}{TP+FP} \tag{8}$$

- **Recall**: Recall represents the proportion of data correctly classified as needing covariation among all the data classified as needing covariation. The definition of the Recall metric is as follows:

$$Recall = \frac{TP}{TP+FN} \tag{9}$$

- **F1-score**: F1-score is the harmonic mean of precision and recall. It evaluates whether the increase in precision (or recall) exceeds the decrease in recall (or precision). The definition of the F1-score metric is as follows:

$$F_1 = \frac{2*Precision*Recall}{Precision+Recall} \tag{10}$$

In addition, there is a trade-off between precision and recall, and the F1-score can strike a balance between them. The higher the evaluation metrics, the better the model's performance.

4.3 Experimental Settings

This paper implemented the proposed model using PyTorch in Python. A pre-trained BERT model served as the encoder, providing context-dependent sentence representations. The commit message encoder, production code change encoder, and test unit code encoder were jointly trained to minimize cross-entropy. The encoded Commit Message, Production Code Change, and Test Unit Code were mapped to 768-dimensional vectors. Adam optimizer was used with a batch size of 1 and 10 training epochs for the MLP with three hidden layers. The hidden layer utilized a pooling rate of 0.2 before computing the final probability. The learning rate of Adam was set to 0.001, and gradient norm was clipped to 2. The model achieving the best performance on the validation set was selected for evaluation.

4.4 Results

In this section, we will analyze the experimental results and provide a detailed description of their performance on the test set. Then, we will conduct a component-based evaluation experiment to demonstrate the role of each data component in the overall performance of the model.

Effectiveness Analysis. After pre-preparation and model training, we can get the performance on the validation set, as shown in Table 2.

Table 2. Performance on the validation set

Metrics	Score
Accuracy	0.934
Precision	0.941
Recall	0.926
F1-score	0.934

After visualizing all training data, the obtained parameter data is shown in the figure below:

According to the above Figs. 8, 9, and 10, it can be seen that this training includes the training results and verification results of 10 epochs. The following are some analyzes of the training results:

1. With increasing epochs, the model showed gradual improvement in performance. Accuracy, precision, recall, and F1 score on both the training set and validation set steadily increased, indicating continuous learning and optimization of the model.
2. The model demonstrates excellent performance with high accuracy, precision, recall, and F1 score above 0.99 on the training set. The validation set also shows favorable performance indicators around 0.93, indicating the model's ability to handle new data effectively.

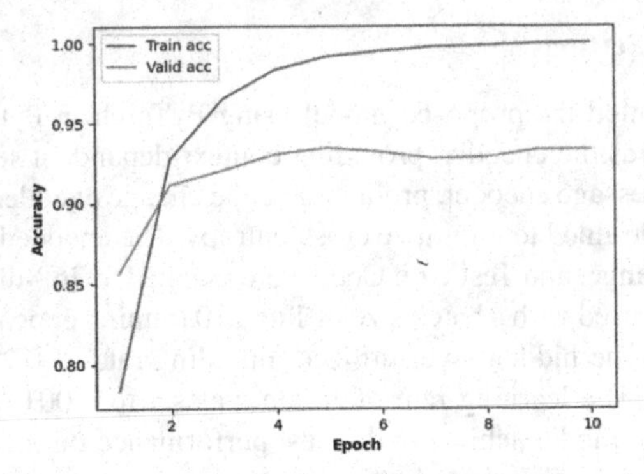

Fig. 8. Accuracy and training loss on training and validation sets

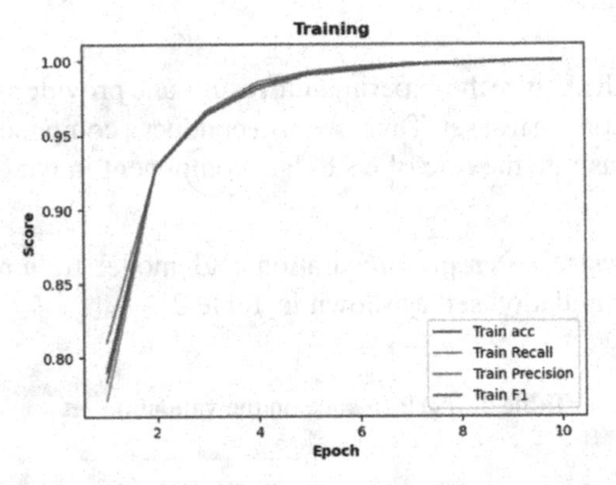

Fig. 9. Changes in various evaluation indicators on the training set

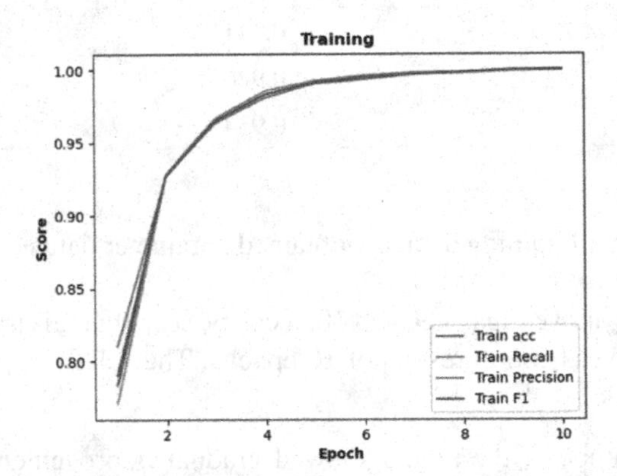

Fig. 10. Changes in various evaluation indicators on the validation set

3. The training loss decreased gradually, indicating convergence of the model. However, the validation loss initially decreased but slightly increased in later epochs, indicating

overfitting. To mitigate overfitting, early stopping was employed to select the best-performing model for testing.

4. After the fifth epoch, the model achieved its peak performance on the validation set. Subsequently, there was a slight decline in performance. This indicates that the optimal performance of the model is attained during the fifth epoch. To evaluate the model's performance, we utilized a test set consisting of data from widely used Internet of Things projects. The test results are shown in the figure below (Table 3).

Table 3. Performance on test sets collected from widely used IoT projects

Metric	Score
test_acc	0.939
test_precision	0.943
test_recall	0.934
test_F1	0.938

In summary, the model performed well during the training process, with good generalization ability and classification performance. However, it is important to avoid overfitting to ensure the model's robustness and reliability.

Component-Wise Evaluation. The key to determining whether test code needs to be collaboratively changed with production code modifications is to effectively capture the relationship and references between the production code changes and the test unit code. To achieve this, the experiment used three encoders, namely the Commit Message Encoder, the Production Code Change Encoder, and the Test Unit Code Encoder, to better represent and link the information between the Commit Message, Production Code Change, and Test Unit Code. To verify the effectiveness of these three encoders, the experiment was evaluated component-wise, evaluating their performance and contribution. Therefore, the experimental model was compared to three other incomplete versions:

- CMG_PCC_Encoder: it retains the Commit Message Encoder and the Production Code Change Encoder. This model does not consider test unit code. It is then trained as a binary classification model by using Commit Message and Production Code Change as input.
- PCC_TUC_Encoder: it retains the Production Code Change Encoder and the Test Unit Code Encoder. This model does not consider Commit Message. It is then trained as a binary classification model by using Production Code Change and Test Unit Code as input.
- TUC_CMG_Encoder: it retains the Test Unit Code Encoder and the Commit Message Encoder. This model does not consider production code changes. It is then trained as a binary classification model by using Test Unit Code and Commit Message as input.
- CMG_PCC_TUC_Encoder: this model includes all components.

After model training, the evaluation results of the above four experimental models are shown in Table 4.

Table 4. Component-Wise Evaluation

	Accuracy	Precision	Recall	F1
CMG_PCC_Encoder	0.861	0.889	0.843	0.864
PCC_TUC _Encoder	0.517	0.855	0.558	0.610
TUC_CMG _Encoder	0.586	0.530	0.591	0.462
CMG_PCC_TUC_Encoder	**0.910**	**0.921**	**0.903**	**0.911**

Based on the experimental results in the table, we can draw the following conclusions:

- Firstly, the CMG_PCC_TUC_Encoder achieved the highest performance in all metrics, demonstrating its superior accuracy, precision, and recall. By incorporating CMG, PCC, and TUC information, this encoder conducts a comprehensive analysis of the data, effectively capturing relevant features and enhancing prediction accuracy.
- Secondly, the CMG_PCC_Encoder showed good precision and recall, but slightly lower accuracy. This may be attributed to its difficulty in accurately predicting negative cases, resulting in lower accuracy. Improving the encoder's handling of negative cases is necessary. Data labels determine the necessity of co-variation, with the test unit code being the object of co-variation. As the Test Unit Code data components were not included in this encoder, it limited the model's ability to learn deeper data correlations, affecting its performance with positive and negative cases.
- The PCC_TUC_Encoder and TUC_CMG_Encoder had the worst model performance.

Based on the training results of each model, it is clear that removing any of the three data components would harm the overall performance of the experimental model in this article. This also verifies the initial assumption of this article, which is that all three encoders can embed useful information from their respective inputs.

5 Threats to Validity

This study has certain limitations that should be addressed. Firstly, the data preparation process involved the removal of diffs containing changes across multiple files. The evaluation data also underwent the same processing steps, resulting in the exclusion of cases involving co-changes produced by multiple code files. Consequently, the current method lacks the ability to handle such scenarios when analyzing unseen data. Additionally, the dataset used in this experiment focused solely on Python projects on GitHub, neglecting projects in other programming languages. Future work should aim to include a broader range of programming languages and cater to a wider audience of designers. Finally, regarding the hyperparameters used to train the models, there are two key parameters to consider: the encoder's embedding size and the MLP hidden layer size. While the size of the hidden layer can be fine-tuned theoretically, exploring the full range of size settings can be computationally expensive and time-consuming.

6 Related Work

With the rapid growth of Internet of Things technology and its widespread adoption in various domains [24, 25], there is an increasing focus on studying the co-variation relationship between production code and test code [16]. Researchers are actively developing advanced techniques and algorithms to effectively match production code with test code, aiming to enhance the quality and reliability of software in the IoT domain.

One approach proposed by researchers is the use of historical test data for test case prediction [13–15], which aims to improve the testing process of IoT applications. By leveraging machine learning and genetic algorithms, models are constructed from historical test data to predict test cases suitable for future testing. This method enables more accurate identification of effective test cases for specific IoT applications. Another line of research focuses on analyzing the co-evolution [20] of test code and production code in Android API [17–19]. A static analysis-based method has been proposed to discover patterns of association between test code and production code, and its effectiveness has been validated through empirical verification. This research provides deeper insights into how test code and production code mutually influence and co-evolve during the development of IoT applications.

Furthermore, researchers have investigated association rule-based approaches [21–23] to uncover covariant patterns between test code and production code. By conducting extensive analyses of open-source software projects, they have successfully identified prevalent covariant patterns and conducted comprehensive studies to validate their findings. This research holds immense importance in enhancing testing efficiency, optimizing testing strategies, and elevating software quality within the realm of IoT.

In all relevant studies [26–30], there is still a gap in leveraging relevant data components for co-variation matching between test code and production code. This area presents an exciting opportunity for further exploration by researchers, and this study contributes to filling this gap, making a valuable contribution to the advancement of the IoT field.

7 Conclusion and Future Work

This paper proposes an automated NLP-based solution to predict the covariant association between production code and test code in IoT development. We extract three data components (commit message, production code change, and test unit code) from the top 1,000 Python repositories on GitHub. Utilizing a neural network model, we learn semantic features and correlations between these data components. Experimental results on widely used IoT projects demonstrate the effectiveness and strong performance of our approach.

Future work includes optimizing the model to enhance accuracy and reliability by increasing training data volume and improving the model architecture. Additionally, we plan to apply this method to diverse software development scenarios such as cloud computing, big data, and artificial intelligence to improve software quality.

References

1. Kour, V.P., Arora, S.: Recent developments of the internet of things in agriculture: a survey. IEEE Access **8**, 129924–129957 (2020)
2. Dang, L.M., Piran, M.J., Han, D., et al.: A survey on internet of things and cloud computing for healthcare. Electronics **8**(7), 768 (2019)
3. Zhou, I., Makhdoom, I., Shariati, N., et al.: Internet of things 2.0: concepts, applications, and future directions. IEEE Access **9**, 70961–71012 (2021)
4. Kouicem, D.E., Bouabdallah, A., Lakhlef, H.: Internet of things security: a top-down survey. Comput. Netw. **141**, 199–221 (2018)
5. Atlam, H.F., Wills, G.B.: IoT security, privacy, safety and ethics. Digit. Twin Technol. Smart Cities, 123–149 (2020)
6. Taivalsaari, A., Mikkonen, T.: On the development of IoT systems. In: 2018 Third International Conference on Fog and Mobile Edge Computing (FMEC), pp. 13–19. IEEE (2018)
7. Taivalsaari, A., Mikkonen, T.: A taxonomy of IoT client architectures. IEEE Softw. **35**(3), 83–88 (2018)
8. Chen, T.Y., Cheung, S.C., You, S.M.: Metamorphic testing: a new approach for generating next test cases. arXiv Preprint arXiv:2002.12543 (2020)
9. Li, W., Le Gall, F., Spaseski, N.: A survey on model-based testing tools for test case generation. In: Itsykson, V., Scedrov, A., Zakharov, V. (eds.) TMPA 2017. CCIS, vol. 779, pp. 77–89. Springer, Cham (2018). https://doi.org/10.1007/978-3-319-71734-0_7
10. Palomba, F., Panichella, A., Zaidman, A., et al.: Automatic test case generation: what if test code quality matters?. In: The 25th International Symposium on Software Testing and Analysis, pp. 130–141 (2016)
11. Lyu, H., Sha, N., Qin, S., et al.: Advances in neural information processing systems. In: Advances in Neural Information Processing Systems, vol. 32 (2019)
12. Taud, H., Mas, J.F.: Multilayer perceptron (MLP). Geomatic Approach. Model. Land Change Scenarios, 451–455 (2018)
13. Noor, T.B., Hemmati, H.: Studying test case failure prediction for test case prioritization. In: The 13th International Conference on Predictive Models and Data Analytics in Software Engineering, pp. 2–11 (2017)
14. Paterson, D., Campos, J., Abreu, R., et al.: An empirical study on the use of defect prediction for test case prioritization. In: 2019 12th IEEE Conference on Software Testing, Validation and Verification (ICST), pp. 346–357. IEEE (2019)
15. Shao, Y., Liu, B., Wang, S., et al.: A novel test case prioritization method based on problems of numerical software code statement defect prediction. Eksploatacja i Niezawodność **22**(3) (2020)
16. Kraut, R.E., Streeter, L.A.: Coordination in software development. Commun. ACM **38**(3), 69–82 (1995)
17. Jiang, Y., Adams, B.: Co-evolution of infrastructure and source code-an empirical study. In: 2015 IEEE/ACM 12th Working Conference on Mining Software Repositories, pp. 45–55. IEEE (2015)
18. Eilertsen, A.M., Bagge, A.H.: Exploring API: client co-evolution. In: The 2nd International Workshop on API Usage and Evolution, pp. 10–13 (2018)

19. Um, S.Y., Yoo, Y.: The co-evolution of digital ecosystems (2016)
20. Zaidman, A., Rompaey, B.V., Deursen, A.V., et al.: Studying the co-evolution of production and test code in open source and industrial developer test processes through repository mining. Empir. Softw. Eng. **16**(3), 325–364 (2011). https://doi.org/10.1007/s10664-010-9143-7
21. Lubsen, Z.A.: Studying Co-evolution of production and test code using association rule mining. Delft Univ. Technol. Softw. Eng. Res. Group (2008). ISSN 1872-5392
22. Lubsen, Z., Zaidman, A., Pinzger, M.: Using association rules to study the co-evolution of production & test code. In: IEEE International Working Conference on Mining Software Repositories. IEEE (2009). https://doi.org/10.1109/MSR.2009.5069493
23. Zaidman, A., Rompaey, B.V., Demeyer, D.S.: Studying the co-evolution of production and test code in open source and industrial developer test processes through repository mining. Empir. Softw. Eng. (2011). https://doi.org/10.1007/s10664-010-9143-7
24. Ploennigs, J., Cohn, J., Stanford-Clark, A.: The future of IoT. IEEE Internet Things Mag. **1**(1), 28–33 (2018)
25. Lee, S.K., Bae, M., Kim, H.: Future of IoT networks: a survey. Appl. Sci. **7**(10), 1072 (2017)
26. Zaidman, A., Rompaey, B.V., Demeyer, S., et al.: Mining software repositories to study co-evolution of production & test code. In: 2008 1st International Conference on Software Testing, Verification, and Validation. IEEE (2008). https://doi.org/10.1109/ICST.2008.47
27. Marsavina, C., Romano, D., Zaidman, A.: Studying fine-grained co-evolution patterns of production and test code. In: 2014 IEEE 14th International Working Conference on Source Code Analysis and Manipulation, pp. 195–204. IEEE (2014)
28. Wang, S., Wen, M., Liu, Y., et al.: Understanding and facilitating the co-evolution of production and test code. In: 2021 IEEE International Conference on Software Analysis, Evolution and Reengineering (SANER), pp. 272–283. IEEE (2021)
29. Vidács, L., Pinzger, M.: Co-evolution analysis of production and test code by learning association rules of changes. In: 2018 IEEE Workshop on Machine Learning Techniques for Software Quality Evaluation (MaLTeSQuE), pp. 31–36. IEEE (2018)
30. Shimmi, S., Rahimi, M.: Patterns of code-to-test co-evolution for automated test suite maintenance. In: 2022 IEEE Conference on Software Testing, Verification and Validation (ICST), pp. 116–127. IEEE (2022)

Cyber-Physical-Social Systems

Fine-Grained Access Control Proxy Re-encryption with HRA Security from Lattice

Jinqiu Hou[1], Changgen Peng[1](✉), Weijie Tan[1,2], Chongyi Zhong[1,3], Kun Niu[1], and Hu Li[1]

[1] State Key Laboratory of Public Big Data, College of Computer Science and Technology, Guizhou University, Guiyang 550025, China
cgpeng@gzu.edu.cn
[2] Key Laboratory of Advanced Manufacturing Technology, Ministry of Education, Guizhou University, Guiyang 550025, China
[3] School of Mathematics and Big data, Guizhou Education University, Guiyang 550025, China

Abstract. With the gradual formation of the cloud computing ecosystem, the value of cloud computing is becoming increasingly evident, and the accompanying information security issues have become core elements. The challenge lies in securely transmitting, computing, and sharing data in a cloud computing environment while keeping privacy. In this paper, we propose an access control proxy re-encryption scheme that supports inner product operations based on lattices in the standard model. First, the ciphertext is linked to the attributes, and condition for transition is associated with the re-encryption key. The proxy may re-encrypt ciphertext that meets attribute conditions instead of all ciphertext, thereby limiting the proxy's conversion permissions. Furthermore, the user outputs the inner product value after decryption. Finally, honestly re-encryption attacks (HRA) security with increased CPA security is employed to better capture the target of proxy re-encryption. In addition, we propose a new method of hiding access policy, which uses differential privacy technology to perturb the attribute values to better safeguard the privacy of users.

Keywords: Lattice-based cryptography · Proxy re-encryption · HRA security · Differential privacy

1 Introduction

Cloud storage has gradually replaced other data storage options as the information age. The volume of data and processing requirements have all changed dramatically. Due to its benefits of comfort, scalability, and flexibility, cloud storage has drawn a lot of attention. Yet, cloud storage has always had issues with data protection and safe sharing.

Supported by the Natural Science Foundation of China (No. 62272124, No. 2022YFB2701400), Youth Science and Technology Talents Cultivating Object of Guizhou Province (No. QJHKY[2022]301), PhD Project of Guizhou Education University (No. 2021BS005) and the Postgraduate Innovation Program in Guizhou Province (No. YJSKYJJ[2021]028).

H. Jin et al. (Eds.): GPC 2023, LNCS 14504, pp. 175–190, 2024.
https://doi.org/10.1007/978-981-99-9896-8_12

Proxy re-encryption (PRE) [2] can successfully address the above-mentioned issues in cloud storage. In a PRE scheme, the proxy can use a re-encryption key to re-encrypt all of the data owner's ciphertext, allowing the delegatee to share all of the owner's data.

But in real-world scenarios, data owners typically only share a portion of their information with partners. The notion of conditional proxy re-encryption (CPRE) [18] was proposed to implement more precise decryption right assignment. In Fig. 1, the delegator can provide access to particular receivers using CPRE, and only those recipients who comply with the requirements can decrypt the data. Traditional CPRE usually limits access to data to those that meet specific requirements, such time or location constraints. In contrast, more intricate and precise access control policies, including attribute-based access control, may be needed in real world. In summary, in terms of access control granularity, PRE provides ciphertext level access control, CPRE adds the function of conditional judgement on the basis of this, and attribute-based CPRE further extends the access control granularity by allowing the definition of access policies based on attributes to achieve more fine-grained and dynamic access control.

Fig. 1. CPRE in a cloud data sharing system.

Furthermore, conventional cryptographic techniques are encountering a growing number of difficulties as a result of the advancement of computing capacity, particularly the development of quantum computing. People are keen to identify a new kind of cryptographic system against quantum assaults in order to address these possible security issues. Due to its great efficiency, verifiable security, and support for homomorphic encryption, lattice cryptography is gaining increasing amounts of attention. Therefore, researching the effective and secure PRE system with fine-grained access control functionality based on lattice is crucial for the advancement and use of cloud computing. The following are the primary contributions of this study.

(1) The inner product function is associated with the private key, and two attribute access policies are embedded in the re-encryption key. Receivers who satisfy the conditions decrypt to get the inner product value about the original plaintext;
(2) Considering the problem of insufficient CPA security for PRE, we construct HRA security ACPRE based on the difficult problem of LWE. Moreover, the security of the ACPRE we designed is strictly proved under the standard model;

(3) To address the issue of user privacy leakage caused by access policies, we propose a method of hiding access policies that satisfies localized differential privacy based on a random response mechanism, protecting user privacy information.

Section 1.1 evaluates the PRE' state of research at present. Next, we provide our proposed scheme in Sect. 2. Our method of access policy hiding is presented in Sect. 3. At last, Sect. 4 concludes this study and creation of the paper.

1.1 Related Works

PRE. PRE [1,6,10,12] is able to address the issue of data transaction in the area of data circulation that is hidden from exchanges and data merchants, as well as facilitate the interchange of data products and services that can be monitored, evaluated, and disseminated by proxy. The traditional PRE allows the proxy to transform all the ciphertext of the delegator, but fails to control the proxy's conversion authority at the fine-grained level. CPRE introduces conditional information into plaintext encryption and re-encryption key generation, which can better control the conversion ability of the proxy. Attribute-based conditional proxy re-encryption (ABCPRE) was originally defined formally and given a security model by Zhao et al. [22] They also constructed a chosen-ciphertext attack (CCA) secure ABCPRE scheme, which supports attribute-based control of decryption delegation. More specifically, in the ABCPRE scheme, the attribute vector $\mathbf{a}$ will be embedded in the ciphertext, and the re-cryption key will be linked to the policy f. Only when $f(\mathbf{a}) = 0$, the proxy can correctly perform the transformation of the ciphertext, which enhances the expression ability and flexibility of the condition. Mao et al. [13] first designed an anonymous ABCPRE scheme based on the linear secret sharing scheme, which ensures that no one can obtain any attribute information connected to the ciphertext except the user who has the corresponding key. Paul et al. [14] designed a unidirectional and single-hop CPRE scheme without pairing operation, and the scheme satisfied CCA security requirements under the assumption of computational Diffie Hellman. Due to its flexible authorisation, CPRE is appropriate for more complicated applications. In the conditional proxy broadcast re-encryption given by Chen et al. [4], the delegator could alter dynamically the user group with which the data was shared. Hörandner et al. [7] used CPRE and secure multi-party computing technology to provide flexible privacy protection for digital twins.

PRE on Lattice. Lattice cryptography is a public key cryptosystem that has attracted much attention and is resistant to quantum computing attacks [3,9]. Scholars have studied PRE based on lattice [8,17,19,20]. Liang et al. [11] based on the difficult problem of LWE on the lattice, constructed a unidirectional, multi-hop ABCPRE scheme using the homomorphic encryption and key exchange technology. To make the scheme more succinct and avoid the occurrence of two-level ciphertext, they divided the ciphertext into two parts, one part about the message ciphertext, and the other part about the attribute ciphertext. Wang et al. [16] put forth a hierarchical identity-based CPRE scheme. In their scheme, hierarchical identity-based encryption in conjunction with PRE made it possible to control the level of user decryption authority.

PRE with HRA Security. The above schemes have been proved to be chosen-plaintext attack (CPA) security under the standard model. However, CPA security can only ensure that the proxy cannot learn delegator's private data from the re-encrypted key. After obtaining an honest re-encrypted ciphertext, the delegatee's security is insufficient to prevent the delegatee from discovering the delegator's private key. Therefore, according to the situation of malicious receivers, Cohen et al. [5] proposed the security of HRA, and pointed out that if the CPA secure PRE has the nature of re-encryption simulation, then the scheme must be HRA secure. Susilo et al. [15] innovatively created HRA secure key-policy attribute-based PRE, which does away with homomorphic encryption technology and linked the key and the ciphertext together with the access policy. Yao et al. [21] gave the formal definition and security model of ABCPRE for adaptive HRA security, and based on the lattice difficulty problem, designed a single-hop, one-way ABCPRE using inner product as the decision condition for re-encryption. However, there is a problem with the above schemes that the exposure of access policies can easily lead to the disclosure of user's privacy, therefore, it is imperative to research on concealable ABCPRE.

In a word, the above schemes are only for data and have not yet considered encryption and decryption control of function calculation. How to deeply integrate PRE and functional encryption to achieve "partial encryption and decryption controllable and on-demand security calculation" is a very meaningful exploration direction. However, the research on this issue is still immature. In addition, there are fewer PRE studies on HRA security. Therefore, an access control proxy re-encryption (ACPRE) scheme is proposed that supports inner product operation based on the lattice difficulty problem to achieve the anti-quantum, flexible, controllable and manageable goal of "data available and invisible" under the open environment of zero trust.

2 Our Scheme

First, we give a thorough description of ACPRE's construction method in 2.1; Second, we examine the scheme's correctness and security in 2.2 and 2.3; Finally, we evaluate the ACPRE scheme with other relevant schemes in 2.4. It is important to note that Table 1 is a list of the main notations used in the paper.

Table 1. Symbol description

Symbols	Definitions
$n \in \mathbb{Z}_q$	Random numbers on integer module q spaces
$\mathbf{a} \in \mathbb{Z}_q^n$	n-dimensional random vectors on integer module q space
$\mathbf{A} \in \mathbb{Z}_q^{n \times m}$	n-row m-column matrices on integer module q space
$\|\mathbf{A}\|$	l_2-norm of a matrix $\mathbf{A}$
Λ	Lattice
Ψ	Gaussian noise distribution
$O(\cdot)$	Asymptotic upper bound

2.1 Construction

Setup$(1^\lambda, \mathscr{F}_\lambda)$: This algorithm takes positive integer $n > \Omega(\lambda)$, prime number $q > 2$, lattice dimension $m > 5n\log q$, security parameters 1^λ, and $\mathscr{F}_\lambda = \left\{f : \{0,1\}^l \to \{0,1\}\right\}$, then performs the algorithm **TrapGen**(n, m, q, σ) to produce two matrices, one is the uniform and random matrix $\mathbf{A}_0 \in \mathbb{Z}_q^{n\times m}$. The other is a small norm matrix $\mathbf{T}_{\mathbf{A}_0} \in \mathbb{Z}_q^{m\times m}$, which is a trapdoor basis for $\Lambda_q^\perp(\mathbf{A}_0)$, where $\|\widetilde{\mathbf{T}_{\mathbf{A}_0}}\| \leq m \cdot \omega\sqrt{\log m}$; Next, chooses random matrices $(\mathbf{A}_1, \mathbf{A}_2, ..., \mathbf{A}_l) \leftarrow \left(\mathbb{Z}_q^{n\times m}\right)^l$, $\mathbf{D} \leftarrow \mathbb{Z}_q^{n\times m}$; Finally, returns the public parameters $pp = \{\mathbf{A}_1, \mathbf{A}_2, ..., \mathbf{A}_l, \mathbf{D}, \mathbf{G}\}$ and the master key $msk = \mathbf{T}_{\mathbf{A}_0}$.

KeyGen$(pp, msk, f, \mathbf{y})$: pp, msk, a function $f \in \mathscr{F}_\lambda$ and a vector $\mathbf{y} \in \mathscr{Y}_\lambda$ are input into this algorithm. Evaluates $\mathbf{A}_f \leftarrow Eval_{pk}\left(f, \{\mathbf{A}_i\}_{i=1}^l\right)$ where $\mathbf{A}_f \in \mathbb{Z}_q^{n\times m}$; Samples the matrix $\mathbf{R}_f \leftarrow$ **SampleBasisRight** $(\mathbf{A}_0, \mathbf{T}_{\mathbf{A}_0}, \mathbf{A}_f, \mathbf{D}, \rho)$ so that $(\mathbf{A}_0|\mathbf{A}_f) \cdot \mathbf{R}_f = \mathbf{D}$, $\mathbf{R}_f \in \mathbb{Z}_q^{2m\times m}$ and $\mathbf{R}_f < \sqrt{2}\rho m$; Finally, outputs the function private key $sk_f = \mathbf{k} = \mathbf{R}_f \cdot \mathbf{y}$.

Encrypt $(pp, \mathbf{a}, \mathbf{x})$: As input, the algorithm accepts pp, $\mathbf{a} = (a_1, a_2, ..., a_l) \in \mathbb{Z}_q^l$ and the massage $\mathbf{x} \in \{0,1\}^m$, then chooses uniformly at random l matrices $\mathbf{S}_i \in \{\pm 1\}^{m\times m}$, $i \in 1, 2, ..., l$, two error vectors $\mathbf{e}_0 \in \Psi_\alpha^m$, $\mathbf{e}_{out} \in \Psi_\alpha^m$ and a vector $\mathbf{s} \in \mathbb{Z}_q^n$; Sets $\mathbf{e} = [\mathbf{I}_m|\mathbf{S}_1|\cdots|\mathbf{S}_l]^T \cdot \mathbf{e}_0 = \left(\mathbf{e}_{in}^T, \mathbf{e}_1^T, ..., \mathbf{e}_l^T\right)^T \in \mathbb{Z}_q^{(l+1)m}$; $\mathbf{H}_\mathbf{a} = (\mathbf{A}_0|a_1\mathbf{G} + \mathbf{A}_1|\cdots|a_l\mathbf{G} + \mathbf{A}_l) \in \mathbb{Z}_q^{n\times(l+1)m}$; Lets $\mathbf{c}_{in} = \mathbf{A}_0^T\mathbf{s} + \mathbf{e}_{in} \in \mathbb{Z}_q^m$; $\mathbf{c}_{out} = \mathbf{D}^T\mathbf{s} + \mathbf{e}_{out} + \lfloor q/K \rfloor \cdot \mathbf{x} \in \mathbb{Z}_q^m$; $\mathbf{c}_\mathbf{a} = \left(\left\{\mathbf{c}_i = (a_i\mathbf{G} + \mathbf{A}_i)^T\mathbf{s} + \mathbf{e}_i\right\}_{i\in\{1,...,l\}}\right) \in \mathbb{Z}_q^{lm}$; $\mathbf{c} = \mathbf{H}_a^T\mathbf{s} + \mathbf{e} = [\mathbf{c}_{in}|\mathbf{c}_\mathbf{a}] \in \mathbf{Z}_q^{(l+1)m}$; $\mathbf{c}_\mathbf{x} = (\mathbf{c}_{in}, \mathbf{c}_{out}) \in \mathbb{Z}_q^{2m}$; At last, outputs the ciphertext $ct = (\mathbf{c}_\mathbf{a}, \mathbf{c}_\mathbf{x}) \in \mathbb{Z}_q^{(l+2)m}$.

ReKeyGen(pp, sk_f, h, g): This algorithm enters pp, sk_f, conversion policy h and new access policy g to return the re-encryption key $rk_{f\xrightarrow{h}g}$ in the manner described below: First, selects an attribute set $\mathbf{b} = (b_1, b_2, ..., b_l)$ such that $g(\mathbf{b}) = 0$; Then, computes $\mathbf{H}_\mathbf{b} = (\mathbf{A}_0|b_1\mathbf{G} + \mathbf{A}_1|\cdots|b_l\mathbf{G} + \mathbf{A}_l) \in \mathbb{Z}_q^{n\times(l+1)m}$; Evaluates $\mathbf{A}_h \leftarrow Eval_{pk}\left(h, \{\mathbf{A}_i\}_{i=1}^k\right)$ where $\mathbf{A}_h \in \mathbb{Z}_q^{n\times m}$; Samples the matrix $\mathbf{R}_h \leftarrow$ **SampleBasisRight** $(\mathbf{A}_0, \mathbf{T}_{\mathbf{A}_0}, \mathbf{A}_h, \mathbf{D}, \rho)$ so that $(\mathbf{A}_0|\mathbf{A}_h) \cdot \mathbf{R}_h = \mathbf{D}$, $\mathbf{R}_h \in \mathbb{Z}_q^{2m\times m}$; Sets $\mathbf{R}_{f\xrightarrow{h}g} = \mathbf{k}\mathbf{e}_h{}^T + \mathbf{R}_h \in \mathbb{Z}_q^{2m\times m}$, where $\mathbf{e}_h \leftarrow \Psi^m$; After that, chooses a uniformly random matrix $\mathbf{R}_1 \leftarrow \mathbb{Z}_q^{2mk\times n}$, and two error matrices $\mathbf{E}_1 \leftarrow \Psi^{2mk\times(l+1)m}$, $\mathbf{E}_2 \leftarrow \Psi^{2mk\times m}$; Sets re-encryption key

$$rk_{f\xrightarrow{h}g} = \begin{bmatrix} \mathbf{R}_1\mathbf{H}_\mathbf{b} + \mathbf{E}_1 & \mathbf{R}_1\mathbf{D} + \mathbf{E}_2 - P2\left(\mathbf{R}_{f\xrightarrow{h}g}\right) \\ \mathbf{0}_{m\times(l+1)m} & \mathbf{I}_{m\times m} \end{bmatrix} \in \mathbb{Z}_q^{(2mk+m)\times(l+2)m}; \quad (1)$$

Finnaly, outputs $rk_{f\xrightarrow{h}g}$ along with the attribute vector $\mathbf{b}$.

ReEncrypt$(pp, rk_{f\xrightarrow{h}g}, ct)$: This algorithm accepts as inputs pp, $rk_{f\xrightarrow{h}g}$ and ct. The proxy then produces the re-encrypted ciphertext ct' in the manner described below: If $h(\mathbf{a}) \neq 0$ or $c_\mathbf{a} = \varnothing$, outputs the symbol $\perp$; Otherwise, do as follows: First, lets $ct = \left(\mathbf{c}_{in}, \{\mathbf{c}_i\}_{i\in[l]}, \mathbf{c}_{out}\right)$; Then, evaluates $\mathbf{c}_h \leftarrow Eval_{ct}\left(\{a_i, \mathbf{A}_i, \mathbf{c}_i\}_{i=1}^{l}, h\right)$;And sets $\mathbf{c}_h{}' = [\mathbf{c}_{in}|\mathbf{c}_h] \in \mathbb{Z}_q^{2m}$; Next, computes $(\mathbf{c_x}{}')^T = (\mathbf{c}_{in}{}'|\mathbf{c}_{out}{}')^T = [BD(\mathbf{c}_h{}')\,|\mathbf{c}_{out}]^T \cdot rk_{f\xrightarrow{h}g} \in \mathbb{Z}_q^{1\times(l+2)m}$, $\mathbf{c_a}{}' = 0$; At last, outputs the re-encryption ciphertext $ct' = (\mathbf{c_a}{}', \mathbf{c_x}{}')$.

Decrypt(pp, sk_f, ct): The user inputs pp, sk_f and ct, then follow the steps below to decrypt: If $f(\mathbf{a}) \neq 0$, outputs the symbol $\perp$; Then sets $\mathbf{c}_f \leftarrow Eval_{ct}\left(\{a_i, \mathbf{A}_i, \mathbf{c}_i\}_{i=1}^{l}, f\right)$; $\mathbf{c}_f{}' = [\mathbf{c}_{in}|\mathbf{c}_f] \in \mathbb{Z}_q^{2m}$; Otherwise, evaluates $\xi' = \mathbf{y}^T\mathbf{c}_{out} - sk_f^T \cdot \mathbf{c}_f{}'$. Finally, outputs $\xi = \langle\mathbf{x}, \mathbf{y}\rangle \in \{0, 1, ..., K\}$ which minimizes $|\lfloor q/K\rfloor \cdot \xi - \xi'|$.

2.2 Correctness

(i) The accuracy of unconverted ciphertext decoding. If the policy f satisfies $f(\mathbf{a}) = 0$, the original ciphertext ct decrypted with the private key sk_f is as follows

$$\mathbf{c}_f{}' = [\mathbf{c}_{in}|\mathbf{c}_f] = \left[(\mathbf{A}_0^T\mathbf{s} + \mathbf{e}_{in})\,|\,(\mathbf{A}_f^T\mathbf{s} + \mathbf{e}_f)\right] = (\mathbf{A}_0|\mathbf{A}_f)^T\mathbf{s} + (\mathbf{e}_{in}|\mathbf{e}_f) \in \mathbb{Z}_q^{2m} \tag{2}$$

$$\begin{aligned}\xi' &= \mathbf{y}^T\mathbf{c}_{out} - sk_f^T \cdot \mathbf{c}_f{}' \\ &= \mathbf{y}^T\left(\mathbf{D}^T\mathbf{s} + \mathbf{e}_{out} + \lfloor q/K\rfloor \cdot \mathbf{x}\right) - (\mathbf{R}_f \cdot \mathbf{y})^T\left[(\mathbf{A}_0|\mathbf{A}_f)^T\mathbf{s} + (\mathbf{e}_{in}|\mathbf{e}_f)\right] \\ &= \lfloor q/K\rfloor\langle\mathbf{x}, \mathbf{y}\rangle + errors\end{aligned} \tag{3}$$

(ii) Decryption correctness of converted ciphertext. For the re-encryption ciphertext ct', the decryption processes are as follows

(a) If $h(\mathbf{a}) = 0$, we need to compute

$$\begin{aligned}BD(\mathbf{c}_h{}')^T \cdot P2\left(\mathbf{R}_{f\xrightarrow{h}g}\right) &= (\mathbf{c}_h{}')^T \cdot \mathbf{R}_{f\xrightarrow{h}g} \\ &= \left((\mathbf{A}_0|h(\mathbf{a})\mathbf{G} + \mathbf{A}_h)^T\mathbf{s} + (\mathbf{e}_{in}|\mathbf{e}_h)\right)^T \cdot \mathbf{R}_{f\xrightarrow{h}g} \\ &= \mathbf{s}^T(\mathbf{A}_0|\mathbf{A}_h)\,\mathbf{ke}_h{}^T + \mathbf{s}^T\mathbf{D} + (\mathbf{e}_{in}|\mathbf{e}_h)^T\mathbf{R}_{f\xrightarrow{h}g}\end{aligned} \tag{4}$$

$$\begin{aligned}(\mathbf{c}_x{}')^T &= \left[BD(\mathbf{c}_h{}')^T|\mathbf{c}_{out}^T\right] \cdot rk_{f\xrightarrow{h}g} \\ &= \left[BD(\mathbf{c}_h{}')^T|\mathbf{c}_{out}^T\right] \cdot \begin{bmatrix}\mathbf{R}_1\mathbf{H_b} + \mathbf{E}_1 & \mathbf{R}_1\mathbf{D} + \mathbf{E}_2 - P2\left(\mathbf{R}_{f\xrightarrow{h}g}\right) \\ \mathbf{0}_{m\times(l+1)m} & \mathbf{I}_{m\times m}\end{bmatrix} \\ &= \left[\overline{\mathbf{s}}^T\mathbf{H_b} + \overline{\mathbf{e}_1}^T|\overline{\mathbf{s}}^T\mathbf{D} + \overline{\mathbf{e}_2}^T + \lfloor q/K\rfloor \cdot \mathbf{x}^T\right]\end{aligned} \tag{5}$$

Where
$\overline{\mathbf{s}}^T = BD(\mathbf{c}_h{}')^T\mathbf{R}_1$, $\overline{\mathbf{e}_1}^T = BD(\mathbf{c}_h{}')^T\mathbf{E}_1$, $\overline{\mathbf{e}_{out}}^T = BD(\mathbf{c}_h{}')^T\mathbf{E}_2 - \mathbf{s}^T(\mathbf{A}_0|\mathbf{A}_h)\,\mathbf{ke}_h{}^T - (\mathbf{e}_{in}|\mathbf{e}_f)^T\mathbf{R}_h + \mathbf{e}_{out}$, $\mathbf{c_x}{}' = \left[\mathbf{H_b}^T\overline{\mathbf{s}} + \overline{\mathbf{e}_1}|\mathbf{D}^T\overline{\mathbf{s}} + \overline{\mathbf{e}_{out}} + \lfloor q/K\rfloor \cdot \mathbf{x}\right] = [\overline{\mathbf{c}_{in}}|\overline{\mathbf{c}_{out}}]$.

(b) Then, evaluate the ciphertext $\mathbf{c}_g \leftarrow Eval_{ct}\left(\{b_i, \mathbf{A}_i, \mathbf{c}_i\}_{i=1}^{l}, g\right)$ and $\mathbf{c}_g{}' = [\mathbf{c}_{in}|\mathbf{c}_g] = (\mathbf{A}_0|g(\mathbf{b})\mathbf{G}+\mathbf{A}_g)^T\mathbf{s}+(\mathbf{e}_{in}|\mathbf{e}_g) \in \mathbb{Z}_q^{2m}$.

(c) Sample the matrix $\mathbf{R}_g \leftarrow SampleRight(\mathbf{A}_0, \mathbf{T}_{\mathbf{A}_0}, \mathbf{A}_g, \mathbf{D}, \rho)$ so that $(\mathbf{A}_0|\mathbf{A}_g)\cdot\mathbf{R}_g = \mathbf{D}$, $\mathbf{R}_h \in \mathbb{Z}_q^{2m\times m}$.

(d) Use the private key sk_g for the following calculation

$$\begin{aligned}\xi' &= \mathbf{y}^T\overline{\mathbf{c}_{out}} - sk_g^T \cdot \mathbf{c}_g{}' \\ &= \mathbf{y}^T\left(\mathbf{D}^T\mathbf{s}+\overline{\mathbf{e}_{out}}+\lfloor q/K\rfloor\cdot\mathbf{x}\right) - (\mathbf{R}_g\cdot\mathbf{y})^T\left[(\mathbf{A}_0|\mathbf{A}_g)^T\mathbf{s}+(\mathbf{e}_{in}|\mathbf{e}_g)\right] \\ &= \lfloor q/K\rfloor\langle\mathbf{x},\mathbf{y}\rangle + errors\end{aligned} \tag{6}$$

Remark 1. In order to decrypt correctly, we need to set the parameter $q > 8KVm\sigma + 8KVm^2\rho\sigma(1+\gamma_{\mathscr{F}})$ so that $\xi = \langle\mathbf{x},\mathbf{y}\rangle \in \{0,1,...,K\}$, which minimizes $|\lfloor q/K\rfloor\cdot\xi - \xi'|$.

2.3 Security Analysis

Theorem 1. *Assume that σ, m, n, α and q are the same as in the previous sentence. In the standard model, ACPRE is IND-sHRA secure under the assumption that the decision* $\mathrm{LWE}_{q,\Psi}$ *problem is difficult.*

Proof. Let $\mathbf{a}^* = (a_1^*, a_2^*, ..., a_l^*)$ be the target attributes set. Then the challenger $\mathscr{C}$ initializes a counter *numCt* to 0, a key-value store *C* and *Derive* to be empty. Next, we construct the simulation algorithms (**Sim.Setup**, **Sim.KeyGen**, **Sim.Enc**, **Sim.ReKeyGen**, **Sim.ReEnc1** and **Sim.ReEnc2**) as below.

- **Sim.Setup**$(1^\lambda, \mathscr{F}_\lambda)$: This algorithm inputs 1^λ, $n > \Omega(\lambda)$, $q > 2$, $m > 5n\log q$ and the function family $\mathscr{F}_\lambda = \left\{f : \{0,1\}^l \rightarrow \{0,1\}\right\}$, then do as follows: First, chooses a random matrix $\mathbf{A}_0 \leftarrow \mathbb{Z}_q^{n\times m}$; Then selects l random matrices uniformly $\mathbf{S}_i^* \leftarrow \{\pm 1\}^{m\times m}$, $i \in [1,l]$; Lets $\mathbf{A}_i = \mathbf{A}_0\mathbf{S}_i^* - a_i^*\mathbf{G}$, $i \in [1,l]$; Randomly chooses a matrix $\mathbf{R}_{\mathbf{a}^*} \leftarrow \mathbf{Z}_q^{(l+1)m\times m}$; Next, creates the equation $[\mathbf{A}_0|a_1^*\mathbf{G}+\mathbf{A}_1\cdots|a_l^*\mathbf{G}+\mathbf{A}_l]\cdot\mathbf{R}_{\mathbf{a}}^* = \mathbf{D}$, that is, $[\mathbf{A}_0|\mathbf{A}_0\mathbf{S}_1^*|\cdots\mathbf{A}_0\mathbf{S}_l^*]\cdot\mathbf{R}_{\mathbf{a}^*} = \mathbf{D}$; Finally, returns $pp = \{\mathbf{A}_1, \mathbf{A}_2, ..., \mathbf{A}_l, \mathbf{D}, \mathbf{G}\}$ and stores $msk = \mathbf{T}_\mathbf{G}$.
- **Sim.KeyGen**$(pp, msk, f, \mathbf{y})$: pp, msk, $f \in \mathscr{F}_\lambda$ and $\mathbf{y} \in \mathscr{Y}_\lambda$ are inputs into this algorithm. Next, the algorithm extracts the function private key as follows: First, evaluates $\mathbf{A}_f \leftarrow Eval_{pk}\left(f, \{\mathbf{A}_i\}_{i=1}^{l}\right)$ where $\mathbf{A}_f \in \mathbb{Z}_q^{n\times m}$; and then evaluates $\mathbf{S}_f^* \leftarrow Eval_{sim}(f, \mathbf{A}_0, \{a_i^*, \mathbf{S}_i^*\}_{i=1}^{l})$; Next, sets $\mathbf{A}_f = \mathbf{A}_0\mathbf{S}_f^* - f(\mathbf{a}^*)\mathbf{G}$; Gets the trapdoor matrix $\mathbf{T}_{(\mathbf{A}_0|\mathbf{A}_f)} \leftarrow$ **SampleBasisLeft**$(\mathbf{A}_0, f(\mathbf{a}^*)\mathbf{G}, \mathbf{T}_\mathbf{G}, \mathbf{S}_f^*)$; Samples the matrix $\mathbf{R}_f \leftarrow$ **SamplePre**$\left((\mathbf{A}_0|\mathbf{A}_f), \mathbf{T}_{(\mathbf{A}_0|\mathbf{A}_f)}, \mathbf{D}, \rho\right)$ so that $(\mathbf{A}_0|\mathbf{A}_f)\cdot\mathbf{R}_f = \mathbf{D}$, $\mathbf{R}_f \in \mathbb{Z}_q^{2m\times m}$; Finally, outputs the secret key as $sk_f = \mathbf{k} = \mathbf{R}_f\cdot\mathbf{y} \in \mathbb{Z}_q^{2m}$.
- **Sim.Encrypt** $(pp, \mathbf{a}, \mathbf{x})$: As input, the algorithm accepts pp, $\mathbf{a}^* = (a_1^*, a_2^*, ..., a_l^*) \in \mathbb{Z}_q^l$ and the massage $\mathbf{x} \in \{0,1\}^m$, then does: First, uniformly selects a vector $\mathbf{s} \in \mathbb{Z}_q^n$ at random and three error vectors $\mathbf{e}_0 \in \Psi_\alpha^m$, $\mathbf{e}_{in} \in \Psi_\alpha^m$, $\mathbf{e}_{out} \in \Psi_\alpha^m$; Lets $\mathbf{e} = \left[\mathbf{I}_m|\mathbf{S}_1^*|\cdots|\mathbf{S}_l^*\right]^T\cdot\mathbf{e}_0 = \left(\mathbf{e}_{in}^T, \mathbf{e}_1^T, ..., \mathbf{e}_l^T\right)^T \in \mathbb{Z}_q^{(l+1)m}$; $\mathbf{H}_{\mathbf{a}^*} =$

$(\mathbf{A}_0|a_1^*\mathbf{G}+\mathbf{A}_1|\cdots|a_l^*\mathbf{G}+\mathbf{A}_l) \in \mathbb{Z}_q^{n\times(l+1)m}$; $\mathbf{c}_{in} = \mathbf{A}_0^T\mathbf{s}+\mathbf{e}_{in} \in \mathbb{Z}_q^m$; $\mathbf{c}_{out} = \mathbf{D}^T\mathbf{s} + \mathbf{e}_{out} + \lfloor q/K \rfloor \cdot \mathbf{x} \in \mathbb{Z}_q^m$; $\mathbf{c}_\mathbf{x}^* = (\mathbf{c}_{in}, \mathbf{c}_{out})$. Computes $\mathbf{c}_\mathbf{a}^* = \left(\left\{\mathbf{c}_i = (a_i^*\mathbf{G}+\mathbf{A}_i)^T\mathbf{s}+(\mathbf{S}_i^*)^T\mathbf{e}_{in}\right\}_{i\in\{1,\ldots,l\}}\right) \in \mathbb{Z}_q^{lm}$ and $\mathbf{c}^* = \mathbf{H}_{a^*}^T\mathbf{s}+\mathbf{e} = [\mathbf{c}_{in}|\mathbf{c}_\mathbf{a}^*] \in \mathbf{Z}_q^{(l+1)m}$; Finally, $ct = (\mathbf{c}_\mathbf{a}^*, \mathbf{c}_\mathbf{x}^*) \in \mathbb{Z}_q^{(l+2)m}$ is produced.

- **Sim.ReKeyGen**(pp, sk_f, h, g): This algorithm enters pp, sk_f, conversion policy h and new access policy g to return the re-encryption key $rk_{f\xrightarrow{h}g}$ in the manner described below: If $f(\mathbf{a}^*) = 0$, $g(\mathbf{a}^*) \neq 0$, outputs $\perp$; If $f(\mathbf{a}^*) = 0$, $g(\mathbf{a}^*) = 0$, chooses two matrices randomly $\mathbf{X}_1 \in \mathbb{Z}_q^{2mk\times n}$, $\mathbf{X}_2 \in \mathbb{Z}_q^{2mk\times m}$ and then simulate $rk_{f\xrightarrow{h}g}$ as follows: $rk_{f\xrightarrow{h}g} = \begin{bmatrix} \mathbf{X}_1 & \mathbf{X}_2 \\ \mathbf{0}_{m\times(l+1)m} & \mathbf{I}_{m\times m} \end{bmatrix} \in \mathbb{Z}_q^{(2mk+m)\times(l+2)m}$; Finally, outputs $rk_{f\xrightarrow{h}g}$ along with the attribute vector $\mathbf{b}$.
- **Sim.ReEnc1**(pp, ct, h, g): This algorithm enters pp, ct, conversion policy h and new access policy g to return the re-encryption ciphertext ct' in the following way: First, judges the equation $f(\mathbf{a}^*) \stackrel{?}{=} 0$, if $f(\mathbf{a}^*) \neq 0$, output $\perp$; Otherwise, selects an attribute set $\mathbf{b} = (b_1, b_2, \ldots, b_l)$ such that $g(\mathbf{b}) = 0$; Then computes $\mathbf{H}_\mathbf{b} = (\mathbf{A}_0|b_1\mathbf{G}+\mathbf{A}_1|\cdots|b_l\mathbf{G}+\mathbf{A}_l) \in \mathbb{Z}_q^{n\times(l+1)m}$; Then, evaluates $\mathbf{A}_h \leftarrow Eval_{pk}\left(h, \{\mathbf{A}_i\}_{i=1}^k\right)$ where $\mathbf{A}_h \in \mathbb{Z}_q^{n\times m}$; Next, takes a matrix $\mathbf{I}' \in \mathbb{Z}_q^{n\times m}$, two Gaussian noise matrices $\mathbf{R}_h' \in \Psi^{n\times 2m}$, $\mathbf{R}'_{f\xrightarrow{h}g} \in \Psi^{2m\times m}$, and sets $(\mathbf{A}_h|\mathbf{I}') = \mathbf{R}_h' - \mathbf{A}_0(\mathbf{R}'_{f\xrightarrow{h}g})^T$; Chooses a uniformly random matrix $\mathbf{R}_1 \leftarrow \mathbb{Z}_q^{2mk\times n}$, and two error matrices $\mathbf{E}_1 \leftarrow \Psi^{2mk\times(l+1)m}$, $\mathbf{E}_2 \leftarrow \Psi^{2mk\times m}$; Constructs a simulated re-encryption keys $rk'_{f\xrightarrow{h}g}$ as follows $rk'_{f\xrightarrow{h}g} = \begin{bmatrix} \mathbf{R}_1\mathbf{H}_\mathbf{b}+\mathbf{E}_1 & \mathbf{R}_1\mathbf{D}+\mathbf{E}_2-P2\left(\mathbf{R}'_{f\xrightarrow{h}g}\right) \\ \mathbf{0}_{m\times(l+1)m} & \mathbf{I}_{m\times m} \end{bmatrix} \in \mathbb{Z}_q^{(2mk+m)\times(l+2)m}$; Following is the breakdown of the re-encrypted ciphertext: (1) Lets $ct = \left(\mathbf{c}_{in}, \{\mathbf{c}_i\}_{i\in[l]}, \mathbf{c}_{out}\right)$; (2) Evaluates $\mathbf{c}_h \leftarrow Eval_{ct}\left(\{a_i, \mathbf{A}_i, \mathbf{c}_i\}_{i=1}^l, h\right)$; (3) Sets $\mathbf{c}_h{}' = [\mathbf{c}_{in}|\mathbf{c}_h] \in \mathbb{Z}_q^{2m}$; (4) Computes $(\mathbf{c}_\mathbf{x}{}')^T = (\mathbf{c}_{in}{}'|\mathbf{c}_{out}{}')^T = [BD(\mathbf{c}_h{}')|\mathbf{c}_{out}]^T \cdot rk'_{f\xrightarrow{h}g} \in \mathbb{Z}_q^{1\times(l+2)m}$ and lets $\mathbf{c}_\mathbf{a}' = 0$. Finally, outputs $ct' = (\mathbf{c}_\mathbf{a}{}', \mathbf{c}_\mathbf{x}{}')$ along with the attribute vector $\mathbf{b}$.
- **Sim.ReEnc2**(pp, ct, h, g): This algorithm enters pp, ct, conversion policy h and new access policy g to return the re-encryption ciphertext ct' for $\mathbf{a} \neq \mathbf{a}^*$ in the following way: $f(\mathbf{a}^*) = 0$, then judges the equation $f(\mathbf{a}) \stackrel{?}{=} 0$, if $f(\mathbf{a}) \neq 0$, output $\perp$; Otherwise, takes the following steps: Since $\mathbf{a} \neq \mathbf{a}^*$, at least one $i \in \{1, 2, \cdots, l\}$ exists for which $(a_i - a_i^*) \neq 0$. Constructs the matrix

$$\begin{aligned} \mathbf{H}_\mathbf{a} &= (\mathbf{A}_0|a_1\mathbf{G}+\mathbf{A}_1|\cdots|a_l\mathbf{G}+\mathbf{A}_l) \\ &= [\mathbf{A}_0|(a_1-a_1^*)\mathbf{G}+\mathbf{A}_0\mathbf{S}_1^*|\cdots|(a_l-a_l^*)\mathbf{G}+\mathbf{A}_0\mathbf{S}_l^*] \in \mathbb{Z}_q^{n\times(l+1)m}; \end{aligned} \tag{7}$$

As $(a_l - a_l^*) \neq 0$ and the matrix $\mathbf{T}_\mathbf{G}$ is the trapdoor of the matrix $\mathbf{G}$, it is also the trapdoor of $(a_l - a_l^*)\mathbf{G}$; Then samples $\mathbf{T}_{(\mathbf{A}_0|(a_l-a_l^*)\mathbf{G}+\mathbf{A}_0\mathbf{S}_l^*)} \leftarrow ExtendLeft(\mathbf{A}_0, (a_l -$

$a_l^*)\mathbf{G}, \mathbf{T}_G, \mathbf{S}_l^*)$ and the matrix $\mathbf{T}_{(\mathbf{A}_0|a_l\mathbf{G}+\mathbf{A}_l|a_1\mathbf{G}+\mathbf{A}_1|\cdots|a_{l-1}\mathbf{G}+\mathbf{A}_{l-1})} \leftarrow$ **SampleBasisRight**$((\mathbf{A}_0|a_l\mathbf{G}+\mathbf{A}_l), \mathbf{T}_{(\mathbf{A}_0|a_l\mathbf{G}+\mathbf{A}_l)}, (a_l\mathbf{G}+\mathbf{A}_l|a_1\mathbf{G}+\mathbf{A}_1|\cdots|a_{l-1}\mathbf{G}+\mathbf{A}_{l-1}))$. We get a matrix $\mathbf{T}_{(\mathbf{A}_0|a_1\mathbf{G}+\mathbf{A}_1|\cdots|a_{l-1}\mathbf{G}+\mathbf{A}_{l-1}|a_l\mathbf{G}+\mathbf{A}_l)}$ by swapping some columns of the matrix $\mathbf{T}_{(\mathbf{A}_0|a_l\mathbf{G}+\mathbf{A}_l|a_1\mathbf{G}+\mathbf{A}_1|\cdots|a_{l-1}\mathbf{G}+\mathbf{A}_{l-1})}$; The matrix $\mathbf{T}_{(\mathbf{A}_0|a_1\mathbf{G}+\mathbf{A}_1|\cdots|a_{l-1}\mathbf{G}+\mathbf{A}_{l-1}|a_l\mathbf{G}+\mathbf{A}_l)}$ is the trapdoor of the matrix $\mathbf{H_a}$; Assumes that $\mathbf{T_{H_a}} = \mathbf{T}_{(\mathbf{A}_0|a_1\mathbf{G}+\mathbf{A}_1|\cdots|a_{l-1}\mathbf{G}+\mathbf{A}_{l-1}|a_l\mathbf{G}+\mathbf{A}_l)}$. Samples the matrix $\mathbf{R}'_f \leftarrow$ **SamplePre**$(\mathbf{H_a}, \mathbf{T_{H_a}}, \mathbf{D}, \sigma)$. Selects an attribute set $\mathbf{b} = (b_1, b_2, ..., b_l)$ such that $g(\mathbf{b}) = 0$, then computes $\mathbf{H_b} = (\mathbf{A}_0|b_1\mathbf{G}+\mathbf{A}_1|\cdots|b_l\mathbf{G}+\mathbf{A}_l) \in \mathbb{Z}_q^{n\times(l+1)m}$; Evaluates $\mathbf{A}_h \leftarrow Eval_{pk}\left(h, \{\mathbf{A}_i\}_{i=1}^k\right)$ where $\mathbf{A}_h \in \mathbb{Z}_q^{n\times m}$; Next, samples the matrix $\mathbf{R}_h \leftarrow$ **SampleBasisRight**$\left(\mathbf{A}_0, \mathbf{T}_{\mathbf{A}_0}, \mathbf{A}_h, \mathbf{D}, \rho\right)$ so that $(\mathbf{A}_0|\mathbf{A}_h)\cdot\mathbf{R}_h = \mathbf{D}$, $\mathbf{R}_h \in \mathbb{Z}_q^{2m\times m}$; Sets $\mathbf{R}_{f\xrightarrow{h}g} = \mathbf{k}'\mathbf{e}_h{}^T + \mathbf{R}_h \in \mathbb{Z}_q^{2m\times m}$, where $\mathbf{k}' = \mathbf{R}'_f \cdot \mathbf{y}$, $\mathbf{e}_h \leftarrow \Psi^m$; Chooses a uniformly random matrix $\mathbf{R}_1 \leftarrow \mathbb{Z}_q^{2mk\times n}$, and two error matrices $\mathbf{E}_1 \leftarrow \Psi^{2mk\times(l+1)m}$, $\mathbf{E}_2 \leftarrow \Psi^{2mk\times m}$; Constructs a simulated $rk'_{f\xrightarrow{h}g}$ as follows

$$rk'_{f\xrightarrow{h}g} = \begin{bmatrix} \mathbf{R}_1\mathbf{H_b}+\mathbf{E}_1 & \mathbf{R}_1\mathbf{D}+\mathbf{E}_2 - P2\left(\mathbf{R}'_{f\xrightarrow{h}g}\right) \\ \mathbf{0}_{m\times(l+1)m} & \mathbf{I}_{m\times m} \end{bmatrix} \in \mathbb{Z}_q^{(2mk+m)\times(l+2)m}; \quad (8)$$

Following is the breakdown of ct': (1) Evaluates $\mathbf{c}_h \leftarrow Eval_{ct}\left(\{a_i, \mathbf{A}_i, \mathbf{c}_i\}_{i=1}^l, h\right)$; (2) Sets $\mathbf{c}_h{}' = [\mathbf{c}_{in}|\mathbf{c}_h] \in \mathbb{Z}_q^{2m}$; (3) Computes $(\mathbf{c_x}')^T = (\mathbf{c}_{in}{}'|\mathbf{c}_{out}{}')^T = [BD(\mathbf{c}_h{}')\,|\mathbf{c}_{out}]^T \cdot rk'_{f\xrightarrow{h}g} \in \mathbb{Z}_q^{1\times(l+2)m}$; lets $\mathbf{c}'_\mathbf{a} = 0$. Finally, outputs $ct' = (\mathbf{c_x}', \mathbf{c_a}')$.

According to the selective security model, we make the premise that a PPT attacker $\mathscr{A}$ with a non-negligible advantage will attempt to attack our system via"game-hopping". The basic idea is to construct three games. The first game is an actual IND-sHRA attack, however the last game is impossible for the attacker to win; Then, based on some difficult assumptions on lattice, it is proved that the first game and the last game are equivalent. We have to demonstrate that these three games are indistinguishable from each other for PPT adversary $\mathscr{A}$ in the following ways.

Game sequence:

- **Game0**: $\mathscr{C}$ and $\mathscr{A}$ are playing a typical IND-sHRA game. And most operations are identical to the real scheme. Besides, $\mathscr{C}$ faithfully responds to different sk_f and $rk_{f\xrightarrow{h}g}$ queries in accordance with the real scheme's algorithms.
- **Game1**: This game changes the generation of pp as in **Sim.Setup**. On the one hand, we know that a uniform distribution results in statistical indistinguishability between $\mathbf{A}_0\mathbf{S}_1^*, \mathbf{A}_0\mathbf{S}_2^*, \cdots, \mathbf{A}_0\mathbf{S}_l^*$ according to the left-over hash lemma. Therefore, $(\mathbf{A}_1, \mathbf{A}_2, \cdots, \mathbf{A}_l)$ are close to uniform as defined in **Sim.Setup**. On the other hand, the leftover hash lemma shows that $[\mathbf{A}_0|\mathbf{A}_0\mathbf{S}_1^*|\mathbf{A}_0\mathbf{S}_2^*|\cdots|\mathbf{A}_0\mathbf{S}_l^*]\cdot\mathbf{R}_{\mathbf{a}^*}$ is statistically equivalent to a uniform distribution. So the matrix $\mathbf{D}$ is statistically indistinguishable with uniform distribution. The following inquiries are possible from $\mathscr{A}$:

Secret Key Query $\mathcal{O}_{sk}$: The attacker $\mathcal{A}$ sends the access policy f to a challenger $\mathcal{C}$ for private key queries. If $f(\mathbf{a}^*) = 0$, the challenger $\mathcal{C}$ returns the symbol $\perp$; Otherwise, according to the parameters in **Sim.KeyGen**, the challenger $\mathcal{C}$ generates sk_f as previously mentioned and gives it back to the attacker $\mathcal{A}$.
Re-encryption Key Query $\mathcal{O}_{rk}$: If $f(\mathbf{a}^*) = 0$, the challenger $\mathcal{C}$ executes the algorithm **Sim.ReKeyGen**; Otherwise, $\mathcal{C}$ first generates sk_f as in $\mathcal{O}_{sk}$. Then, the challenger $\mathcal{C}$ replies with $rk_{f \xrightarrow{h} g}$ by running **ReKeyGen**(pp, sk_f, h, g); The re-encryption key $rk_{f \xrightarrow{h} g}$ is sent by the the challenger $\mathcal{C}$ to the attacker $\mathcal{A}$. The attacker $\mathcal{A}$ can query multiple times.
Encryption Query $\mathcal{O}_{Enc}$: $\mathcal{A}$ can also make some queries $\mathcal{O}_{Enc}$ to $\mathcal{C}$. $\mathcal{C}$ only executes the algorithm **Sim.Encrypt** faithfully; $\mathcal{C}$ increments $numCt$ and adds ct to the set C with key $(\mathbf{a}, numCt)$.
Re-encryption Query $\mathcal{O}_{ReEnc}$: After entering parameters f, h, g and $(\mathbf{a}, k)$ where $k \leq numCt$, then $\mathcal{C}$ follows these actions: (i) If $\mathbf{a} = \mathbf{a}^*$ and C with key $(\mathbf{a}^*, k)$ has no value, the symbol $\perp$ is produced. Otherwise, sets $ct = (\mathbf{c}_a, \mathbf{c}_x)$ to that value in C and uses **Sim.ReEnc1** to generate the re-encrypted ciphertext; (ii) Symbol $\perp$ is printed if $\mathbf{a} \neq \mathbf{a}^*$ and C with key $(\mathbf{a}, k)$ has no value. Otherwise, sets $ct = (\mathbf{c}_a, \mathbf{c}_x)$ be that value in C. Additionally, there are two cases to consider: If $f(\mathbf{a}^*) = 0$, execute the algorithm **Sim.ReEnc2** to produce the re-encryption ciphertext ct'; If $f(\mathbf{a}^*) \neq 0$, the challenger $\mathcal{C}$ first generates $rk_{f \xrightarrow{h} g}$ as in $\mathcal{O}_{rk}$. Then, the challenger $\mathcal{C}$ replies with the re-encryption ct' by running the algorithm **ReEncrypt**$(pp, rk_{f \xrightarrow{h} g}, ct)$.
Finally, the attacker $\mathcal{A}$ receives the re-encryption ciphertext ct' back from the challenger $\mathcal{C}$. And the settings of other parameters are identical to **Game0**.

- **Game2**: The production of the challenge ciphertext ct^* distinguishes this game from **Game1**. In **Game1**, the **Sim.Encrypt** $(pp, \mathbf{a}, \mathbf{x})$ algorithm creates ct^*. However, in **Game2**, the ct^* is a uniformly random and independent matrix from $\mathbb{Z}_q^{(l+2)m}$. The rest of the settings are the same as **Game1**. In this case, the advantage of the attacker $\mathcal{A}$ is zero.

If **Game1** is indistinguishable from **Game0**, and **Game2** is indistinguishable from **Game1**, our ACPRE is IND-sHRA secure in the standard model.

Game transfer:

- **Game0** to **Game1**: We now prove that **Game0** and **Game1** are indistinguishable through the Lemma 1.

Lemma 1. *If* $mn > (n+1)\log_2 q + \omega(\lg n)$*, then* ***Game1*** *and* ***Game0*** *are indistinguishable, and the answer to the attacker* $\mathcal{A}$*'s inquiries becomes indistinguishable from the true scheme.*

Proof. As a result of the public parameters and query responses being statistically similar to those in **Game0**, from the Lemma 1, we can see that the $\mathcal{A}$'s advantage in **Game1** is equal to that in **Game0**. Therefore, **Game0** and **Game1** are indistinguishable.

- **Game1** to **Game2**: Assuming that in the selective security model, an attacker $\mathscr{A}$ can discriminate between **Game1** and **Game2** with a non-negligible advantage by HRA, the challenger $\mathscr{C}$ will create an algorithm $\mathscr{B}$ intended to effectively solve the decision LWE problem. Following is the reduction's progression:
  - **Target.** The challenger $\mathscr{C}$ receives a statement from the $\mathscr{A}$ announcing the target attribute set $\mathbf{a}^* = (a_1^*, a_2^*, \cdots, a_l^*)$ to attack.
  - **Instance.** The $\mathscr{B}$ begins by getting an LWE challenge consisting of two random matrices $\mathbf{A}_0, \mathbf{D}\ in \mathbb{Z}_q^{n\times m}$, and two vectors $\mathbf{c}_{in}, \mathbf{c}_{out} \in \mathbb{Z}_q^m$. Here, we take into account two cases about $\mathbf{c}_{in}$ and $\mathbf{c}_{out}$. One is to choose at random from the $\mathbb{Z}_q^m$, and the other is to figure out $\mathbf{c}_{in} = \mathbf{A}_0^T \mathbf{s} + \mathbf{e}_0$ and $\mathbf{c}_{out} = \mathbf{D}^T \mathbf{s} + \mathbf{e}_{out}$, where $\mathbf{s} \leftarrow \mathbb{Z}_q^n$, $\mathbf{e}_0 \leftarrow \Psi_\alpha^m$, $\mathbf{e}_{out} \leftarrow \Psi_\alpha^m$. What's more, the objective of $\mathscr{B}$ is to differentiate these two circumstances applying $\mathscr{A}$ with non-negligible advantage.
  - **Setup.** The following is how $\mathscr{B}$ sets up the public parameters once it has obtained the target attribute set $\mathbf{a}^*$: The central authority chooses the appropriate system parameters n, m, q, σ; Selects l random matrices uniformly $\mathbf{S}_i^* \leftarrow \{\pm 1\}^{m\times m}$, $i \in [1,l]$; Lets $\mathbf{A}_i = \mathbf{A}_0 \mathbf{S}_i^* - a_i^* \mathbf{G}$, $i \in [1,l]$; The remaining parameters are the same as **Game1** settings. Finally, $\mathscr{B}$ transmits $pp = \{\mathbf{A}_1, \mathbf{A}_2, ..., \mathbf{A}_l, \mathbf{D}, \mathbf{G}\}$.

Remark 2. The above settings have the correct distribution.

- **Phase 1.** The attacker $\mathscr{A}$ can ask $\mathscr{B}$ for all key questions and $\mathscr{B}$ answers $\mathscr{A}$ as in **Game1**:
- **Challenge.** In order to indicate that it is open to a challenge, $\mathscr{A}$ selects two messages $\{\mathbf{x}_0, \mathbf{x}_1\} \leftarrow \mathbb{Z}_q^m$ of equal-length. A message $\mathbf{x}_\varphi$ ($\varphi \leftarrow \{0,1\}$) is selected at random by $\mathscr{B}$ for encryption. $\mathscr{B}$ replies with $ct^* = (\mathbf{c}^*, \mathbf{c}_{out}^*) \in \mathbb{Z}_q^{(l+2)m}$ built as follows:

$$\begin{aligned} \mathbf{c}^* &= [\mathbf{I}_m | \mathbf{S}_1^* | \cdots | \mathbf{S}_l^*]^T \cdot \mathbf{c}_{in} \in \mathbb{Z}_q^{(l+1)m}; \\ \mathbf{c}_{out}^* &= \mathbf{c}_{out} + \lfloor q/K \rfloor \mathbf{x}_\varphi \in \mathbb{Z}_q^m \end{aligned} \tag{9}$$

  Next, $\mathscr{B}$ sends the challenge ciphertext ct^* to $\mathscr{A}$. Meanwhile, increases $numCt$ and adds $numCt$ to the set $Derive$. At last, $\mathscr{B}$ stores the value ct^* to the C with the key $(\mathbf{a}^*, numCt)$. Here are two cases for challenging ciphertext.
  (i) The $\mathscr{B}$ randomly selects a bit $\varphi \leftarrow \{0,1\}$, if $\varphi = 0$, $ct^* = (\mathbf{c}^*, \mathbf{c}_{out}^*)$ that will be returned to the attacker $\mathscr{A}$ as challenge ciphertext. Assuming that $\mathbf{c}_{in} = \mathbf{A}_0^T \mathbf{s} + \mathbf{e}_0$, $\mathbf{c}_{out} = \mathbf{D}^T \mathbf{s} + \mathbf{e}_{out}$ are constructed based on LWE samplings. It can be seen from the algorithm **Sim.Encrypt** that

$$\mathbf{H}_{a^*} = [\mathbf{A}_0 | a_1^* \mathbf{G} + \mathbf{A}_1 | \cdots | a_l^* \mathbf{G} + \mathbf{A}_l] = [\mathbf{A}_0 | \mathbf{A}_0 \mathbf{S}_1^* | \cdots | \mathbf{A}_0 \mathbf{S}_l^*] \in \mathbb{Z}_q^{n\times(l+1)m} \tag{10}$$

$$\mathbf{e} = [\mathbf{I}_m | \mathbf{S}_1^* | \cdots | \mathbf{S}_l^*]^T \cdot \mathbf{e}_0 \tag{11}$$

$$\mathbf{c}^* = [\mathbf{I}_m | \mathbf{S}_1^* | \cdots | \mathbf{S}_l^*]^T \cdot (\mathbf{A}_0^T \mathbf{s} + \mathbf{e}_0) = \mathbf{H}_{a^*}^T \mathbf{s} + \mathbf{e} \tag{12}$$

  It is easy to see that the calculation results for $\mathbf{c}^*$ and Game1 are the same. Furthermore,

$$\mathbf{c}_{out}^* = \mathbf{D}^T \mathbf{s} + \mathbf{e}_{out} + \lfloor q/K \rfloor \mathbf{x}_\varphi \tag{13}$$

To sum up, the challenge ciphertext $ct^* = (\mathbf{c}^*, \mathbf{c}^*_{out})$ is a legitimate ciphertext of the message $\mathbf{x}_\varphi$ with the attribute $\mathbf{a}^*$.
(ii) If $\varphi = 1$, $\mathbf{c}_{in}$ and $\mathbf{c}_{out}$ are randomly chosen from $\mathbb{Z}_q^m$, we know that $\mathbf{c}^*$ is also random in $\mathbb{Z}_q^{(l+1)m}$ according to the left over hash lemma. Therefore, ct^* as a challenge ciphertext will be uniformly random in $\mathbb{Z}_q^{(l+2)m}$ as in **Game2** and finally returns to the attacker $\mathscr{A}$.

- **Phase 2.** The simulator $\mathscr{B}$ operates in a similar manner to **Phase 1**, the attacker $\mathscr{A}$ is able to query secret keys multiple times, and the set of attributes $\mathbf{a}^*$ that have done so have not yet met the access structure f. In addition, $\mathscr{O}_{ReEnc}$ will output the symbols $\perp$ if $h(\mathbf{a}^*) \neq 0 \wedge g(\mathbf{a}^*) \neq 0 \wedge k \in Derive$.
- **Guess.** After enough questioning, the $\mathscr{A}$ outputs his guess of φ as φ'. Next, $\mathscr{B}$ returns the $\mathscr{A}$'s guess as the response to the corresponding LWE challenge.
In the attacker $\mathscr{A}$'s view, the $\mathscr{B}$'s behavior is close to that of a real, adaptive security experiment. In this game, the $\mathscr{A}$ has the advantage since $\varepsilon = |Pr[\varphi' = \varphi] - \frac{1}{2}|$.
Therefore, the advantages of the LWE predictor are as follows:
(i) In a pseudo-random sampler, an attacker $\mathscr{A}$ has the advantageous value ε. In this case, $\mathscr{O} = \mathscr{O}_\$$, $Pr[\varphi' = \varphi \mid \mathscr{O} = \mathscr{O}_\$] = \frac{1}{2} + \varepsilon$ and the $\mathscr{C}$'s advantage is

$$Pr[\mathscr{O}' = \mathscr{O} \mid \mathscr{O} = \mathscr{O}_\$] = \frac{1}{2} + \varepsilon; \tag{14}$$

(ii) In a true random predictor, an attacker $\mathscr{A}$ has an advantage of 0. In this case, $\mathscr{O} = \mathscr{O}_s$, $Pr[\varphi' = \varphi \mid \mathscr{O} = \mathscr{O}_s] = \frac{1}{2} + \varepsilon$, the $\mathscr{B}$'s advantage is

$$Pr[\mathscr{O}' = \mathscr{O} \mid \mathscr{O} = \mathscr{O}_s] = \frac{1}{2}. \tag{15}$$

Therefore, assuming that an attacker $\mathscr{A}$ guesses the correct probability is $Pr[\varphi' = \varphi] \geq \frac{1}{2} + \varepsilon$, the $\mathscr{B}$ has the advantage

$$\begin{aligned} &\frac{1}{2}(Pr[\mathscr{O}' = \mathscr{O} \mid \mathscr{O} = \mathscr{O}_\$] + Pr[\mathscr{O}' = \mathscr{O} \mid \mathscr{O} = \mathscr{O}_s]) - \frac{1}{2} \\ &= \frac{1}{2}(\frac{1}{2} + \varepsilon + \frac{1}{2}) - \frac{1}{2} = \frac{\varepsilon}{2} \end{aligned} \tag{16}$$

to solve the decision $\mathrm{LWE}_{\mathrm{q},\Psi}$ problem.

As a result, **Game2** and **Game1** cannot be distinguished under the assumption of the decision $\mathrm{LWE}_{\mathrm{q},\Psi}$ problem.

In conclusion, the security of our designed ACPRE scheme is compactly reduced to the decision $\mathrm{LWE}_{q,\Psi}$ difficulty assumption by the use of the three equivalent games mentioned above. Additionally, under the standard model, our ACPRE is IND-sHRA secure. The proof is now complete.

2.4 Comparison

In this part, we compare our ABCPRE with other relevant schemes ([4,6,11,14,15,19, 21]) in terms of access policy, support for inner product operations, security, resistance to quantum attacks, support for proxy control, and the size of secret key. And Table 2 displays the comparison's findings.

Table 2. Comparison with previous related work

Schemes	Access Policy	Inner Product	Security	Anti Quantum	Proxy Control	Secret Key Size
[6]	Attribute-based	✗	Semantic Security	✗	✗	—
[14]	Identity-based	✗	CCA(RO)	✗	✓	—
[4]	Identity-based	✗	CPA	✗	✓	—
[19]	Identity-based	✗	CPA	✓	✗	$O((2m \times l)log_2q)$
[11]	Attribute-based	✗	CPA	✓	✓	$O(2(m \times m)log_2q)$
[15]	Attribute-based	✗	HRA	✓	✗	$O((2m \times m)log_2q)$
[21]	Attribute-based	✗	HRA	✓	✓	$O((ml \times ml)log_2q)$
Ours	Attribute-based	✓	HRA	✓	✓	$O(2mlog_2q)$

As can be seen from the Table 2, with regard to access policies, our scheme makes use of attribute-based encryption, which can decide whether users have access privileges based on their unique attributes. This allows for more granular access management and is more flexible than simply using identity information to make decisions. More significantly, it is challenging to implement fine-grained ciphertext computation because current proxy re-encryption schemes cannot support inner product operations. Literature [14] achieves CCA security, whereas literature [6] only gets semantic security, but it has been shown to be secure under the random oracle model. In contrast to conventional conditional proxy re-encryption systems based on number theory ([4,14]), our scheme is based on lattice difficulty problems and is hence resistant to quantum attacks. Compared with PRE constructed based on lattice theory, [15,19] cannot control the conversion permissions of the proxy. In addition, among the PRE schemes constructed based on LWE ([11,15,19,21]), our proposed ACPRE scheme has the smallest secret key size.

In summary, the scheme we proposed is well suited for scenarios involving fine-grained ciphertext sharing across numerous users since it is secure against quantum attacks, restricts the proxy's capacity to convert, and supports inner product operations.

3 Access Policy Hidden

Users typically utilize attribute-based encryption to establish fine-grained ciphertext restriction to guarantee the security and controllability of data in the cloud. Although their access policy is posted to the cloud in plaintext along with ciphertext, this can result in the leakage of user-sensitive data if the access strategy is revealed. We employ differential privacy technology to conceal access policies and improve user privacy security by addressing the ambiguity of attribute replies in access policies in order to address the same problems as those discussed above. The specific process is as follows.

Firstly, encode the user attributes, installing the attribute set be $\mathbf{a} = (attribute_1, at-tribute_2, ..., attribute_l)$, the total number of attributes be l, and the value of each attribute $attribute_i$ be 1 or 0. Assuming that the user has an attribute set of $\{nationality, gender, age, ethnicity, research\ direction, workplace, marriage\}$. An existing access policy is $W = (t, l) = (3,7)$, which means that a user's attributes must meet at least three attributes in the access policy $\{Chinese, male, under\ 35\ years\ old, Han, cryptography, Beijing, married\}$ in order to have access permissions. Then, represent the access policy with $\{0,1\}$, ensuring that the number of 1 is at least 3, such as $\{1,0,0,1,1,0,1\}$. Secondly, based on the binary random response mechanism, the access policy is randomly perturbed. After differential perturbation, the number of 1 remains unchanged, achieving policy hiding and protecting the user's personal privacy. Finally, in a hardware environment of Inter Core i5-7300HQ 2.50GHz, operating system of Windows 10, and system memory of 16g, experiments were conducted using python language to verify the hiding effect of access policies under different privacy budgets. Among them, the experiment used the publicly available data set-Adult. The experimental results obtained are the average value after running the same algorithm five times. To clarify the error between the encoded perturbed data and the real data, we use Hamming distance as a measure of the error between the perturbation score and the actual score, and the specific experimental results are shown in the Figure. 2.

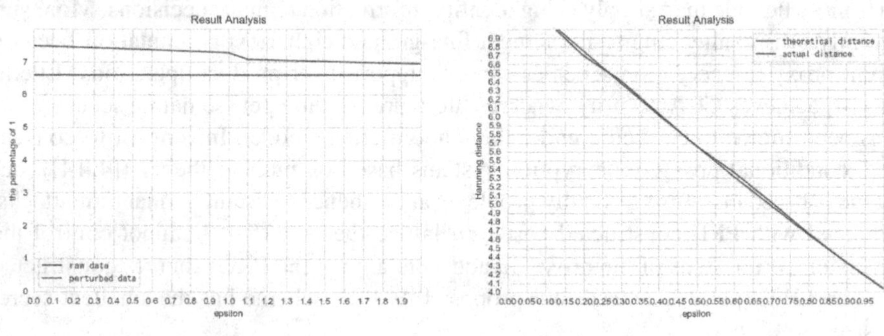

(a) Ratio of 1 under different privacy budgets

(b) Hamming distance under different privacy budgets

Fig. 2. The results of binary random response

Figure. 2-(a) shows that the proportion of 1 in the original data remains unchanged, which means that when the privacy budget is between 0 and 1, the privacy protection effect is better. When the epsilon value is 1.1, the proportion of 1 tends to stabilize, approaching the true value, and can achieve policy hiding. From Figure. 2-(b), it is clear that utilizing differential privacy can achieve policy hiding while ensuring data utility since the Hamming distance of the perturbed data approaches the Hamming distance of the real data.

4 Conclusion

Based on the lattice difficulty problem, we construct an access control inner product function PRE scheme, which can not only resist quantum attacks, but also consider the fine-grained control problem from three aspects. First of all, in order to control the authority of the proxy and delegatee, attribute-based encryption is used to generate ciphertext and re-encrypt key. Secondly, for the proxy, only the ciphertext satisfying the conversion strategy can be re-encrypted. Finally, the delegatee can only obtain the inner product value of the plaintext after decrypting with the function key, which not only protects the privacy of the delegator, but also improves the efficiency of the ciphertext calculation. In addition, the scheme achieves IND-sHRA security in the standard model and has high security. Moreover, we adopt localized differential privacy technology based on random response mechanism to hide access policies to ensure the privacy of user information. In the future, we are going to construct ACPRE based on the RLWE problem on lattice to increase the effectiveness of the scheme.

Acknowledgements. The authors are willing to express our appreciation to the reviewers for their constructive comments which significantly enhanced the presentation of the study.

References

1. Agyekum, K.O.O., Xia, Q., Sifah, E.B., Cobblah, C.N.A., Xia, H., Gao, J.: A proxy re-encryption approach to secure data sharing in the internet of things based on blockchain. IEEE Syst. J. **16**(1), 1685–1696 (2022). https://doi.org/10.1109/JSYST.2021.3076759
2. Blaze, M., Bleumer, G., Strauss, M.: Divertible protocols and atomic proxy cryptography. In: Nyberg, K. (ed.) EUROCRYPT 1998. LNCS, vol. 1403, pp. 127–144. Springer, Heidelberg (1998). https://doi.org/10.1007/BFb0054122
3. Cao, Y., Xu, S., Chen, X., He, Y., Jiang, S.: A forward-secure and efficient authentication protocol through lattice-based group signature in vanets scenarios. Comput. Netw. **214**, 109149 (2022). https://doi.org/10.1016/j.comnet.2022.109149
4. Chen, Z., Chen, J., Meng, W.: A new dynamic conditional proxy broadcast re-encryption scheme for cloud storage and sharing. In: 2020 IEEE International Conference on Dependable, Autonomic and Secure Computing, International Conference on Pervasive Intelligence and Computing, International Conference on Cloud and Big Data Computing, International Conference on Cyber Science and Technology Congress (DASC/PiCom/CBDCom/CyberSciTech), pp. 569–576 (2020). https://doi.org/10.1109/DASC-PICom-CBDCom-CyberSciTech49142.2020.00101
5. Cohen, A.: What about bob? the inadequacy of CPA security for proxy reencryption. In: Lin, D., Sako, K. (eds.) PKC 2019. LNCS, vol. 11443, pp. 287–316. Springer, Cham (2019). https://doi.org/10.1007/978-3-030-17259-6_10
6. Ge, C., Susilo, W., Baek, J., Liu, Z., Xia, J., Fang, L.: A verifiable and fair attribute-based proxy re-encryption scheme for data sharing in clouds. IEEE Trans. Dependable Secur. Comput. **19**(5), 2907–2919 (2022). https://doi.org/10.1109/TDSC.2021.3076580
7. Hörandner, F., Prünster, B.: Armored twins: Flexible privacy protection for digital twins through conditional proxy re-encryption and multi-party computation. In: di Vimercati, S.D.C., Samarati, P. (eds.) Proceedings of the 18th International Conference on Security and Cryptography, SECRYPT 2021, 6-8 July 2021, pp. 149–160. SCITEPRESS (2021). https://doi.org/10.5220/0010543301490160

8. Hua, N., Li, J., Zhang, K., Zhang, L.: A novel deterministic threshold proxy re-encryption scheme from lattices. Int. J. Inf. Secur. Priv. **16**(1), 1–17 (2022). https://doi.org/10.4018/ijisp.310936
9. Kumari, S., Singh, M., Singh, R., Tewari, H.: A post-quantum lattice based lightweight authentication and code-based hybrid encryption scheme for iot devices. Comput. Netw. **217**, 109327 (2022). https://doi.org/10.1016/j.comnet.2022.109327
10. Li, W., Jin, C., Kumari, S., Xiong, H., Kumar, S.: Proxy re-encryption with equality test for secure data sharing in internet of things-based healthcare systems. Trans. Emerg. Telecommun. Technol. **33**(10) (2022). https://doi.org/10.1002/ett.3986
11. Liang, X., Weng, J., Yang, A., Yao, L., Jiang, Z., Wu, Z.: Attribute-based conditional proxy re-encryption in the standard model under LWE. In: Bertino, E., Shulman, H., Waidner, M. (eds.) ESORICS 2021. LNCS, vol. 12973, pp. 147–168. Springer, Cham (2021). https://doi.org/10.1007/978-3-030-88428-4_8
12. Luo, F., Al-Kuwari, S., Susilo, W., Duong, D.H.: Chosen-ciphertext secure homomorphic proxy re-encryption. IEEE Trans. Cloud Comput. **10**(4), 2398–2408 (2022). https://doi.org/10.1109/TCC.2020.3042432
13. Mao, X., Li, X., Wu, X., Wang, C., Lai, J.: Anonymous attribute-based conditional proxy re-encryption. In: Au, M.H., Yiu, S.M., Li, J., Luo, X., Wang, C., Castiglione, A., Kluczniak, K. (eds.) NSS 2018. LNCS, vol. 11058, pp. 95–110. Springer, Cham (2018). https://doi.org/10.1007/978-3-030-02744-5_7
14. Paul, A., Selvi, S.S.D., Rangan, C.P.: A provably secure conditional proxy re-encryption scheme without pairing. J. Internet Serv. Inf. Secur. **11**(2), 1–21 (2021). https://doi.org/10.22667/JISIS.2021.05.31.001
15. Susilo, W., Dutta, P., Duong, D.H., Roy, P.S.: Lattice-based HRA-secure attribute-based proxy re-encryption in standard model. In: Bertino, E., Shulman, H., Waidner, M. (eds.) ESORICS 2021. LNCS, vol. 12973, pp. 169–191. Springer, Cham (2021). https://doi.org/10.1007/978-3-030-88428-4_9
16. Wang, C., Han, Y., Duan, X., Guo, K.Y.: Hierarchical identity-based conditional proxy re-encryption scheme based RLWE and NTRU variant. In: Zeng, J., Qin, P., Jing, W., Song, X., Lu, Z. (eds.) ICPCSEE 2021. CCIS, vol. 1452, pp. 240–259. Springer, Singapore (2021). https://doi.org/10.1007/978-981-16-5943-0_20
17. Wang, X., Wang, Y., Wang, M.: Lattice-based revocable identity-based proxy re-encryption with re-encryption verifiability. In: Wang, L., Segal, M., Chen, J., Qiu, T. (eds.) WASA 2022, Part I. LNCS, vol. 13471, pp. 535–544. Springer, Cham (2022). https://doi.org/10.1007/978-3-031-19208-1_44
18. Weng, J., Deng, R.H., Ding, X., Chu, C., Lai, J.: Conditional proxy re-encryption secure against chosen-ciphertext attack. In: Li, W., Susilo, W., Tupakula, U.K., Safavi-Naini, R., Varadharajan, V. (eds.) Proceedings of the 2009 ACM Symposium on Information, Computer and Communications Security, ASIACCS 2009, Sydney, Australia, 10–12 March 2009, pp. 322–332. ACM (2009). https://doi.org/10.1145/1533057.1533100
19. Wu, L., Yang, X., Zhang, M., Wang, X.A.: IB-VPRE: adaptively secure identity-based proxy re-encryption scheme from LWE with re-encryption verifiability. J. Ambient. Intell. Humaniz. Comput. **13**(1), 469–482 (2022). https://doi.org/10.1007/s12652-021-02911-9
20. Yang, N., Tian, Y., et al.: Identity-based unidirectional collusion-resistant proxy re-encryption from u-lwe. Sec. Commun. Netw. **2023** (2023)
21. Yao, L., Weng, J., Wang, B.: Conditional attribute-based proxy re-encryption and its instantiation. Cryptology ePrint Archive (2022)
22. Zhao, J., Feng, D., Zhang, Z.: Attribute-based conditional proxy re-encryption with chosen-ciphertext security. In: Proceedings of the Global Communications Conference, GLOBECOM 2010, Miami, Florida, USA, 6–10 December 2010, pp. 1–6. IEEE (2010). https://doi.org/10.1109/GLOCOM.2010.5684045

A Smart Glasses-Based Real-Time Micro-expressions Recognition System via Deep Neural Network

Siyu Xiong[1], Xuan Huang[3], Kiminori Sato[2], and Bo Wu[2](✉)

[1] Graduate School of Bionics, Computer and Media Sciences, Tokyo University of Technology, Hachioji 192-0982, Tokyo, Japan
g212302911@edu.teu.ac.jp

[2] School of Computer Science, Tokyo University of Technology, Hachioji 192-0982, Tokyo, Japan
{satohkmn,wubo}@stf.teu.ac.jp

[3] Metaverse Research Institute, Waseda University, Tokorozawa 359-1192, Saitama, Japan
x.huang6@kurenai.waseda.jp

Abstract. The rapid development of new communication technologies such as social software and email has brought convenience, but it has also increased the distance between people. This research aims to capture micro-expressions in real-time through smart glasses and uses deep learning technology to develop a real-time emotion recognition system based on RGB color values to improve interpersonal communication. We trained a multi-layer fully deep neural network (DNN) model using the CASME2 dataset to effectively learn the association between facial expressions and emotions. The experimental results of model show that the emotion classification accuracy of the system reaches 95% on several test samples. The system is adapted to run on smart glasses, the emotion recognition results are immediately fed back to the screen of the smart glasses and displayed to the user in the form of an emotion label. The experimental results of system show that the system can achieve accurate emotion recognition, help users better understand the psychological state of the communication object, and improve the communication environment and quality. In addition, the system can also store communication records and emotion analysis results to help users understand their own communication performance and changes in others' emotions, thus improving communication skills and abilities.

Keywords: Smart Glasses · MediaPipe · Micro expression Analysis · Deep Learning · Improving Communication Skills · Real-Time Emotion Labels

1 Introduction

The rapid development of the new communication technology [1–3] such as social software as well as email has indeed brought people a convenient life [4], but at the same time it has also brought and increased the distance between people [5]. The ability to communicate between people is reduced, people's emotional intelligence and ability

H. Jin et al. (Eds.): GPC 2023, LNCS 14504, pp. 191–205, 2024.
https://doi.org/10.1007/978-981-99-9896-8_13

to perceive emotions cannot be exercised, and even psychological disorders such as social phobia and other social disorders have emerged [6]. In this context, how to use the existing technological means to improve interpersonal communication skills becomes crucial [7].

By observing the facial expressions and emotional changes of others, individuals can gain a deeper understanding of their psychological state [8]. This heightened awareness allows for appropriate responses, fostering positive interpersonal relationships. Micro expressions [9] are instantaneous flashes of facial expression that reveal a person's true feelings and emotions. Although a subconscious expression may only last for a fraction of a second, it can easily reveal emotions [10]. When the face makes a certain expression, these very short expressions can suddenly flash and sometimes express opposite emotions. By observing micro expressions, we gain a more complete understanding of others' emotional states. This knowledge provides new avenues to enhance interpersonal communication.

Simultaneously, the advancement of Internet of Things (IoT) sensors has paved the way for a plethora of innovative wearable devices. These devices include motion capture devices [11, 12], eye trackers [13], and smart glasses [14]. They not only enhance convenience in people's daily lives but also introduce intelligence to their lifestyles. One prime example is wearable smart glasses, which are capable of capturing real-time facial expressions [15]. This feature enables functions like instant photography and video recording. And has an independent development system, can develop a variety of programs according to real-time feedback, users can see the surrounding situation through the screen, you can see the program feedback information.

Therefore, this study focuses on capturing the micro-expressions using wearable smart glasses and attempts to improve communication by developing a real-time emotion recognition system based on the user's micro-expressions using Red, Green, Blue (RGB) color values to determine each other's emotions. The main purpose of using RGB values is to extract color information from key points on the face, which can add more distinguishing features to the model. To be specific, according to build a facial micro expression recognition system, at first, we used a set of micro expressions videos from the CASME2 dataset [16] as training and validation data. The MediaPipe's FaceMesh model [17] is used to transfer the video data to the facial feature data (key facial points), which will be used to train the recognition system by using the deep neural network (DNN).

We designed the recognition system via a smart glass with front camera. According to the video frames data from the glasses via MediaPipe's FaceMesh model, we continuously capture and process facial images and extract facial feature vectors in real-time. At the same time, based on the trained DNN model, the system can realize the purpose of real-time recognition of micro-expressions. Using the system can let the user better understand communicatee's psychological state, to improve the communication environment and the quality of communication.

This study has important research significance and practical application value. By accurately identifying micro-expressions, we can gain a deeper understanding of other people's mental states, including those that are hidden or suppressed, thereby establishing deeper and more authentic human interactions. In addition, the constructed facial micro expression recognition system is real-time and accurate and can perform micro expression analysis in real-time video without interfering with the observer. In conclusion, this study provides a new approach and practical application for facial micro expression recognition system to improve interpersonal communication, making our facial micro expression recognition system more accurate and practical.

2 Related Works

In this section, we present research on the use of smart glasses and emotion related techniques that can demonstrate accurate and efficient communication through micro-expressions using wearable devices.

2.1 Micro Expression Emotion Recognition

Bin Xia et al. proposed a method for recognizing micro-expressions using macro-expression samples as a guide. An expression identity separation network was also proposed to extract expression substrates from expression images without identity related information. Extensive experimental expressions were performed on three more public spontaneous micro-expression datasets (i.e., SMIC, CASME2, and SAMM), this framework due to advanced micro-expression recognition system based on manual or deep features [18].

Also, a comprehensive survey and analysis of various micro-expression recognition system by Madhumita Takalkar et al. analyzed the general framework of micro-expression recognition system and decomposed it into basic components, i.e., face detection, preprocessing, facial feature detection and extraction, dataset, and classification. The roles of these elements are also discussed and the models and emerging trends they follow in their design are highlighted. In addition, a comprehensive analysis of micro-expression recognition systems is presented by comparing their performance [19].

Kamlesh Mistry's core research method is a feature selection method based on micro genetic algorithms embedded with particle swarm optimization algorithms for intelligent facial emotion recognition. The method uses a PSO algorithm for feature selection, which is performed separately for each emotion category to identify the distinguishing features of each expression. At the same time, neural networks, multi class support vector machines and integrated classifiers are used for emotion recognition. When tuning the classifier, they use trial and error and grid search methods to find the optimal parameters. In addition, they used methods such as small population concepts and sub-dimensional search strategies to speed up the evolution and improve the search efficiency. Experimental results show that their proposed method is significantly superior to other PSO variants and traditional methods in global optimal search and discriminative feature selection [20].

2.2 Judging Emotions Through Smart Glasses

Lin Zixiang et al. When using smart glasses, they can be used to improve communication through their features, namely the ability to follow the wearer's gaze and microphone, and the available augmented reality (AR) capabilities. The emotion recognition system on smart glasses is designed and constructed. Cameras and microphones capture the facial expressions, voice emotions, and intonation of individuals to enhance communication. These captured data are then analyzed for emotion recognition and analysis [21].

YINGYING ZHAO's team uses deep learning methods to implement relationship analysis. They designed a smart glasses system, called EMOShip, equipped with a deep neural network, Emoship-Net, to extract emotion-related features and emotional cues in visual attention and fuse them together to identify emotions and quantify emotional relationships more accurately. Specifically, they used convolutional neural networks (CNNS) and gated cycle units (GRUs) for feature extraction, and multi-task learning for simultaneous emotion recognition and image labelling. They also conducted field trials to verify the performance and usability of EMOShip [22].

Mukhriddin Mukhiddinov's core research method is a mask facial emotion recognition method based on facial markers and deep learning. The researchers used the AffectNet dataset for training and evaluation and applied a specific facial signature model (MediaPipe face mesh method) to identify facial emotions. They created a convolutional neural network (CNN) model with feature extraction, full join and SoftMax classification layers, and used Mish activation functions in each convolutional layer. Using this method, they achieved an accuracy of 69.3% and an average confusion matrix of 71.4%. However, the method has certain limitations in the case of different orientations and cannot correctly identify the emotion of multiple faces present in the same image. To further improve the classification model and image datasets, the researchers plan to explore methods such as semi-supervised and self-supervised learning and investigate how to generate the attentional parts of faces without facial markers. They also plan to work on the hardware side of smart glasses, creating a prototype of a device that can help visually impaired people identify people, places, and objects in near environments [23].

3 Methods

3.1 Face Point Detection and Feature Extraction

As shown in Fig. 1, face key point detection and feature extraction is a process of extracting facial feature points from face images by using the FaceMesh module of the MediaPipe framework to process each frame in the video.

The steps of feature extraction are as follows. For each frame in the video, the position of the facial feature points is detected by MediaPipe's FaceMesh module to obtain the (x, y) coordinates of each feature point in the image.

The positional information of these feature points is utilized to acquire the RGB color values of their respective pixel points. RGB color values consist of three components, namely red, green, and blue, which depict the color information for each pixel in the image. Each feature point's RGB color value is considered as a feature vector, and all feature vectors are merged to form a comprehensive set of features.

The feature set extracted from each video frame is stored in a data set for subsequent training and evaluation of the emotion classification model.

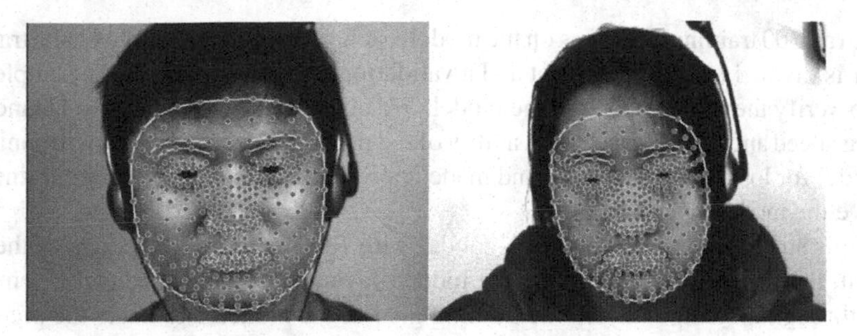

Fig. 1. MediaPipe's FaceMesh module

3.2 Deep Learning Model

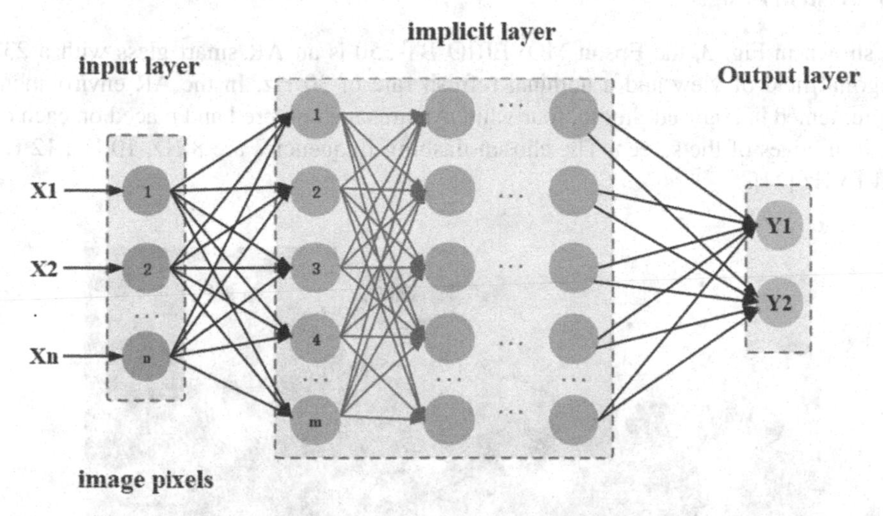

Fig. 2. Deep neural network (DNN)

As shown in Fig. 2, we design a multi-layer fully connected neural network (DNN) model to extract emotional features from color information of facial key points. Specifically, the output layer is encoded with RGB color and mood labels for facial key points. And a sequential model with multiple hidden layers is used. First, we use a 64-node hidden layer with ReLU (modified linear unit) as the activation function to capture the initial features. Next, we add two 32-node hidden layers that can further extract abstract representations of emotion from the color information. To increase the expressiveness of the model, we again use a 64-node hidden layer in the middle of the network. Finally, we take an output layer with the same number of nodes as the number of emotion classes and use the SoftMax activation function for multi-class emotion classification.

In the training process of the model, we use the Adam optimizer to adjust the weight of the model. We chose binary cross entropy (binary_crossentropy) as the loss function to measure the difference between the model prediction and the true label. We also set the learning rate of the optimizer to 0.001 to control the step size of each weight update.

We ran 300 training iterations on the model, each with a batch size of 64. The training dataset is divided into a training set and a validation set, where 20% of the samples are used to verify the performance of the model. We chose these parameters to balance the training speed and the generalization ability of the model. In each iteration, we monitored the model for losses and accuracy, and made appropriate adjustments during training to improve the model's performance.

In this study, we introduce a DNN model with RGB color of key points of the face as input, learns the features through the hidden layer, and finally predicts the emotion labels through the output layer. The training goal of the model is to make the predicted emotion labels as close to the real ones as possible, so that the model can accurately recognize facial micro-expressions and predict the corresponding emotions.

3.3 System Design

As shown in Fig. 3, the Epson MOVERIO BT-350 is an AR smart glass with a 23° diagonal field of view and a nominal refresh rate of 30 Hz. In the AR environment implemented in Android Studio, four white squares are rendered and placed on each of the four edges of the screen. The chosen flashing frequencies are 8 Hz, 10 Hz, 12 Hz and 15 Hz [24].

Fig. 3. Epson MOVERIO BT-350 Smart Glasses and BT-350 Glasses Exclusive Controller

As shown in Fig. 4, the contained face is first obtained by the data input module, and then facial features are extracted by the face feature extraction module. Then, the deep neural network-based emotion recognition module analyses the extracted features and predicts people's emotional states. The emotion recognition results are fed back to the smart glasses screen in real time and displayed to the system user in the form of an "emotion label". At the same time, users can provide feedback, which is considered and used to improve the system through user feedback and customization modules.

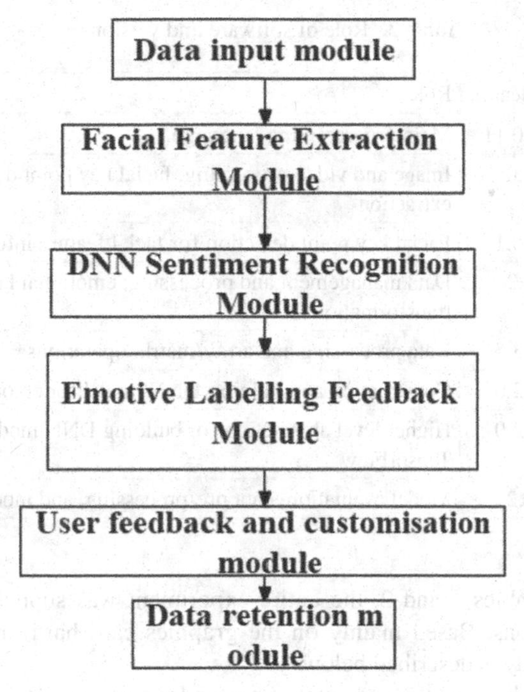

Fig. 4. The structure chart of this study

4 Model Training

4.1 Training Environment and Dataset

This study uses the CASME2 dataset, which contains rich micro-expression data with higher temporal resolution (200 fps) and spatial resolution (approximately 280 × 340 pixels in the facial region). Participants' facial expressions were recorded in a well-controlled laboratory environment with appropriate lighting (e.g., removal of light flicker). From nearly 3000 facial movements, 247 micro-expressions were selected as a database and tagged with movement units and emotions.

Table 1. Hardware

Hardware	Version
CPU	I5 13400F
Memory	XPG DDR4 3200 MHz 8 * 2
Solid state drive	Kingston NV2 500 GB PCIE4.0
Graphics card	PALIT RTX 3060 TI 8 GB
Operating system	WIN10 64 bit

Table 2. Role of software and versions

Software Name	Releases	Role
Python	3.10.11	Main programming language
OpenCV	4.7.0	Image and video processing, facial key point detection and feature extraction
MediaPipe	0.10.1	Facial key point detection for facial feature information
Pandas	2.0.2	Data management and processing, emotional label processing and transformation
NumPy	1.23.5	Data processing and array/matrix operations
TensorFlow	2.12.0	Constructing and training DNN model for emotion recognition
Keras	2.12.0	Higher level abstraction for building DNN models with TensorFlow
Scikit-learn	1.2.2	Model evaluation, data pre-processing, and model optimization

As shown in Tables 1 and 2, the entire experiment was supported by the above software and versions. Based mainly on the graphics card hardware, the role of all software in this study is described below.

In this study, Python was the main programming language used to develop and implement the system, which used OpenCV and MediaPipe for image and video processing, facial key point detection and feature extraction. In addition, Pandas and NumPy were used for dataset management, emotional label processing and data manipulation, while TensorFlow and Keras were used to construct and train a DNN model for emotion recognition. Scikit-learn was used for model evaluation, data pre-processing and optimization tasks.

4.2 Data Preprocessing

The program first prepares the data by reading the video files in the video folder and the associated emotion tags. The video file path is stored in a variable, while the emotion tag is retrieved from the document file.

Next, the program uses the MediaPipe framework for sentiment analysis and sentiment detection. To do this, it calls the FaceMesh module for face key point detection.

The program then reads the video file and processes it frame by frame. It analyses each frame; extracts face key point coordinates and stores RGB color values in a data point dictionary. At the same time, it adds emotional labels to the data points that correspond to the detected facial features. Finally, the extracted facial key features and corresponding emotional labels are stored in the form file.

4.3 Judging Method of Emotion Label

In the judging method of emotion label, we use the supervised learning method. First, we prepared a data set containing the tagged emotion label data and the RGB color value information of the facial feature points extracted by the FaceMesh module. Each sample consists of RGB color value information for facial feature points and corresponding emotion labels.

As shown in Fig. 5, to enable deep learning models to automatically recognize associations between facial expressions and emotions, we map emotion labels to discrete emotion categories. We classify emotions into seven categories based on the labels in the dataset: "happiness", "disgust", "depression", "surprise", "fear", "sadness "and "other". This mapping allows the model to classify seven emotions based on facial expressions.

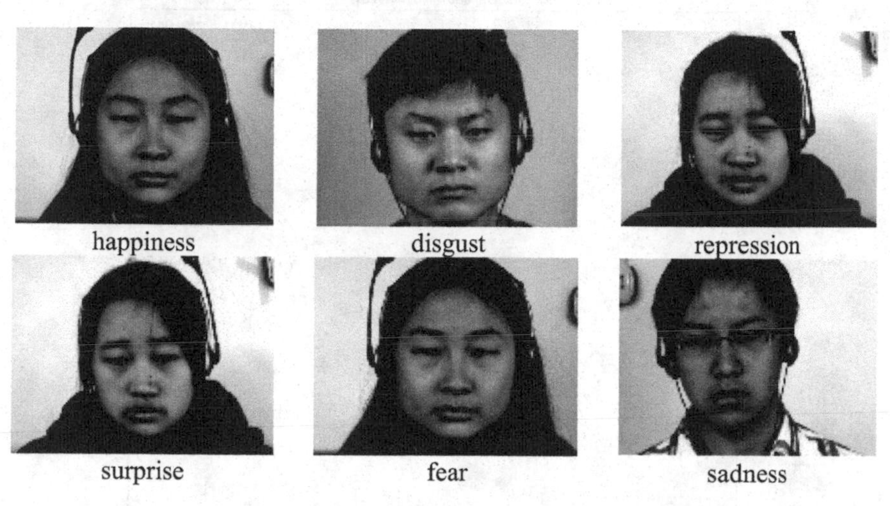

Fig. 5. Emotion Recognition

Finally, we output the extracted RGB color value information of the facial feature points and the corresponding emotion labels and save them as CSV files for subsequent model training and evaluation.

4.4 Model Training and Optimization

In the model training and optimization phase, we use the CASME2 dataset to train the deep learning model. We chose the cross-entropy loss function as the loss function of the model, which measures the difference between the model's predictions and the true emotion label. At the same time, we use the Adam optimizer to optimize the model parameters to improve the model's convergence speed and performance.

To further improve the performance of the model, we use a learning rate scheduler. The learning rate scheduler can automatically adjust the learning rate and dynamically adjust the learning rate according to the performance of the model during the training process, which helps the model to converge faster and avoid falling into the local optimal solution.

Through continuous iteration and optimization, the deep learning model gradually learns the correlation between facial expressions and emotions and can accurately predict the corresponding emotion category based on the RGB color value information of facial feature points.

The goal of model training and optimization is to enable the model to perform emotion classification of facial expressions in real-time scenes and achieve accurate and stable prediction results. At the same time, the optimized model has good generalization ability and can adapt to the changes of different faces and emotional expressions, to achieve better results in practical applications.

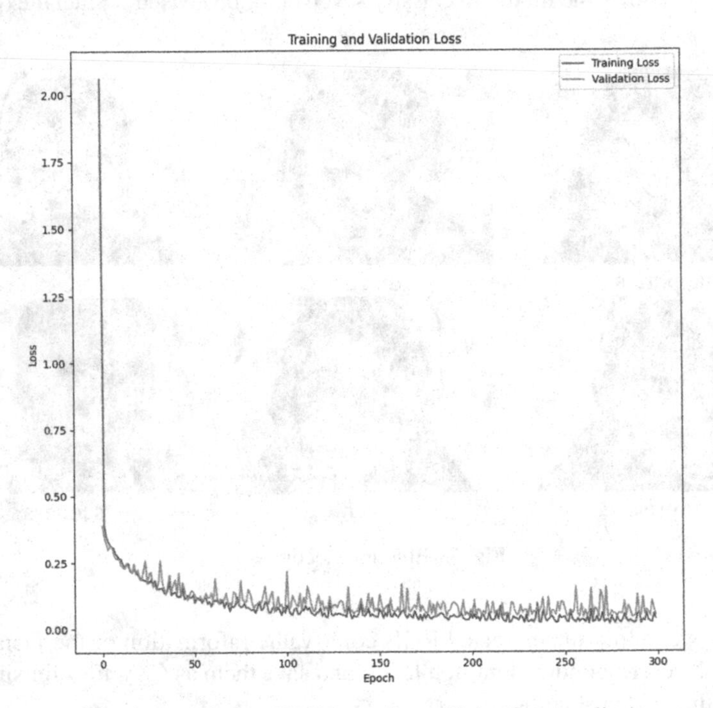

Fig. 6. Training and Validation Loss (Color figure online)

As shown in Fig. 6, the accuracy, precision and recall of this emotion classification model on the CASME2 dataset are all 95%. In the two graphs, the red line represents the change in the performance index (loss and accuracy) of the training set data with the training rounds, while the blue line represents the change in the performance index of the validation set data with the training rounds.

For the Training Loss and Validation Loss curves, the red line represents the Training Loss, i.e., the loss value of the model on the training set, which measures the predictive effect of the model on the training set, i.e., the difference between the prediction result of the model and the actual label of the training set. The blue line represents the validation loss, the value of the model's loss on the validation set, which measures the model's ability to generalize on the validation set, i.e., the model's adaptability to new data. As training progresses, the red line gradually decreases, indicating that the model is gradually optimized on the training set, learning better features and laws. At the same time, the blue line also decreases, indicating that the model is gradually generalizing on the validation set and can predict the previously unseen data. However, if the blue line bounces or oscillates at a later stage, it may indicate that the model is beginning to overfit, remembering too much of the noise in the training set, leading to poor performance on the validation set.

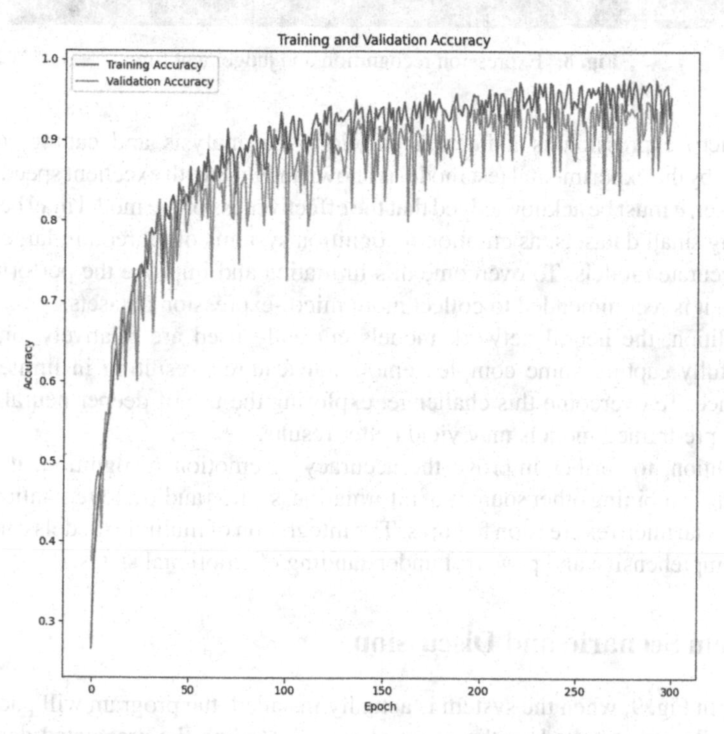

Fig. 7. Training and Validation Accuracy (Color figure online)

As shown in Fig. 7, for the Training Accuracy and Validation Accuracy curves, the red line represents Training Accuracy, the model's prediction accuracy on the training set, which represents the proportion of training set samples that the model correctly predicts. The blue line represents Validation Accuracy, which is the accuracy of the model's predictions on the validation set and represents the proportion of the model that correctly predicts the validation set sample. As training progresses, the red line gradually rises, indicating that the accuracy of the model on the training set continues to improve. At the same time, the blue line also rises, indicating that the accuracy of the model on the verification set is gradually increasing. If the blue line drops or flattens at a later stage, this may indicate that the model is beginning to overfit, learning too much about the details of the training set, resulting in poor performance on the validation set.

By using the deep neural network, the complex feature representation is effectively learned, the classification performance of the model is improved, and the emotion recognition system achieves a high accuracy rate. The system uses the MediaPipe framework for face key point detection, which ensures efficient extraction of face features and provides valuable input for emotion recognition.

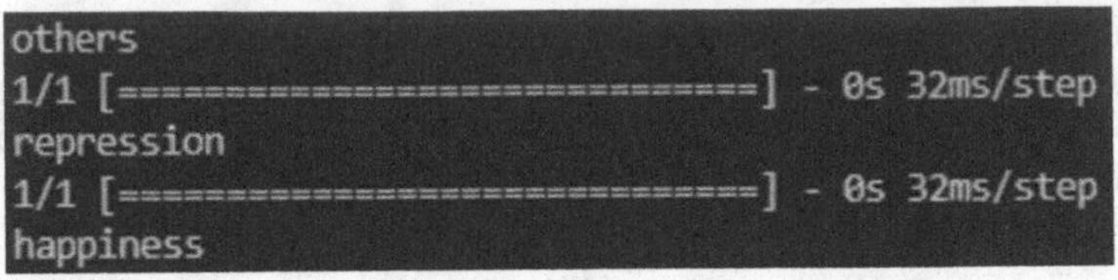

```
others
1/1 [==============================] - 0s 32ms/step
repression
1/1 [==============================] - 0s 32ms/step
happiness
```

Fig. 8. Expression recognition and judgement time

As micro expression is a fleeting emotion, the analysis and capture of micro expression by the experimental test model is shown in Fig. 8 with excellent speed (32 ms).

However, it must be acknowledged that the effectiveness of the model may be reduced when using small datasets, as emotion recognition systems often require large datasets to train accurate models. To overcome this limitation and improve the performance of the model, it is recommended to collect more micro-expression datasets.

In addition, the neural network models currently used are relatively simple and may not fully capture some complex emotional features, resulting in limited model performance. To overcome this challenge, exploring the use of deeper neural network models or pre-trained models may yield better results.

In addition, to further improve the accuracy of emotion recognition, it is worth considering combining other sources of information, such as audio and semantic features, with key facial micro expression features. The integration of multiple models can provide a more comprehensive and powerful understanding of emotional states.

5 System Scenario and Discussion

As shown in Fig. 9, when the system is actually installed, the program will packaged as an ".apk" file and installed on the smart glasses, including the associated dependency libraries and the emotion recognition model files. While chatting with each other, system user can use the smart glasses and activate a mood analysis program so that the smart glasses capture each other's facial expressions through the camera. Through the system, sentiment analysis program can analyze the micro-expressions in real time and translates them into sentiment labels that visualizes the results on the display of the smart glasses (as shown in Fig. 10). The system can optimization and adjustment improve the communication environment in real-time, which can make it easier for both sides to understand and trust each other and achieving better communication results.

On the other hand, smart glasses also can store communication records and sentiment analysis results for users to view and analyze later. Users can understand their own communication performance and the other party's mood changes according to the stored records, thus improving communication skills and enhancing communication ability.

At the same time, the user can provide feedback on the interface for each recognition result, such as marking which recognition is correct and which is incorrect or providing additional information such as the correct emotional label. Feedback data is then collected from ethical users, which may include user markups, comments, or other forms of feedback. Incorporate the user feedback data into the model's training data. The model is retrained using new data that incorporates user feedback. This allows the model to

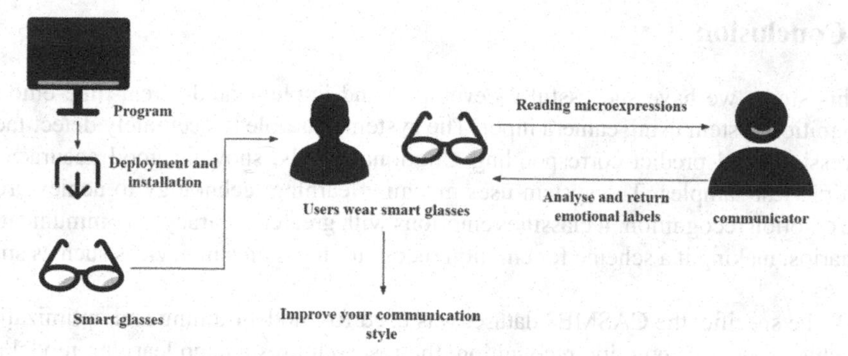

Fig. 9. Experimental environment

learn from user feedback and improve its performance over time. After the model has been retrained, evaluate the performance of the model to see if there is a significant improvement.

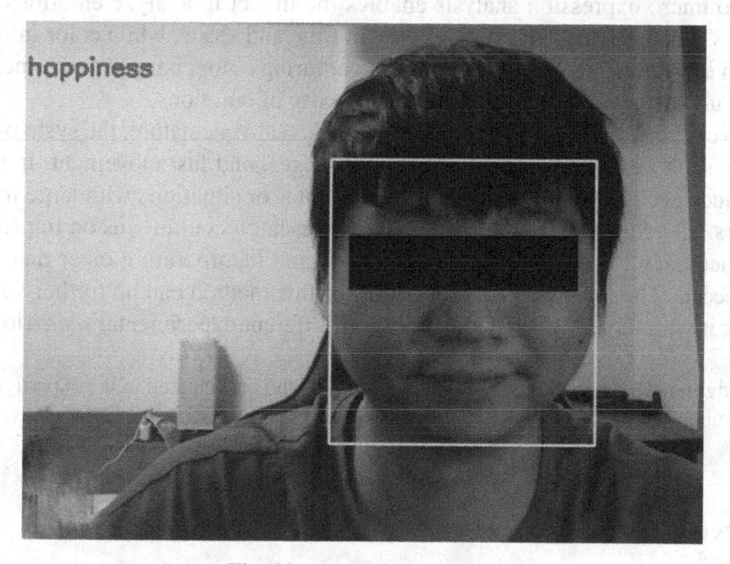

Fig. 10. Real-time interface

In conclusion, the implemented emotion recognition system shows good performance, but to achieve higher accuracy and effectiveness, it is necessary to further expand the dataset, explore more advanced model architectures, and combine complementary information from different modes.

6 Conclusion

In this study, we have successfully developed and implemented a real-time emotion recognition system using camera input. The system was able to accurately detect facial expressions and predict corresponding emotional labels, showing good accuracy on multiple test samples. The system uses machine learning technology to achieve real-time emotion recognition. It classifies emotions with greater accuracy in communication scenarios, making it a scheme for emotion recognition systems in devices such as smart glasses.

To be specific, the CASME2 dataset was used for model training and optimization. To achieve accurate emotion recognition, the research uses a deep learning model and designs a multi-layer fully connected neural network (DNN) model. After experimental verification, the emotion recognition system performs well, and the emotion classification accuracy rate for multiple test samples reaches 95%. By applying this approach to devices such as smart glasses, it can enhance interpersonal communication, thereby improving efficiency in chat environments.

Compared to traditional emotion classification methods, the introduction of RGB values into micro expression analysis enables the model to analyze emotions in more dimensions. Micro expressions tend to vary in time and space, while color information can add an extra dimension to the model. By capturing color changes over time, models can better understand the dynamics and complexity of emotions.

However, there are some limitations. Due to its real-time nature, the system can face challenges such as facial occlusion, lighting changes, and fast movement. In addition, it may reduce accuracy for complex emotional states or situations with large individual differences. To address these limitations, future research could focus on improving the system's accuracy for complex emotional states and incorporating other data, such as tone of speech. The feasibility and reliability of this method can be further verified by evaluating it with larger affective data sets and different experimental scenarios.

Acknowledgments. Appreciations for all the workers who participated in the experiments. This work was supported by JSPS KAKENHI Grant Number 21K11876.

References

1. Zhou, X., Li, S., Li, Z., Li, W.: Information diffusion across cyber-physical-social systems in smart city: a survey. Neurocomputing **444**, 203–213 (2021). https://doi.org/10.1016/j.neucom.2020.08.089
2. Zhou, X., Jin, Q.: A heuristic approach to discovering user correlations from organized social stream data. Multimedia Tools Appl. **76**(9), 11487–11507 (2017). https://doi.org/10.1007/s11042-014-2153-5
3. Zhou, X., Wang, W., Jin, Q.: Multi-dimensional attributes and measures for dynamical user profiling in social networking environments. Multimedia Tools Appl. **74**(14), 5015–5028 (2015). https://doi.org/10.1007/s11042-014-2230-9
4. Sevük, T.: The influence of Facebook on interpersonal communication. Eastern Mediterranean University (EMU)-Doğu Akdeniz Üniversitesi (DAÜ) (2013)

5. Safina, A.M., Leontyev, G.D., Gaynullina, L.F., et al.: Dialectics of freedom and alienation in the space of the internet. Revista ESPACIOS **39**, 27 (2018)
6. Weinstein, A., Dorani, D., Elhadif, R., et al.: Internet addiction is associated with social anxiety in young adults. Ann. Clin. Psychiatry **27**(1), 4–9 (2015)
7. Zeng, Z., Pantic, M., Roisman, G.I., Huang, T.S.: A survey of affect recognition methods: audio, visual, and spontaneous expressions. IEEE Trans. Pattern Anal. Mach. Intell. **31**(1), 39–58 (2009)
8. Shiota, M.N., Campos, B., Keltner, D., et al.: Positive emotion, and the regulation of interpersonal relationships. Regulat. Emotion **68** (2004)
9. Li, X., Pfister, T., Huang, X., Zhao, G., Pietikäinen, M.: A spontaneous micro-expression database: inducement, collection, and baseline. In: FG 2013, pp. 1–6. IEEE (2013)
10. Frank, M.G., Svetieva, E.: Microexpressions and deception. In: Mandal, M., Awasthi, A. (eds.) Understanding Facial Expressions in Communication: Cross-cultural and Multidisciplinary Perspectives, pp. 227–242. Springer, New Delhi (2015). https://doi.org/10.1007/978-81-322-1934-7_11
11. Wu, B., Wu, Y., Dong, R., et al.: Behavioral analysis of mowing workers based on hilbert-huang transform: an auxiliary movement analysis of manual mowing on the slopes of terraced rice fields. Agriculture **13**(2), 489 (2023)
12. Wu, B., Wu, Y., Nishimura, S., Jin, Q.: Analysis on the subdivision of skilled mowing movements on slopes. Sensors **22**(4), 1372 (2022)
13. Wu, B., Zhu, Y., Yu, K., Nishimura, S., Jin, Q.: The effect of eye movements and culture on product color selection. HCIS **10**(48), 2020 (2020)
14. Cirani, S., Picone, M.: Wearable computing for the internet of things. IT Prof. **17**(5), 35–41 (2015)
15. Khan, S., Javed, M.H., Ahmed, E., et al.: Facial recognition using convolutional neural networks and implementation on smart glasses. In: 2019 International Conference on Information Science and Communication Technology (ICISCT), pp. 1–6. IEEE (2019)
16. Yan, W.J., Li, X., Wang, S.J., et al.: CASME II: an improved spontaneous micro-expression database and the baseline evaluation. PLoS ONE **9**(1), e86041 (2014)
17. Lugaresi, C., Tang, J., Nash, H., et al.: Mediapipe: a framework for building perception pipelines. arXiv preprint arXiv:1906.08172 (2019)
18. Xia, B., Wang, W., Wang, S., et al.: Learning from macro-expression: a micro-expression recognition framework. In: Proceedings of the 28th ACM International Conference on Multimedia, pp. 2936–2944 (2020)
19. Takalkar, M., Xu, M., Wu, Q., et al.: A survey: facial micro-expression recognition. Multimedia Tools Appl. **77**(15), 19301–19325 (2018)
20. Mistry, K., Zhang, L., Neoh, S.C., et al.: A micro-GA embedded PSO feature selection approach to intelligent facial emotion recognition. IEEE Trans. Cybern. **47**(6), 1496–1509 (2016)
21. Lin, T.H., Huang, L.A., Hannaford, B., et al.: Empathics system: application of emotion analysis AI through smart glasses. In: Proceedings of the 13th ACM International Conference on PErvasive Technologies Related to Assistive Environments, pp. 1–4 (2020)
22. Zhao, Y., Chang, Y., Lu, Y., et al.: Do smart glasses dream of sentimental visions? Deep emotionship analysis for eyewear devices. Proc. ACM Interact. Mob. Wearable Ubiquit. Technol. **6**(1), 1–29 (2022)
23. Mukhiddinov, M., Djuraev, O., Akhmedov, F., et al.: Masked face emotion recognition based on facial landmarks and deep learning approaches for visually impaired people. Sensors **23**(3), 1080 (2023)
24. Arpaia, P., De Benedetto, E., De Paolis, L., et al.: Highly wearable SSVEP-based BCI: performance comparison of augmented reality solutions for the flickering stimuli rendering. Meas. Sens. **18**, 100305 (2021)

Pervasive and Green Computing

Genetic-A* Algorithm-Based Routing for Continuous-Flow Microfluidic Biochip in Intelligent Digital Healthcare

Huichang Huang[1,3], Zhongliao Yang[1,3], Jiayuan Zhong[1,3], Li Xu[2], Chen Dong[1,3](✉), and Ruishen Bao[1,3]

[1] College of Computer and Data Science, Fuzhou University, Fuzhou 350116, China
dongchen@fzu.edu.cn
[2] College of Computer and Cyber Security, Fujian Normal University, Fuzhou 350007, China
[3] Fujian Key Laboratory of Network Computing and Intelligent Information Processing, Fuzhou University, Fuzhou, China

Abstract. In the field of intelligent digital healthcare, Continuous-flow microfluidic biochip (CFMB) has become a research direction of widespread concern. CFMB integrates a large number of microvalves and large-scale microchannel networks into a single chip, enabling efficient execution of various biochemical protocols. However, as the scale of the chip increases, the routing task for CFMB becomes increasingly complex, and traditional manual routing is no longer sufficient to meet the requirements. Therefore, this paper proposes an automatic routing framework for CFMB based on Genetic algorithm (GA) and A* algorithms. Specifically, we adopt a two-stage A* algorithm to design the routing between modules, using the routing results obtained from the A* algorithm as the basis for evaluating the quality of solutions in the GA algorithm. Then, the GA algorithm is used to search for the optimal approximate solution in the solution space. Experimental results show that this method can reduce routing length and minimize routing crossings, thereby improving the parallel transmission speed of reagents on CFMB. This approach provides a feasible solution for large-scale automated routing of CFMB in the field of intelligent digital healthcare.

Keywords: Continuous-flow microfluidic biochip · A* algorithms · Routing · Genetic algorithm · Intelligent Digital Healthcare

1 Introduction

The past decade has witnessed rapid development in flow-based microfluidic technology, which has had a profound impact on biochemical experiments in traditional laboratories. This technology integrates large-scale microvalves and microchannel networks into a single chip [1], enabling the integration of various complex biochemical experiments. Microfluidic technology has evolved from

H. Jin et al. (Eds.): GPC 2023, LNCS 14504, pp. 209–223, 2024.
https://doi.org/10.1007/978-981-99-9896-8_14

simple basic components such as channels [2], valves and pumps [3] to chips with integrated large-scale components [1,4,5]. CAD technology, the same technology used in the design and manufacturing of microelectronic circuits, is employed for the design and fabrication of these devices.

Continuous-flow microfluidic biochip (CFMB) is a technology that allows for the manipulation of fluids and suspended objects at the nanoscale in micro-scale channels. Due to its advantages of automation, low cost, portability, and efficiency, CFMB has been widely applied in the field of intelligent digital healthcare, such as disease diagnosis [6], real-time DNA sequencing [7], and antigen detection [8]. Furthermore, several studies have been conducted on compound-protein interactions by deep learning [9]. This technology has had a profound impact on advancements in the field of intelligent digital healthcare.

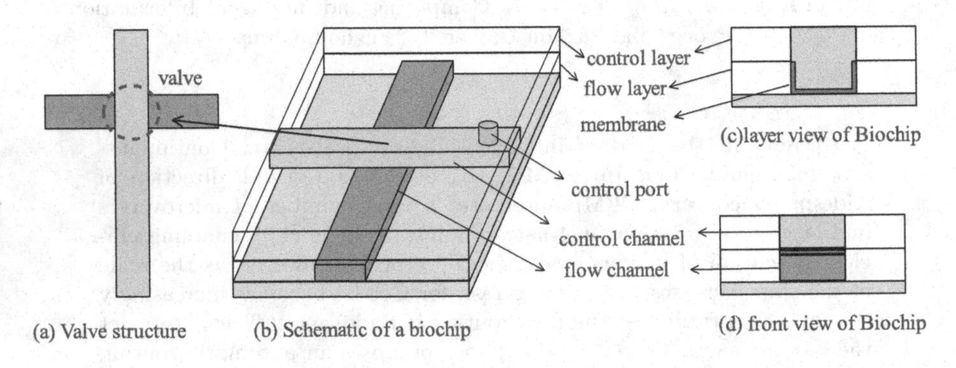

Fig. 1. An overview of Continuous-flow microfluidic biochip.

The basic microfluidic device consists of a two-layer elastomeric structure, with each layer containing its own network of channels, as shown in Fig. 1. The channels in the control layer are colored yellow, while the channels in the flow layer are colored blue. There are overlapping regions between the two layers of channels, forming a flexible membrane, known as a microvalve. The biochemical fluids (also known as the reaction liquid) flow in the flow layer, while the control layer prevents further flow of fluid by applying pressure to bend the membrane valves into the flow channels. The control layer can be located above or below the flow layer and is called a push-up valve or a push-down valve, respectively. These channels are connected by an external pressure pump, and the power provided by the pressure pump is used to drive the liquid flow and the movement of the membrane valve [10]. Connections to external ports, including fluidic ports and pressure sources, are made through holes on the chip, forming fluidic layer pins and control layer pins, enabling access to the fluidic and control layers. External tubes can be connected to the chip via pins to connect to external fluid reservoirs, pumps, or pressure sources [4]. All input ports of the control layer and flow layer channels were connected to external pumps. By combining multiple basic structures, more complex components such as mixers, micropumps, multiplexers, etc. can be constructed, and hundreds of components can be accommodated on a single chip. This approach is known as microfluidic large-scale integration [1,4].

However, as the complexity of biochips continues to increase, traditional manual design methods [11] for pipeline routing are no longer able to meet the demands of chip design. This approach requires a significant amount of manpower and expertise. Additionally, due to the unique physical structure of CFMB itself, the routing methods used in ultra-large-scale integrated circuits cannot be directly applied to CFMB. To address the aforementioned issues, this paper designs a routing method suitable for CFMB modules based on the physical structure characteristics of CFMB. The main contributions of this paper are as follows:

1. A two-stage A* algorithm for CFMB routing is proposed. By improving the A* algorithm, a unique routing algorithm suitable for CFMB is obtained, which considers minimizing the number of crossover points during the routing stage.
2. A Genetic algorithm (GA) algorithm is proposed to find better routing results. By using the routing results obtained from the two-stage A* algorithm as the basis for evaluating the quality of particle solutions in the GA algorithm, better solutions can be discovered.
3. A CFMB routing design framework is proposed. By combining the GA algorithm with the A* algorithm to form the GA* algorithm, the routing design stage is implemented, and its effectiveness is verified through experiments.

The rest of this paper is organized as follows: Sect. 2 provides a review of related literature in the field. Section 3 provides a detailed description of the routing problem in CFMB. Section 4 presents the proposed method in detail. Section 5 presents the experimental results of the algorithm and provides an analysis. Finally, Sect. 6 concludes the paper.

2 Related Works

In the field of physical design research, [12] addressed congestion issues in biochemical protocols by proposing a feasibility-based flow routing algorithm. This algorithm considered fluid constraints and minimized the analysis completion time on PMD, effectively resolving fluid constraint problems on PMD while minimizing the experimental completion time and ensuring feasibility. [13] addressed the gap between the layout and routing stages by introducing an innovative physical design method. This method significantly reduced the number of crossover points and microvalves on the chip, thereby improving chip performance and reliability and achieving a close connection between the routing and layout stages. Additionally, [14] utilized a powerful solving engine and appropriate pruning strategies to traverse the physical design solution space as much as possible, aiming to obtain approximate optimal solutions. [15] proposed a time-delay-driven flow layer channel network construction scheme to enhance real-time efficiency.

In terms of the control layer, [16] also proposed a top-down synthesis approach that generates control layer solutions in two steps. Firstly, the control logic is determined, which is the driving sequence for each valve. Then, considering the compatibility of the driving sequences between valves, multiple valves share a control port to minimize the number of control ports required in the control layer. Taking into account the inherent physical phenomenon in the control layer, specifically the varying lengths of control pipes in the valves sharing a control port, which results in inconsistent arrival times of pressure, potentially leading to erroneous execution of reagents in the flow layer. Therefore, [10] imposed certain constraints on propagation delay during the routing process to achieve synchronized state changes in valve combinations. [17] proposed a minimum spanning tree-based algorithm to address this issue.

According to the principle of the bucket effect mentioned in [18], in a chip, if a valve malfunctions, it is highly likely to cause the entire chip to lose functionality or even produce incorrect results. Therefore, several valve switching modes were designed [19] with the aim of reducing the maximum activation frequency of valves. Iterative designs were carried out in [20–22] using a synthesis tool called Columba, which employed multiplexer synthesis to achieve efficient and reconfigurable valve control, enabling the realization of designs ranging from simple to complex.

Similar to some path planning problems, [23] proposes an improved deep reinforcement learning navigation strategy, which enables robots to learn navigation and collision avoidance strategies more accurately. [24, 25] Some heuristic algorithms also show their superiority in the path-planning process. In addition, the security of chips has been extensively studied [26]. Due to the specific characteristics of CFMB's physical structure, the traditional VLSI routing method is no longer applicable. Therefore, in this paper, we propose and design a routing method suitable for CFMB modules. This method minimizes the number of crossover points in the pipelines while shortening the routing length as much as possible.

3 Preliminaries

This section will describe the routing problem of CFMB physical design in detail and formulate the problem.

3.1 Problem Description

The design task of CFMB mainly involves two key stages: high-level synthesis and physical design. The goal is to generate a synthesis architecture that meets the requirements for executing biochemical protocols, while adhering to certain constraints and optimization objectives. Among them, the physical routing of CFMB is an important part of the physical design stage and is closely related to the component allocation and scheduling tasks in high-level synthesis.

During the high-level synthesis design process, once the binding relationships between components and operations, as well as the scheduling order of operations, are determined, we can derive the connectivity between components. Based on these connectivity relationships and the layout solution, the task in the routing stage is to concretize the connectivity between components onto the routing of each flow channel.

During the routing process, it is essential to minimize the occurrence of crossover points between flow channels. This is because an increase in the number of crossovers will inevitably require the use of switches to control the flow direction at those points, resulting in an increased number of valves and added complexity in the design of the control layer. On the other hand, it is also necessary to reduce the length of the routing mesh as much as possible. As the length of the flow channels increases, the transmission time also increases, thereby reducing the efficiency of executing biochemical protocols. Based on the binding and scheduling solutions depicted in Fig. 2(a) and the layout solution shown in Fig. 2(b), we can deduce that five pairs of connections need to be routed. One possible routing scheme is shown in Fig. 2(c).

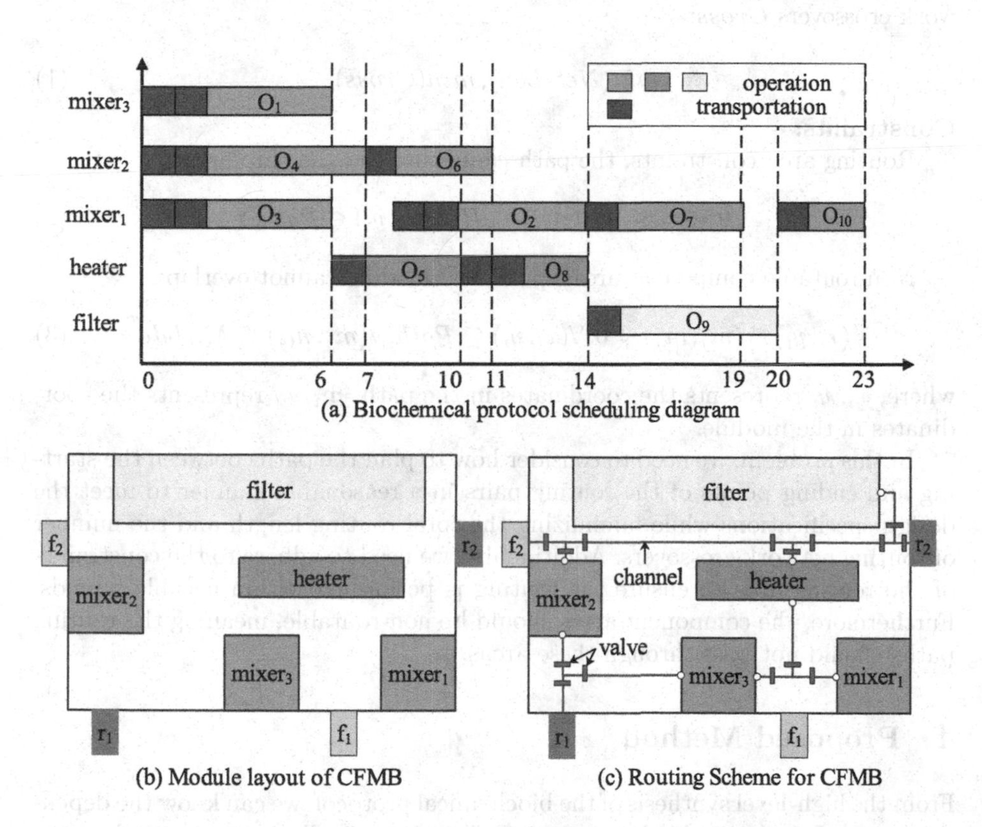

Fig. 2. Routing tasks for Continuous-Flow microfluidic biochip.

3.2 Problem Formulation

The physical routing problem between CFMB resource modules can be defined as follows:

Input:

Routing pairs information: Includes the starting and ending points of each routing pair (s_i, t_i); The number of transmissions for each pair during the scheduling process in Fig. 2(a).

Layout design solution: Contains the positional information of each component in Fig. 2(b).

Design specifications: Chip size constraints $B_{W \times H}$; maximum duration T_{max}, etc.

Output:

Detailed routing information for the routing pairs between components.

Objective:

Minimize the total routing length $NetsLen$ and the number of routing network crossovers $Cross$:

$$min(NetsLen), min(Cross) \tag{1}$$

Constraints:

Routing area constraints, the path cannot exceed the chip area:

$$0 \leq x_i \leq W, 0 \leq y_i \leq H; \forall(x_i, y_i) \in Path \tag{2}$$

Non-routable component areas, path and module cannot overlap:

$$(x_i, y_i) \cap (m_j, m_j) = 0; \forall(x_i, y_i) \subseteq Path, \forall(m_j, m_j) \subseteq Module \tag{3}$$

where, x_i, y_i represents the coordinates in the path. m_i, $m_i$represents the coordinates in the module.

In this problem, we need to consider how to plan the paths between the starting and ending points of the routing pairs in a reasonable manner to meet the design specifications while minimizing the total routing length and the number of routing network crossovers. Additionally, we need to adhere to the constraints of the routing area to ensure the routing is performed within feasible bounds. Furthermore, the component areas should be non-routable, meaning the routing paths should not pass through these areas.

4 Proposed Method

From the high-level synthesis of the biochemical protocol, we can know the dependency transfer relationship between operations. Specifically, in the directed acyclic graph representing the biochemical protocol, the starting point of an arrow represents an output operation, while the endpoint represents an input operation.

Therefore, the results of the output operations need to be transferred to the corresponding input operations as inputs. From the binding and scheduling solutions in high-level synthesis, the correspondence between components and operations can be determined. Consequently, we can deduce the connectivity between components and identify the locations where routing needs to be performed. Once the layout design determines the position of each component, the routing design phase involves detailed routing of the routing pairs between components.

The following subsections will describe the steps of the two-stage A* algorithm and the GA* algorithm, respectively.

4.1 Two-Stage A* Algorithm Steps

Since one of the optimization objectives in the routing phase is to minimize the total bus length in order to reduce overall transmission time, this paper adopts the classical heuristic pathfinding algorithm, A* algorithm.

Another optimization objective is to minimize the number of crossover points because an increase in crossover points in the flow layer pipelines will require the use of switches to control the direction of fluid flow. This also means introducing more valves to the control layer. A pipeline crossing usually requires four valves (A, B, C, D) to be controlled, which significantly increases the complexity of the control layer design. Another reason is that when fluid passes through intersecting pipelines, the other pipelines connected to the crossover point cannot be utilized due to one side being closed by the valves. For example, when fluid flows in a specific direction at the intersection shown in Fig. 3, the valve states at the intersection are AB (valves open) and CD (valves closed). This results in two pipelines connected to CD being unable to perform transmission tasks due to the principle of pressure. Additionally, valve states usually remain unchanged during transmission tasks. Therefore, the CFMB routing phase needs to minimize the number of crossover points as much as possible to improve system efficiency and enable high-concurrency execution of multiple transmission tasks.

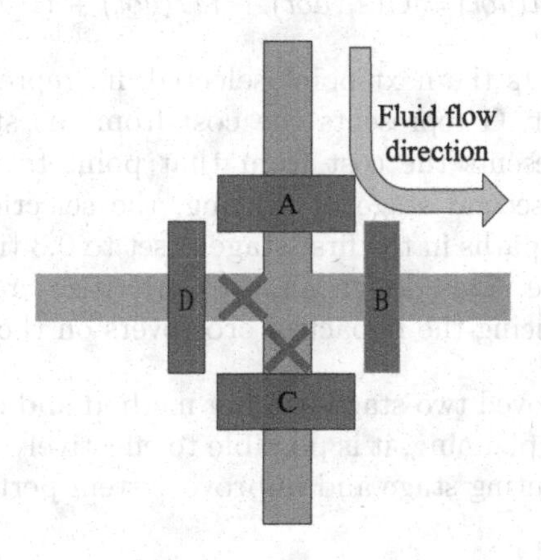

Fig. 3. Example of pipeline crossing.

To achieve this objective, a series of strategies are employed in the CFMB routing phase, including optimizing the heuristic pathfinding process of the A* algorithm to consider crossover point minimization. At the same time, during the path planning process, it is necessary to consider the limitations of the routing area and design constraints. Through these measures, the number of crossover points in the routing can be effectively reduced, improving the execution efficiency of CFMB and enabling efficient concurrent execution of multiple transmission tasks.

Given the objective of minimizing the number of crossover points in the routing phase, an improvement is made to the A* algorithm by adopting a two-stage routing method [13]. The overall algorithm steps are as follows:

(1) Initialize m routing pairs, determine the number of randomly generated solutions n, and set the regions where the components are located as obstacles.
(2) Initialize the optimal solution as empty, and set the quality of the optimal solution to infinity. Perform the following steps n times:
   A. Generate a random routing solution Rt.
   B. Perform the following steps for each routing pair in Rt:a) Use the A* algorithm for routing, not allowing passage through obstacles.b) If the routing is successful, set the path of the route as an obstacle and record relevant information; otherwise, add the routing pair to the failure queue.
   C. Check the failure queue. If it is not empty, set a higher cost for the successful routes' paths and perform the following steps for each failed routing pair:a) Use the A* algorithm for routing, not allowing bypassing obstacles but allowing crossover with a cost.b) Set a certain cost selection for the successfully routingd paths.
   D. Calculate the quality of the current routing solution Rt which A* algorithm evaluates the cost of a point *dot*dot as shown in formula (4) and decide whether to update the global optimal solution.
(3) Output the global optimal solution.

$$Cost(dot) = G(s_i, dot) + fix(dot) + H(dot, t_i) \tag{4}$$

where, dot represents the next point selected, fix represents the penalty for selecting that point, G represents the cost from the starting point to that point, and H represents the cost from that point to the endpoint. Set to 0 initially. In the second stage of routing, the selection cost of points on successfully routed paths in the first stage is set to 0.3 times the length of the corresponding route. This design aims to prioritize crossovers with shorter paths, thereby reducing the impact of crossovers on the system.

By using this improved two-stage routing method and combining it with the A* algorithm for path planning, it is possible to effectively minimize the number of crossovers in the routing stage and improve system performance.

4.2 GA* Algorithm Steps

To search for a globally approximate optimal solution in the solution space, the global optimization in the solution space is required. In this paper, the GA* algorithm, which combines the genetic algorithm and the A* algorithm, will be used to achieve routing routing design. The GA algorithm has characteristics such as strong global optimization capabilities and suitability for nonlinear, non-convex, and multimodal optimization problems. In the GA algorithm, we will use the two-stage routing routing results obtained from the A* algorithm as the basis for evaluating the quality of particle solutions in the GA algorithm.

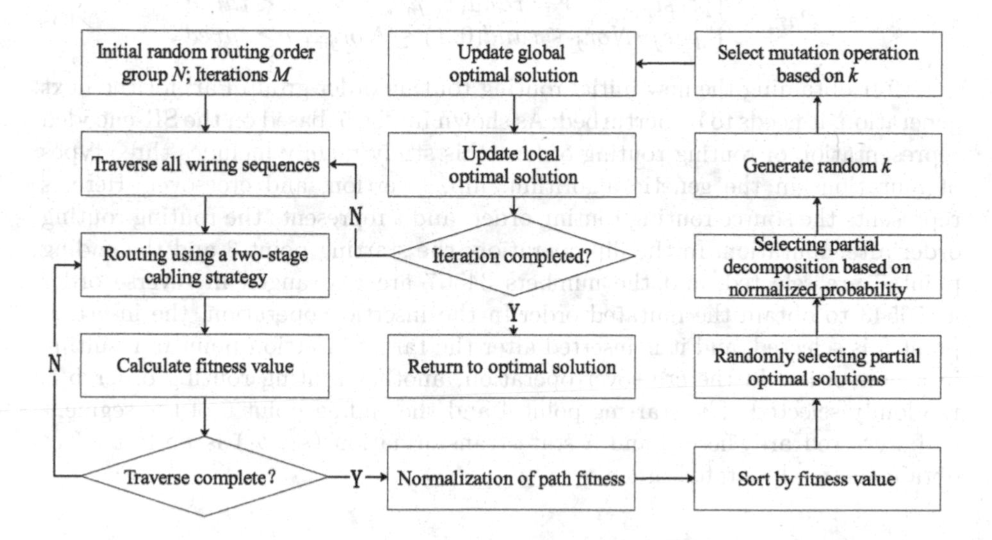

Fig. 4. GA* algorithm flowchart.

Figure 4 illustrates the process of the GA* algorithm. To better integrate the mutation strategy in the genetic algorithm, a coding method called SR encoding is employed in this paper. For a set of routing routing pairs $Path = (p_1(s_1, t_1), ..., p_n(s_n, t_n))$, in the initial stage of the algorithm, a set of random routing routing orders $Rt = (r_1, ..., r_n)$ will be generated. The SR encoding follows the following rules:1. s_i and t_i represent the starting and ending points of routing routing pair p_i, respectively;2. r_i in Rt corresponds to p_j in $Path$ have a one-to-one correspondence; 3. RtRt represents the routing routing order, i.e., the order in which the routing routing pairs appear from left to right.

Normalization is performed according to the fitness value, as shown in formula (5):

$$Nor_y = \frac{\sum_{i=1}^{y} fit_i}{\sum_{j=1}^{y_{max}} fit_j}, y \in (1, y_{max}) \tag{5}$$

where y represents the yth routing routing order, y_{max} denotes the maximum scale of routing routing solutions, and fit_i represents the fitness value of the ith routing routing order, which represents the quality of the routing routing result. After normalization, the value of Nor_y is obtained, ranging from 0 to 1, for further processing.

After normalization, two strategies are used to select the initial population for the next generation. One strategy is to select a certain number of routing routing orders corresponding to the local historical best solutions, *pbest*, from a set of *thred* routing routing orders. The other strategy is to probabilistically select routing routing orders from the previous generation based on the normalized fitness values (ranging from 0 to 1), as shown in formula (6):

$$Par_i^{'} = \begin{cases} pbest_r & r = rand(1, y_{max}) \quad i < thred \\ par_j & Nor_j \leq rand(0,1) \leq Nor_{j+1} \; i \geq thred \end{cases} \tag{6}$$

After obtaining the new initial routing routing order group Par' for the next generation, it needs to be perturbed. As shown in Fig. 5, based on the SR-encoded representation of routing routing orders, this study mainly includes three types of mutations in the genetic algorithm: flip, insertion, and crossover. Here, s represents the source routing routing order, and t represents the routing routing order after mutation. In the flip operation, the starting point 3 and the ending point 7 are selected, and the numbers 34567 are rearranged in reverse order as 76543 to obtain the mutated order.In the insertion operation, the insertion point 3 is selected, and it is inserted after the target insertion point 6, resulting in a new order. In the crossover operation, another routing routing order b is randomly selected. The starting point 4 and the ending point 6 of the segment to be crossed are chosen, and a state transformation (s_1, b_1) is performed to obtain the final mutated order.

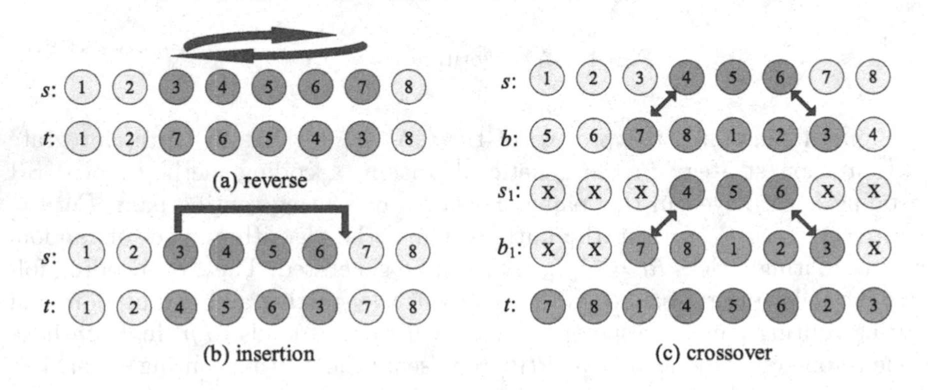

Fig. 5. Mutation Operation of Genetic Algorithm.

5 Experimental Results and Discussion

This section will introduce the experimental description of CFMB, the simulation experiment scheme, and the corresponding simulation experiment results and analysis in detail.

5.1 Experiment Setup

The experiment is conducted on a Windows 10 operating system with an Intel i5-10500 processor running at a frequency of 3.10 GHz and 8GB of memory. Programs are written in C++ and run on the Microsoft Visual Studio Community 2022 platform.

To validate the effectiveness of the proposed algorithm, in this section, the six standard biochemical protocol test cases listed in Table 1 are used for architecture synthesis.

Table 1. Biochemical protocol information and number of allocated components [27].

	Biochemical protocol					
	PCR	IVD	EA	Syn1	Syn3	Syn4
operand	7	12	10	10	30	40
(Mixer, detector, filter, heater)	(3,0,0,0)	(2,3,0,0)	(3,0,1,1)	(2,1,1,1)	(5,1,2,3)	(6,3,2,2)
Edges	6	6	9	9	29	39

In this experiment, the population size for routing is set to twice the number of operations in the biochemical protocol. Since the size of the solution space is greatly influenced by the scale of the biochemical protocol, the size of the routing order group is dynamically generated in the initial stage of the algorithm and needs to be larger than the scale of the biochemical protocol. Additionally, with selection probabilities of 0.4, 0.4, and 0.2, the three mutation strategies of flip, insertion, and crossover are chosen. This choice of probabilities helps maintain the diversity and exploratory capability of the algorithm to search for globally approximate optimal solutions in the solution space.

5.2 Experimental Results and Analysis

During actual operation, the transmission delay of the chip is directly proportional to the routing length obtained during the routing stage. Therefore, optimizing the overall routing length can improve the chip's operational efficiency. Additionally, in CFMB, there is a control layer that changes the state of valves to alter the state of the fluid layer's pipelines. Unlike circuit crossings in VLSI which can lead to short circuits, CFMB allows crossing between flow layer pipes to enable successful routing of all paths in one flow layer.

The CFMB routing stage uses a two-stage routing A* algorithm combined with a GA strategy. The two-stage routing is used to generate routing solutions, which include two optimization objectives: pipeline cross count Cross and total routing length NetsLen. The formula for calculating the pipeline Cross is as follows:

$$Cross = \sum(x_i, y_i), (x_i, y_i) = 2 \tag{7}$$

where (x_i, y_i) represents the coordinates of the grid point in the routing area that is covered the most times, and the formula for calculating the total routing length NetsLen is as follows:

$$NetLen = \sum_{i=1}^{Rt} length_i \tag{8}$$

where $length_i$ represents the actual length of the i-th routing, and Rt represents the number of channels. The evaluation function for assessing the quality of the routing solution is as follows:

$$Engy = \omega \times Cross + \psi \times NetsLen \tag{9}$$

where ω and Ψ are balance parameters. Since the number of crossovers has a different order of magnitude relationship with the routing lengths, in this experiment, the balance parameters ω and Ψ are set to 15 and 3, respectively.

Table 2. Routing experiment results.

Biochemical protocol	GA * algorithm		Two-stage A * algorithm		Decline rate(%)	
	NetsLen	Cross	NetsLen	Cross	NetsLen	Cross
PCR	19	0	19	0	0	0
IVD	62	2	62	2	0	0
EA	92	3	109	4	15.6	25
Syn1	113	3	153	4	26.1	25
Syn3	769	51	1026	73	25	30.1
Syn4	1241	93	1486	125	16.5	25.6
Avg	–	–	–	–	13.9	17.6

To validate the effectiveness of the two-stage A* algorithm with the inclusion of a genetic algorithm, we also implement a baseline two-stage A* algorithm for routing routing design as a comparative experiment. In order to ensure the reliability of the experimental results, for each standard test case, 30 routing designs were carried out, and the average value was obtained, as shown in Table 2. we observe that for the PCR and IVD protocols, which involve fewer components and routing routing pairs, both algorithms find the optimal routing routing solutions for the given layouts. However, for other more complex biochemical protocols, the genetic algorithm combined with the A* algorithm outperforms the basic two-stage A* algorithm. This indicates that the genetic algorithm can extensively search the solution space, resulting in an average reduction of 13.9% in routing length and a 17.6% reduction in the number of pipeline crossovers.

Figure 6 shows the results of the EA test case, where both resource binding and scheduling solutions and module placements are manually set. Both routing routing algorithms achieve optimal routing effects for the same input.

In Fig. 6(a), we have the solution obtained from the basic two-stage A* algorithm, while in Fig. 6(b), we have the optimal solution obtained from the genetic algorithm combined with the A* algorithm. It can be observed that due to the combination of the genetic algorithm, the genetic algorithm combined with A* algorithm exhibits stronger optimization capabilities, fewer pipeline crossovers, and less impact on the operation of input-output pipeline crossovers, resulting in higher quality solutions.

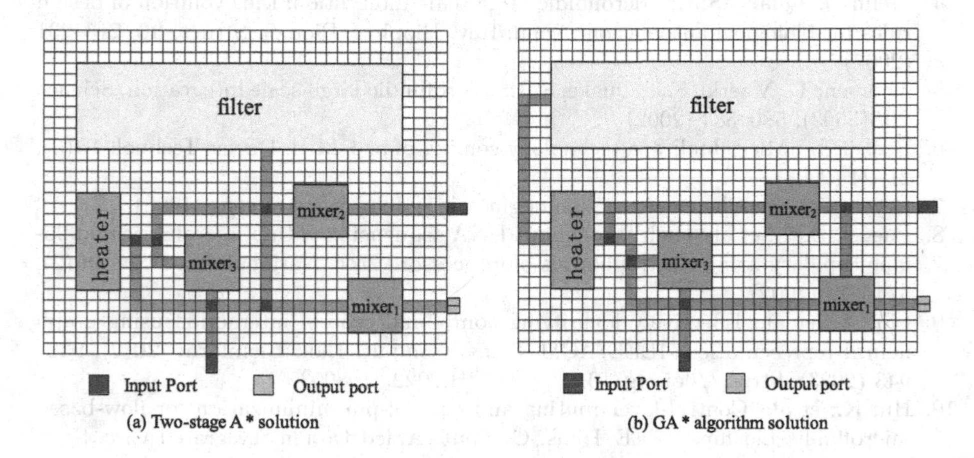

Fig. 6. Comparison of routing results for EA test cases.

6 Conclusion

CFMB utilizes the principles of microfluidics, microchannels, and Microelectromechanical Systems technology to control the handling and reactions of biological samples, enabling detection and separation of biomolecules and cells. It is widely used in the field of intelligent digital healthcare. However, the rapid growth in the number of components on CFMB has led to increased complexity in the routing process. To address this issue, this paper proposes an automatic routing algorithm for CFMB based on the GA* algorithm, which combines the GA with the A* algorithm. Through a series of benchmark experiments, the results demonstrate the effectiveness of the GA* algorithm in CFMB routing. It significantly reduces the routing length, thereby lowering the chip cost. Additionally, the GA* algorithm reduces the number of crossover points in the routing, enhancing the execution efficiency of CFMB's biochemical protocols. This is crucial for the automated development of CFMB construction, accelerating the application of intelligent digital healthcare, and effectively safeguarding people's lives and health.

Acknowledgements. This work is supported by the fund of Fujian Province Digital Economy Alliance, the National Natural Science Foundation of China (No. U1905211), and the Natural Science Foundation of Fujian Province (No. 2020J01500).

References

1. Araci, I.E., Quake, S.R.: Microfluidic very large scale integration (mVLSI) with integrated micromechanical valves. Lab Chip **12**(16), 2803 (2012)
2. Duffy, D.C., et al.: Rapid prototyping of microfluidic systems in Poly (Dimethylsiloxane). Analy. Chem. **70**(23), 4974–4984 (1998)
3. Unger, M.A., et al.: Monolithic microfabricated valves and pumps by multilayer soft lithography. Science **288**(5463), 113–116 (2000)
4. Melin, J., Quake, S.R.: Microfluidic large-scale integration: the evolution of design rules for biological automation. Annu. Rev. Biophys. Biomol. Struct. **36**, 213–231 (2007)
5. Thorsen, T., Maerkl, S.J., Quake, S.R.: Microfluidic large-scale integration. Science **298**(5593), 580–584 (2002)
6. Becker, H.: Microfluidics: a technology coming of age. Med. Device Technol. **19**(3), 21–24 (2008)
7. Levenspiel, O.: Chemical reaction engineering. John wiley & sons (1998)
8. Paegel, B.M., et al.: High throughput DNA sequencing with a microfabricated 96-lane capillary array electrophoresis bioprocessor. Proc. National Acad. Sci. **99**(2), 574–579 (2002)
9. Lin, X., et al.: Effectively identifying compound-protein interaction using graph neural representation. IEEE/ACM Trans. Comput. Biol. Bioinform. **20**(2), 932–943 (2023). https://doi.org/10.1109/TCBB.2022.3198003
10. Hu, K., et al.: Control-layer routing and control-pin minimization for flow-based microfluidic biochips. IEEE Trans. Comput. Aided Design Integrated Circ. Syst. (2017)
11. Chakraborty, S., Das, C., Chakraborty, S.: Securing module-less synthesis on cyber-physical digital microfluidic biochips from malicious intrusions'. In: 2018 31st International Conference on VLSI Design and 2018 17th International Conference on Embedded Systems (VLSID), pp. 467–468.IEEE (2018)
12. Yi-Siang, S., Ho, T.-Y., Lee, D.-T.: A routability-driven flow routing algorithm for programmable microfluidic devices. In: 2016 21st Asia and South Pacific Design Automation Conference (ASP-DAC), pp. 605–610. IEEE (2016)
13. Wang, Q.: Sequence-pair-based placement and routing for flowbased microfluidic biochips. In: 2016 21st Asia and South Pacific Design Automation Conference (ASP-DAC), pp. 587–592. IEEE (2016)
14. Grimmer, A.: Close-to-optimal placement and routing for continuous flow microfluidic biochips. In: 2017 22nd Asia and South Pacific Design Automation Conference (ASP-DAC), 530–535. IEEE (2017)
15. Huang, X., et al.: Timing-driven flow-channel network construction for continuous-flow microfluidic biochips. IEEE Trans. Comput.- Aided Design Integrated Circ. Syst. **39**(6), 1314–1327 (2019)
16. Minhass, W.H., et al.: Control synthesis for the flow-based microfluidic large-scale integration biochips. In: Design Automation Conference (2013)
17. Yao, H., Ho, T.Y., Cai, Y.: PACOR: practical control-layer routing flow with length-matching constraint for flow-based microfluidic biochIPs. In: Design Automation Conference (2015)
18. Li, X., et al.: a potential information capacity index for link prediction of complex networks based on the cannikin law. Entropy **21**(9), 863 (2019)
19. Wang, Q., et al.: Hamming-distance-based valve-switching optimization for control-layer multiplexing in flow-based microfluidic biochips. In: Design Automation Conference (2017)

20. Tseng, T.M., et al.: Columba: co-layout synthesis for continuous-flow microfluidic biochips. In: Proceedings of the 53rd Annual Design Automation Conference, pp. 1–6 (2016)
21. Tseng, T.M., et al.: Columba S: a scalable co-layout design automation tool for microfluidic large-scale integration. In: the 55th Annual Design Automation Conference (2018)
22. Tseng, T.M., et al.: Columba 2.0: a co-layout synthesis tool for continuous-flow microfluidic biochips. IEEE Trans. Comput.- Aided Design Integrated Circ. Syst. **37**(8), 1588–1601 (2018)
23. Zheng, Q., et al.: An improved deep reinforcement learning for robot navigation. In: Third International Conference on Machine Learning and Computer Application (ICMLCA 2022), vol. 12636. SPIE, pp. 171–176 (2023)
24. Fan, X., et al.: Combine discussion mechanism and chaos strategy on particle swarm optimization algorithm. In: 2019 IEEE 10th International Conference on Software Engineering and Service Science (ICSESS). IEEE, pp. 642–645 (2019)
25. Dong, C., et al.: Dual-search artificial bee colony algorithm for engineering optimization. IEEE Access **7**, 24571–24584 (2019)
26. Dong, C., et al.: A cost-driven method for deep-learning-based hardware trojan detection. Sensors **23**(12), 5503 (2023)
27. Minhass, W.H., et al.: Architectural synthesis of flow-based microfluidic large-scale integration biochips. In: International Conference on Compilers, p. 181 (2012)

A Cloud-Based Sign Language Translation System via CNN with Smart Glasses

Siwei Zhao[1(✉)], Jun Wang[1], Kiminori Sato[2], Bo Wu[2], and Xuan Huang[3]

[1] Graduate School of Bionics, Computer and Media Sciences, Tokyo University of Technology, Hachioji, Tokyo 192-0982, Japan
{g21230217b,g2123002c8}@edu.teu.ac.jp

[2] School of Computer Science, Tokyo University of Technology, Hachioji, Tokyo 192-0982, Japan
{satohkmn,wubo}@stf.teu.ac.jp

[3] Metaverse Research Institute, Waseda University, Tokorozawa, Saitama 359-1192, Japan
x.huang6@kurenai.waseda.jp

Abstract. In situations where ordinary people and the hearing-impaired person need to communicate, it is possible that the average person does not know sign language and thus communication may be impaired, which means that a technology or device to assist communication is needed. Therefore, this study develops a new cloud sign language translation system on smart device based on the Browser/Server architecture, so that when the hearing-impaired person makes a sign language movement in front of the user who using the system with a smart device (e.g., smart glasses), the screen of the smart device will display the subtitle of the sign languages. We use MediaPipe to recognize and collect sign language action data from WLASL dataset and provide it to TensorFlow's 1D-CNN deep learning model for training, so as to realize the sign language translation function. In the test phase, we invited five experimenters to test the sign language translation system ten times for each person, and the final average accuracy rate was 72%. Through such an interpreting system, it brings a convenient and efficient communication experience to the hearing-impaired person and people who need to communicate with the hearing-impaired person, and at the same time, it can also provide support for broader sign language research and application.

Keywords: Smart Glasses · Sign Language Translation · TensorFlow · Deep Learning · MediaPipe

1 Introduction

Sign language is a language used by the hearing-impaired person to communicate and interact with signs and finger movements. Sign language expresses feelings, intentions, and thoughts through finger movements, and body movements [1, 2]. In the situation that ordinary people and the hearing-impaired person need to communicate, it is possible that the average person does not know how to use sign language, and thus communication can be impaired, a man-carried translation system with low-cost smart devices which can aid communication is needed.

H. Jin et al. (Eds.): GPC 2023, LNCS 14504, pp. 224–237, 2024.
https://doi.org/10.1007/978-981-99-9896-8_15

In recent years, with the development of computer vision related technologies, sign language translation techniques based on video streams and IoT sensors have been widely studied. For example, there are many existing technical solutions for sign language translation such as Kinect [3], Leap Motion [4, 5] and wireless hand gesture recognition glove [6].

However, existing solutions such as Leap Motion-based sign language Translators usually require special devices to implement, which creates a significant inconvenience for users that users need to carry these devices and additional settings, this may increase the burden and limit flexibility [4, 5].

On the other hand, by using the MediaPipe, as a multimedia processing framework developed as Google [7], it possible to achieve multiple effects such as body movement recognition including hand movements, object recognition and expression recognition by analyzing video data from commercially available webcams. For example, in an existing study, S. Shriram et al. developed a visually based hand tracking system for non-face-to-face interaction by tracking hand movements to operate on a virtual whiteboard. They used MediaPipe to analyze hand movements and allowed users to create basic drawing, writing, and deleting operations on the virtual whiteboard, enabling non-face-to-face interaction. This study demonstrated that hand data generated using the MediaPipe framework have enough precision that can be used for sign language translation as well by extracting features [8].

Therefore, in this study, based on the using of a trained CNN recognition model, we try to design a new translating system on smart glasses for deaf mute to change their sign language to an available well-understood subtitles displaying on the screen.

To be specific, we use the MediaPipe framework to recognize and collect sign language action data from sign language videos in the WLASL (World Level American Sign Language) dataset [9]. Then, we provide this data to TensorFlow 1D-CNN for deep learning and generating recognition models [10]. After that, the Browser/Server architecture is used to develop a front-end application based on the MediaPipe framework and a local smart device, whose acquired hand movement data will be passed to our cloud-based, sign language translation system using the Sanic high concurrency web framework for recognition and feedback.

This system combines computer vision technology as well as modern Internet technology, and the relevant results can provide an accurate and efficient translation solution for language communication. The design of the system is characterized by its support for different smart devices and requires only browser software available on the smart device to realize the translation function [11].

Compared with the traditional local mode translation system, the advantage of the cloud mode can lower the threshold of terminal smart devices, users are no longer limited to specific devices and applications, and can enjoy the sign language translation function anytime and anywhere through a simple network connection. At the same time, this system can support wider sign language research and application while bringing a convenient and efficient communication experience for the hearing-impaired person and people who need to communicate with the hearing-impaired person.

The rest of this thesis is as follows: The second part is the correlation study. The third part is the experimental design. The fourth part is the experimental procedure, test results. The fifth part gives the conclusion of this study and future work.

2 Related Works

2.1 Research Related to Sign Language Translation Technology

Sign language recognition, although it has been explored for many years, is still a challenging problem in practical applications [1, 2]. For example, Ryota Takahashi et al. used deep learning techniques for sign language recognition in their study 3D CNN convolutional networks [12]. They utilized machine learning to construct a recognition model for understanding sign language grammar, which is of great the orifical significance for the development and application of sign language translation technology. In addition, BiyiFang et al. proposed an approach based on DeepASL and Leap Motion to capture hand gestures [4]. They used Leap Motion (an infrared light-based sensing device that extracts skeletal joint information from fingers, palms, and forearms) to non-invasively capture gestures from the hearing-impaired person and collect hand skeletal joint information. For machine learning, bidirectional recurrent neural networks (RNN) and long short-term memory (LSTM) models were used to analyze signers' hand movements and then translate them into text. As well as Lih-Jen et al. implemented sign language translation using wireless hand and smartphone in their research portable sign language translation system [6].

2.2 Research Related to the Use of Smart Glasses Technology

Smart glasses are used as a wearable device with translator augmented reality (AR) capabilities [13]. In the field of industrial production, the use of smart glass-es has revolutionized many traditional processes. For example, Roberto Pierdicca et al. realized that workers wearing smart glasses can receive detailed work instructions in real time, which greatly improves work efficiency and accuracy in industrial related research. In some dangerous places, workers approaching some dangerous equipment prompt the danger of the current equipment, which ensures the safety of the workers' work [14]. In addition, Stefan Mitrasinovic et al. in their research related to the field of healthcare industry have realized that surgeons are able to pre-pare a simulated image of the patient's surgery and the expected outcome of the completed surgery before the surgery through smart glasses. This image was displayed on a prism display to overlay the patient to produce augmented reality [15]. As well as in the study by Ali Samini et al. the augmented reality solution was provided, which provided a valuable reference for hardware selection in this study [16].

According to previous studies, Leap Motion [4] and wireless hand [6] devices can realize sign language interpretation, but users carrying these devices may increase the burden and limit the flexibility. Therefore, we choose smart glasses as the data collection client for sign language users, it has a transparent and borderless screen, users will not block the peripheral vision because of the content displayed on the screen, and the cloud-based sign language translation is real-time Translation, which will not lead to inefficient communication due to the operation of the software. Moreover, Macmini, as a cloud-based sign language translation server, has a compact design compared to traditional 1U2U size servers, and can be easily deployed in a variety of scenarios. This makes Macmini an ideal lightweight server choice for providing translation capabilities to smart devices.

3 Methodology

This chapter describes the role of MediaPipe and Convolutional Neural Network (CNN) in this study.

3.1 MediaPipe

MediaPipe is a multimedia application with real-time processing capabilities [5]. It supports multiple platforms, including mobile devices, desktop systems and embedded devices, and provides cross-platform APIs and toolchains. This makes it possible to achieve multiple effects such as body movement recognition including hand movements, object recognition and expression recognition by analyzing video data from commercially available webcams using MediaPipe.

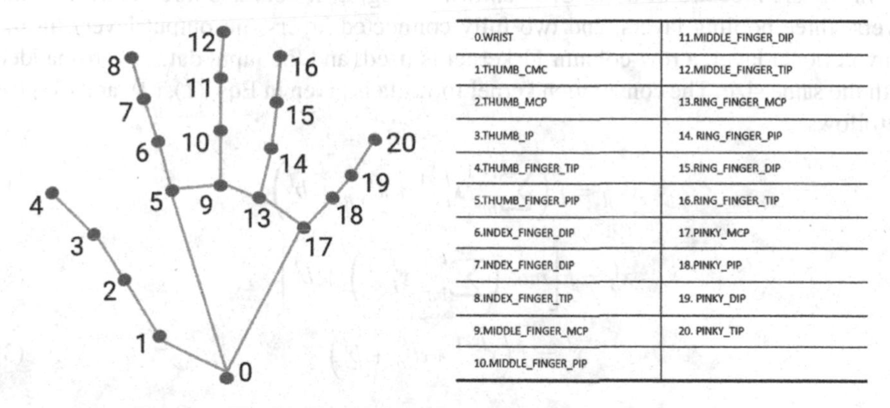

0.WRIST	11.MIDDLE_FINGER_DIP
1.THUMB_CMC	12.MIDDLE_FINGER_TIP
2.THUMB_MCP	13.RING_FINGER_MCP
3.THUMB_IP	14. RING_FINGER_PIP
4.THUMB_FINGER_TIP	15.RING_FINGER_DIP
5.THUMB_FINGER_PIP	16.RING_FINGER_TIP
6.INDEX_FINGER_DIP	17.PINKY_MCP
7.INDEX_FINGER_DIP	18.PINKY_PIP
8.INDEX_FINGER_TIP	19. PINKY_DIP
9.MIDDLE_FINGER_MCP	20. PINKY_TIP
10.MIDDLE_FINGER_PIP	

Fig. 1. Hand joint point marking and sorting.

As shown in Fig. 1, we use the hand analysis module to store 21 hand bone feature points into a Mysql database. Mysql database has good scalability and can handle up to tens of thousands of concurrent connections. This is important for the large-scale hand feature points required by our cloud sign language translation system.

3.2 Convolutional Neural Network (CNN)

We choose the TensorFlow deep learning framework [10, 17, 18]. It can be designed for multiple servers for deep learning, thus improving learning efficiency. It also has a visualization tool, TensorBoard, for presenting deep learning model evaluation metrics data. In convolutional training, we use 1D-CNN to train hand movements because in the data samples, the hand feature points we recognized are time-series data and 1D-CNN model is suitable to be used to learn them.

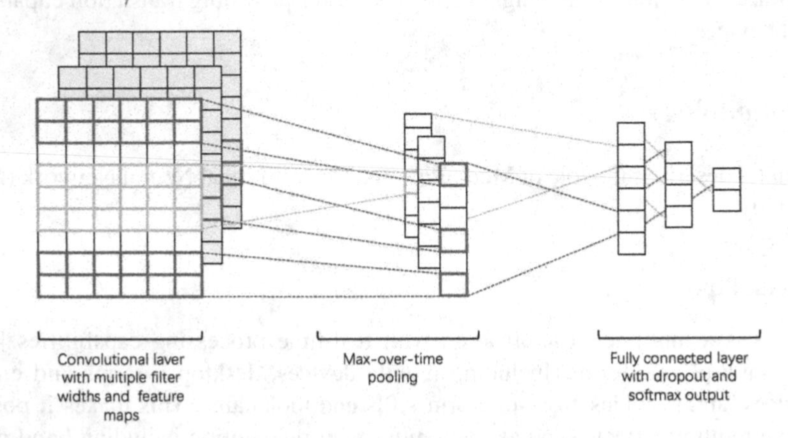

Fig. 2. Deep Learning Model Design

In the architecture of a 1D CNN shown in Fig. 2, it contains three convolutional layers, three pooling layers, and two fully connected layers (no output layer). In the convolutional layer, a row-column 11 kernel is used, and the input data is zero-padded with the same size. The convolution kernel formula is given in Eqs. (1), (2), and (3) [19] as follows.

$$x_{0,fl}^{l} = f\left(\sum_{im}^{l-1} x_i^{l-1} * k_{io,fl}^{l} + b^l\right) \tag{1}$$

$$x_o^l = f\left[max\left(\sum_{im}^{l-1} x_i^{l-1}\right) + b^l\right] \tag{2}$$

$$x_o^l = f\left(x_i^{l-1} * d_{io}^l + b^l\right) \tag{3}$$

Each convolution kernel corresponds to each gesture feature. After the first convolutional layer, the Relu (Rectified Linear Unit) activation function is used, and the Relu formula is shown in Eq. (4) [20] below.

$$f(x)\{x; x \geq 00; x < 0\} \tag{4}$$

In the pooling layer, to reduce overfitting, sampling is done using maximum pooling with a default step size of 1.

After converting the multidimensional data to 1-dimensional data using Flatten prior to the fully connected layer, it is input to the fully connected layer. After each nerve is fully connected in the fully connected layer, it is activated using Relu. Finally, the output layer is fully connected to the fully connected layer and outputs 6 neurons and uses SoftMax to calculate the probability of each neuron and predicts the current sign language meaning with the highest probability.

To prevent overfitting during network training, the Dropout algorithm is used in the convolutional layer, and the fully connected layer is set to 0.5.

Table 1. 1D CNN Summary.

Layer(type)	Output Shape	Param
conv1d (Conv1D)	(None, 63, 256)	70656
max_pooling1d (MaxPooling1D)	(None, 32, 256)	0
dropout (Dropout)	(None, 32, 256)	0
conv1d_1 (Conv1D)	(None, 32, 128)	360576
max_pooling1d_1 (MaxPooling1D)	(None, 16, 128)	0
dropout_1 (Dropout)	(None, 16, 128)	0
conv1d_2 (Conv1D)	(None, 16, 64)	90176
max_pooling1d_2 (MaxPooling1D)	(None, 8, 64)	0
dropout_2 (Dropout)	(None, 8, 64)	0
flatten (Flatten)	(None, 512)	0
dense (Dense)	(None, 128)	65664
dropout_3 (Dropout)	(None, 128)	0
dense_1 (Dense)	(None, 64)	8256
dropout_4 (Dropout)	(None, 64)	0
dense_2 (Dense)	(None, 6)	390

Table 1 summarizes the 1D CNN model. The first convolutional layer has 70656 parameters, the second convolutional layer has 360576 parameters, the third convolutional layer has 90176 parameters, the first fully connected layer has 65664 parameters, and the second fully connected layer has 8256 parameters.

We used 183 dataset samples and trained the 1D CNN neural network. In the training process, we adopt the setting of Batch size 4096 and optimize the network with Adam algorithm. To get the probability of the output result, we add a SoftMax activation function to the final fully connected layer of the network. The loss is calculated according to Eq. (5) [20], and through 1000 iterations of training, we successfully stabilize the accuracy of the training set at about 80%. After training, we saved the model file for later use. Such a training configuration makes our lightweight neural network a feasible and efficient solution to be applied in various smart devices to provide them with powerful sign language translation services.

$$loss = -\sum_{i}^{c} y'_i log(y_i) \tag{5}$$

3.3 System Design

This chapter describes the hardware, software, experimental design, and cloud server architecture design.

Fig. 3. EPSON MOVERIO BT-40 Smart Glasses & Mac Mini 2021

As shown in Fig. 3, we use EPSON MOVERIO BT-40 Smart Glasses for translator hearing-impaired person sign language. We use a Mac Mini as a deep learning server as well as a cloud-based architecture sign language translation server.

Table 2. List of software used in this study.

Software Name	Version	Role
MacOS	13.4	OS
Anaconda	1.11.2	Python Scientific Computing Library
TensorFlow	2.13.0	Deep Learning Framework
Sanic	23.3	Web Framework
Nginx	1.25	Web Server

As shown in Table 2, We use Macmini as the sign language translation cloud server. In the configuration of Macmini, we choose MacOS as the operating system of the server, because MacOS has high stability and security. We then installed the Anaconda runtime, which has a large library of scientific computing and provides a GPU version of TensorFlow for Macmini M1 chip. Deep Learning Framework We have chosen TensorFlow, which can support various deep learning models. For the Web framework we chose Sanic which is lightweight and high performance. Finally, we will use Nginx as our network entry point, which not only provides a Web service, but also a reverse proxy that forwards data to our backend services [21].

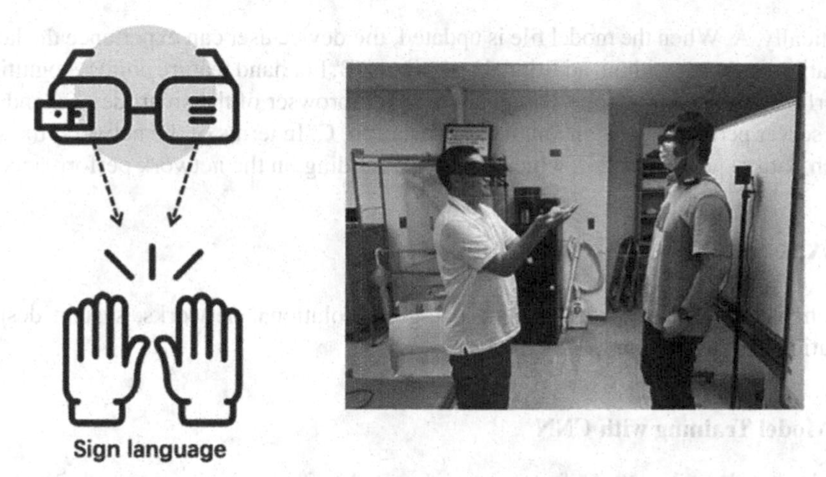

Fig. 4. Experimental schematic

As shown in Fig. 4, the user who needs to translate the sign language needs to wear glasses, the lenses of the smart glasses contain a display, and the user holds a controller which is equipped with a camera. When the signer on the opposite side uses sign language, the controller captures and records the gesture movements. Then, the controller transmits the sign language movement to a server, which translates the sign language content and finally displays the sign language translation result on the display of the smart glasses.

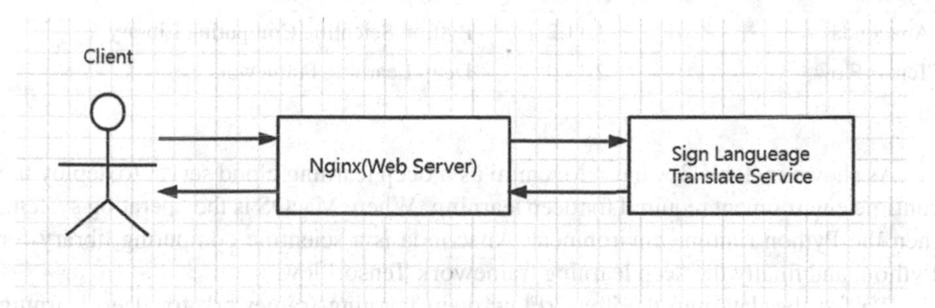

Fig. 5. Server Architecture Diagram

Figure 5 is the server architecture and the data forwarding process. We use Nginx as the web portal for all web requests and forward them based on URL matching. It is a high-performance Web server that supports features such as reverse proxy. On the other hand, we also use Nginx to process the html request from MediaPipe as well as forward the gesture data to the sign language translation service, and then return the translation result to the original way back to the smart glasses to present it in the form of text on the smart glasses. The design is characterized by the fact that there is no need to consider the limited CPU/GPU computing power of the end device. Therefore, we use a cloud server to realize sign language translation based on the Browser/Server architecture.

Specifically, A. When the model file is updated, the device user can experience the latest translation accuracy without additional operations. B. For hand feature point recognition, we perform hand feature point computation in the browser of the smart device, and the cloud server performs the sign language translation. C. In terms of the network, the size of each data request is 325bit, which is less demanding on the network performance.

4 Experiments and Disscution

This chapter describes model training using convolutional networks, system design, evaluation and discussion.

4.1 Model Training with CNN

This chapter describes the software, samples and training results required for model training. The book, Drink, Computer, Before, Chair and Go words, totaling 183 data samples, were selected from the MediaPipe preprocessed data, and then the model was trained using Convolutional Neural Networks and Fully Connected Neural Networks.

Table 3. Software needed for deep learning.

Software Name	Version	Role
MacOS	13.4	Operating System
Python	3.10.12	Python Environment
Anaconda	1.11.2	Python Scientific Computing Library
TensorFlow	2.13.0	Deep Learning Framework

As shown in Table 3, we use Macmini as a deep learning cloud server to deploy the runtime environment required for deep learning. Where MacOS is the operating system, then the Python runtime environment, Anaconda is a scientific computing library for Python, and finally the deep learning framework TensorFlow.

We fed the data into the TensorFlow deep learning framework for deep learning and performed 1000 training iterations to generate the model files. According to Fig. 6, the accuracy increases steadily with the number of iterations, which indicates good performance in deep learning. Similarly, the loss value in Fig. 7 should gradually decrease and stabilize. The result shows an accuracy of 0.8 and loss finger of 0.4. Indicating that our deep learning model performs satisfactorily to some extent.

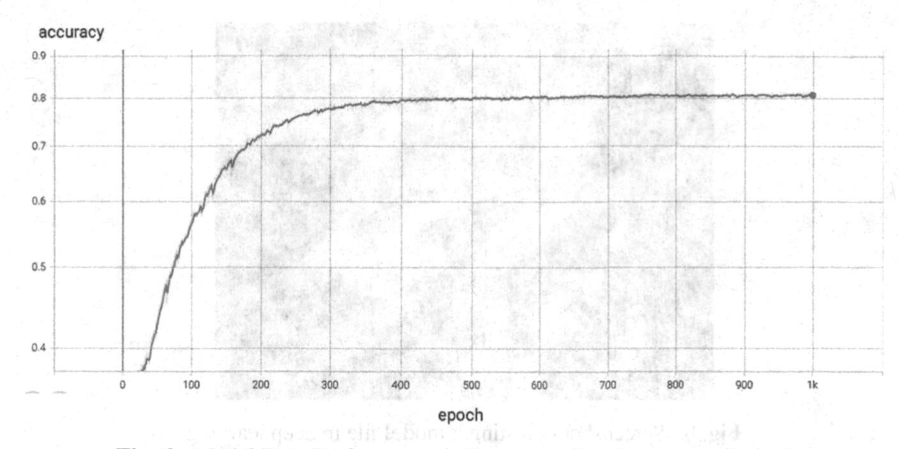

Fig. 6. Model Data Performance in Deep Learning (Accuracy Curves)

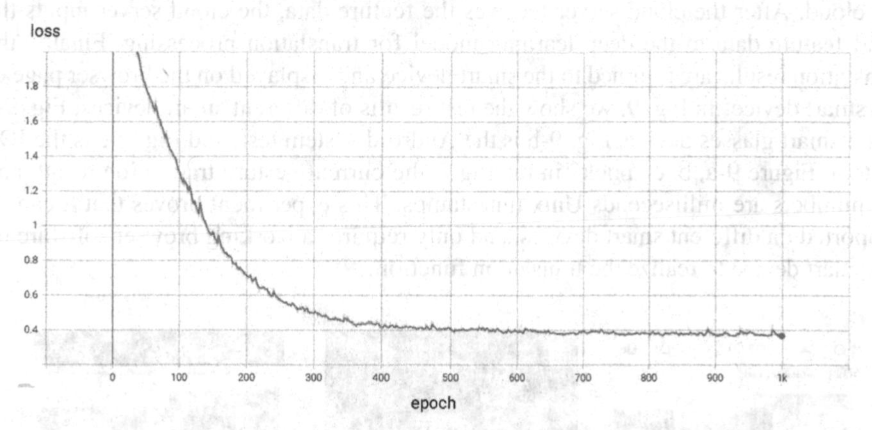

Fig. 7. Loss rate performance in deep learning (loos curves)

4.2 Evaluation and Discussion

This chapter an evaluation and discussion of the model and the cloud-based sign language translation service.

As shown in Fig. 8, in our deep learning experiment, we tested the model files generated by training. We asked sign language users to make hand movements in front of the camera. With the MediaPipe hand recognition module, we can detect the hand movements and input the detected hand feature data to the model for translation processing. Finally, the translated results are output on the video screen. As shown in Fig. 9, "computer" is to show the meaning of the text represented by the current gesture.

As shown in Fig. 9, we conducted an experiment combining multiple smart devices and cloud-based translation function tests. A browser is opened on the smart device and the MediaPipe page is visited. The sign language user makes a hand movement in front of the camera. MediaPipe detects the hand movement and maps the hand feature points. The detected feature data is then sent to a sign language translation server in

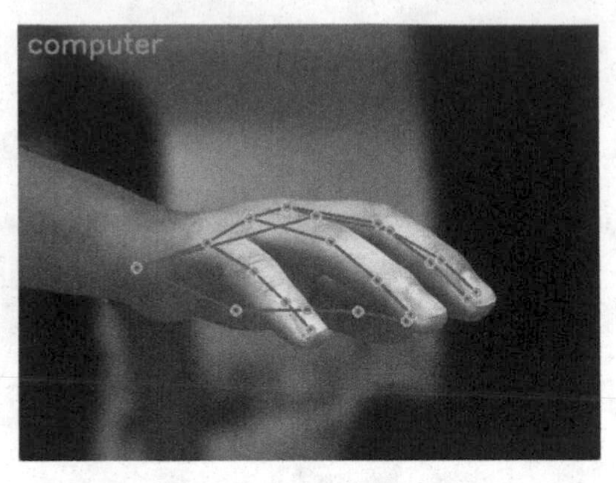

Fig. 8. Screenshot of testing a model file in deep learning

the cloud. After the cloud server receives the feature data, the cloud server inputs the hand feature data to the deep learning model for translation processing. Finally, the translation results are returned to the smart device and displayed on the browser page of the smart device. In Fig. 9, we show the test results of different smart devices, Fig. 9-a is the smart glasses device, Fig. 9-b is the Android system test, and Fig. 9-c is the IOS system. Figure 9-a, b, c "book" in the Fig is the current gesture translation result, and the numbers are milliseconds Unix timestamps. This experiment proves that it can be supported on different smart devices, and only requires a working browser software on the smart device to realize the translation function.

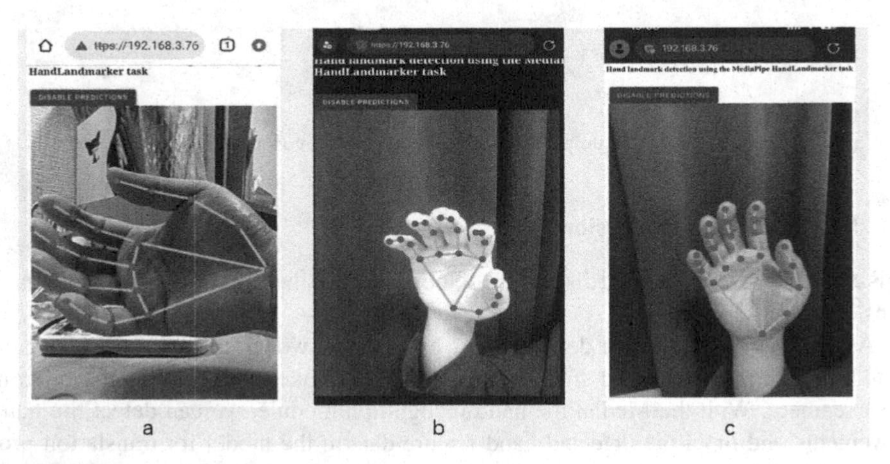

Fig. 9. Different smart devices + cloud translation function test

In addition, we conducted a multi-person test of the accuracy of the sign language translation system.

As shown in Fig. 10, after completing the sign language translation system, we recruited five experimentalists for testing. During the test, we taught the experimenter

Fig. 10. Photograph taken by the experimenter during the test of sign language.

the sign language movement represented by "computer" and asked the experimenter to perform the sign language movement in front of the sign language translation system in the computer, and the sign language system captured the sign language movement of the experimenter and then translated it. To obtain the results, a total of 10 runs were performed for each experimenter.

Table 4. Sign language recognition results test list

	time 1	time 2	time 3	time 4	time 5	time 6	time 7	time 8	time 9	time 10
experimenter 1	√	√	×	√	√	√	√	√	×	√
experimenter 2	√	×	√	√	×	×	√	√	×	√
experimenter 3	√	√	×	×	√	√	√	×	√	√
experimenter 4	×	×	√	√	√	×	√	√	×	√
experimenter 5	√	√	√	√	√	√	√	√	√	×

As shown in Table 4, the number of tests performed by the five participants and whether the test was correct or not. After the sign language system translated the hand movements of the participants, we obtained a correct rate of 72% according to the number of correct tests performed by each participant. It indicates that the sign language translation system has some accuracy in translation.

5 Conclusion

In order to assist sign language users to communicate more efficiently, we designed and implemented a sign language translation system based on MediaPipe and TensorFlow. MediaPipe is used to identify and collect sign language action data of WLASL dataset and provide it to the 1D-CNN deep learning model of TensorFlow for training, so as to realize the sign language translation function. In addition, we opened the browser through the smart device, visited the page of MediaPipe, and used the cloud translation server to input the hand feature data into the deep learning system for translation processing. Finally, the translation results were returned to the smart device and displayed on the browser page of the smart device. The system can be supported on smart glasses, Android and IOS devices, and the translation function can be realized only by using the browser software. In the test phase, we invited five experimenters to test the sign language translation system ten times for each person, and the final average accuracy rate was 72%. Our sign language translation system is prepared and can help sign language users to communicate efficiently.

At the same time, we found that some testers' sign language movements could not be accurately translated during the testing phase, which may be caused by insufficient samples of the trained hand data. Therefore, in the future, we will introduce more training data and combine expressions, speech and hand movements, as well as optimize the existing server architecture to improve the accuracy and stability of sign language translation.

Acknowledgments. Appreciations for all the workers who participated in the experiments. This work was supported by JSPS KAKENHI Grant Number 21K11876.

References

1. Zhou, X., Yen, N.Y., Jin, Q., Shih, T.K.: Enriching user search experience by mining social streams with heuristic stones and associative ripples. Multimedia Tools Appl. **63**(1), 129–144 (2013). https://doi.org/10.1007/s11042-012-1069-1
2. Wu, B., Zhu, Y., Yu, K., Nishimura, S., Jin, Q.: The Effect of eye movements and culture on product color selection. HCIS **10**(48), 2020 (2020)
3. Aliyu, S., Mohandes, M., Deriche, M., Badran, S.: Arabie sign language recognition using the Microsoft Kinect. In: 2016 13th International Multi-Conference on Systems, Signals & Devices (SSD), pp. 301–306 (2016)
4. Fang, B., Co, J., Zhang, M.: Deepasl: enabling ubiquitous and non-intrusive word and sentence-level sign language translation. In: Proceedings of the 15th ACM Conference on Embedded Network Sensor Systems, pp. 1–13 (2017)
5. Liang, W., Zhou, X., Huang, S., Hu, C., Xu, X., Jin, Q.: Modeling of cross-disciplinary collaboration for potential field discovery and recommendation based on scholarly big data. Futur. Gener. Comput. Syst. **87**, 591–600 (2018). https://doi.org/10.1016/j.future.2017.12.038
6. Kau, L.J., Su, W.L., Yu, P.J., Wei, S.J.: A real-time portable sign language translation system. In: 2015 IEEE 58th International Midwest Symposium on Circuits and Systems (MWSCAS), pp. 1–4 (2015)
7. Lugaresi, C., et al.: MediaPipe: a framework for building perception pipelines. arXiv preprint arXiv:1906.08172 (2019)

8. Shriram, S., Nagaraj, B., Jaya, J., Shankar, S., Ajay, P.: Deep learning-based real-time AI virtual mouse system using computer vision to avoid COVID-19 spread. J. Healthcare Eng. (2021). https://www.hindawi.com/journals/jhe/2021/8133076/
9. WLASL (World Level American Sign Language) Videos. www.kaggle.com/datasets/gazquez/wlasl-processed. Accessed 7 Jul 2023
10. Abadi, M., et al.: Tensorflow: large-scale machine learning on heterogeneous distributed systems. arXiv preprint arXiv:1603.04467 (2016)
11. Wu, B., Wu, Y., Nishimura, S., Jin, Q.: Analysis on the subdivision of skilled mowing movements on slopes. Sensors **22**(4), 1372 (2022)
12. Takahashi, R., Saito, H.: Sign language recognition by 3D CNN transformer. In: Proceedings of the 36th National Conference of Japanese Society for Artificial Intelligence. Japanese Society for Artificial Intelligence, pp. 4C1GS703–4C1GS703 (2022)
13. Rauschnabel, P.A., Hein, D.W., He, J., Ro, Y.K., Rawashdeh, S., Krulikowski, B.: Fashion or technology? A fashnology perspective on the perception and adoption of augmented reality smart glasses. i-com **15**(2), 179–194 (2016)
14. Pierdicca, R. et al.: Augmented reality smart glasses in the workplace: safety and security in the fourth industrial revolution era. In: De Paolis, L., Bourdot, P. (eds.) Augmented Reality, Virtual Reality, and Computer Graphics, AVR 2020. LNCS, vol. 12243, pp. 231–247. Springer, Cham (2020). https://doi.org/10.1007/978-3-030-58468-9_18
15. Mitrasinovic, S., et al.: Clinical and surgical applications of smart glasses. Technol. Health Care **23**(4), 381–401 (2015)
16. Samini, A., Palmerius, K.L., Ljung, P.: A review of current, complete augmented reality solutions. In: 2021 International Conference on Cyber-worlds (CW), pp. 49–56 (2021)
17. Zhou, X., Wang, W., Jin, Q.: Multi-dimensional attributes and measures for dynamical user profiling in social networking environments. Multimedia Tools Appl. **74**(14), 5015–5028 (2015). https://doi.org/10.1007/s11042-014-2230-9
18. Liu, P., Pan, J., Zhu, H., Li, Y.: A wearable fall detection sys-tem based on 1D CNN. In: 2021 2nd International Conference on Artificial Intelligence and Computer Engineering (ICAICE), pp. 200–203 (2021)
19. Wu, B., Wu, Y., Dong, R., et al.: Behavioral analysis of mowing workers based on Hilbert–Huang transform: an auxiliary movement analysis of manual mowing on the slopes of terraced rice fields. Agriculture **13**(20), 489 (2023)
20. Ozcanli, A.K., Baysal, M.: Islanding detection in microgrid using deep learning based on 1D CNN and CNN-LSTM networks. Sustain. Energy Grids Networks **32**, 100839 (2022)
21. Nedelcu, Clément. Nginx Http Server. Packt Publishing Ltd. (2010)

TBSA-Net: A Temperature-Based Structure-Aware Hand Pose Estimation Model in Infrared Images

Hongfu Xia[1], Yang Li[2,3], Chunyan Liu[1], and Yunlong Zhao[1](✉)

[1] College of Computer Science and Technology, Nanjing University of Aeronautics and Astronautics, Nanjing, China
zhaoyunlong@nuaa.edu.cn

[2] Unmanned Aerial Vehicles Research Institute, Nanjing University of Aeronautics and Astronautics, Nanjing, China

[3] Key Laboratory of Advanced Technology for Small and Medium-Sized UAV, Ministry of Industry and Information Technology, Nanjing, China

Abstract. In recent years, numerous researchers have conducted in-depth studies and made significant progress in 2D Hand Pose Estimation (HPE) tasks on RGB images. However, the field of HPE in the context of infrared images has received limited attention. Due to the limited channel information and high correlation with temperature, models designed for RGB images may suffer from insufficient accuracy in infrared images. Our experiments reveal that the temperature distributions of the human hand in infrared images exhibit significant regularity. In this paper, we propose the Temperature-Based Hand Judgement Model (TB-HJM) that leverages this characteristic. During the training phase, a higher penalty is given when the predicted pose's temperature distribution does not align with the actual temperature distribution, and vice versa. In the testing phase, TB-HJM is utilized to select a hand proposal that closely matches the temperature distribution as the final output. Additionally, to address the lack of visual information in infrared images, we use PBN-Head and GCN Refine Module to merge structural information into the network to ensure model accuracy. Experimental results demonstrate that our model outperforms the benchmark model (HRNet) by 1.72% in terms of AUC and achieves an improvement of 0.6448 by reducing the EPE from 3.02 to 2.38, achieving state-of-the-art performance on our infrared hand dataset.

Keywords: Infrared Image · Hand Pose Estimation · GCN

1 Introduction

Hand pose estimation (HPE) is the task of finding the joints of the hand from an image or set of video frames. This field has experienced rapid development in recent years [17,23,24] and has found widespread applications in various domains such as augmented reality (AR), virtual reality (VR), and smart homes. Due to

H. Jin et al. (Eds.): GPC 2023, LNCS 14504, pp. 238–250, 2024.
https://doi.org/10.1007/978-981-99-9896-8_16

the similarity between hand pose estimation and human pose estimation tasks, many hand pose estimation models are often derived from human pose estimation [4]. Numerous scholars have made remarkable contributions to the field of hand pose estimation [3,5,10,12,14,20–22,25,26]. Currently, significant progress has been achieved in 2D human pose estimation tasks, with the highest accuracy reaching 66.9% on COCO whole body dataset [6]. However, existing HPE models primarily focus on HPE task using RGB images. RGB images, with their rich color and shading information, enable these models to achieve high levels of performance in RGB image.

Infrared refers to electromagnetic waves with wavelengths between microwaves and visible light, typically ranging from 760 nm to 1 mm [15]. Infrared images are obtained through imaging with infrared radiation. Compared to RGB images, infrared images are lack of color and shading, have lower resolution, and lack depth, resulting in limited optical information. Current human pose estimation models designed for RGB images overly rely on visual information and neglect the structural information of the hand. Additionally, the grayscale values in infrared images are related to temperature, which is a useful characteristic for hand pose estimation tasks. Consequently, existing hand pose estimation models that achieve good performance on regular RGB images cannot effectively perform in infrared images.

Hand pose estimation in infrared images can serve as a supplementary task to RGB image-based hand pose estimation. In scenarios with poor visibility, such as nighttime or low-light conditions, performing keypoint detection using RGB images can be challenging. In such cases, the consideration of hand keypoint detection in infrared images becomes valuable. Fields like AR, VR smart homes, and specialized scenarios like combat operations can benefit from this approach. In situations with inadequate lighting, the use of infrared cameras, which are not dependent on ambient light, enables enhanced keypoint detection accuracy through the utilization of infrared images.

This paper utilizes the distinctive feature of the grayscale distribution of 21 key hand keypoints in infrared images to design the TB-HJM model. The TB-HJM model evaluates the rationality of the generated Hand Proposal by inputting the grayscale values corresponding to the hand keypoints in infrared images. The TB-HJM model is used to enhance the performance of hand pose estimation tasks in infrared images. Additionally, this paper incorporates the skeletal structure information of the human hand into the keypoint detection model using GCN and PBNHead, further improving the network's performance. Experimental results demonstrate that the proposed model achieves higher accuracy in hand keypoint detection tasks in infrared images, surpassing existing models and reaching the state-of-the-art level.

In summary, the contributions of this work are 1) We created a hand objects and keypoints dataset for infrared images, named the InfraredHands Dataset. All the experiments conducted in this paper were performed on this dataset. 2) we introduce a hand keypoint detection model named TBSA-Net, which demonstrates high accuracy in infrared images. To address the lack of visual information in infrared images, we introduce skeletal structural information into

TBSA-Net through the GCNRefine module and the PBNHead module. 3) We have constructed a temperature-based hand judgement model (TB-HJM) that incorporates information about the potential temperature distribution of hand keypoints. This discriminative model allows TBSA-Net to generate results that better align with the underlying temperature distribution of the hand.

2 Related Works

Given the similarity between hand pose estimation tasks and general human pose estimation tasks, and the scarcity of dedicated hand pose estimation, we will discuss the related works for 2D human pose estimation in general. Human pose estimation is an upstream and fundamental problem in computer vision. Currently, the majority of 2D human pose estimation tasks can be divided into two categories based on the different output representations: regression-based HPE and heatmap-based HPE [29]. Next, we will introduce the related work for each of these two types of HPE methods.

Regression-Based HPE

Regression-based HPE networks take a optical image like RGB image as input and output the coordinate representation of the human body. Regression-based HPE models typically have lower accuracy compared to heatmap-based HPE models. However, they have smaller computational overhead and are relatively easier to implement. Due to the accuracy limitations, this approach is commonly found in early HPE models. For instance, DeepPose [21], as the first deep learning-based method for single-person human pose estimation, adopts a multi-stage regression approach using CNN. It directly regresses the 2D coordinates of human skeletal keypoints, with coordinates as the optimization target. Subsequent stages of optimization are applied to achieve higher accuracy. The principle reason of low precision of Regression-based HPE is because of lack supervisory information. Li et al. [8] introduce the residual log-likelihood (RLE), which utilizes the normalizing flows to capture the underlying output distribution and makes regression-based methods match the accuracy of state-of-the-art heatmap-based methods.

Heatmap-Based HPE

Unlike Regression-based HPE, Heatmap-based HPE produces N (N is the number of keypoints, set to 21 in this paper) heatmaps as an intermediate representation, and then employs post-processing techniques to obtain direct coordinate representations from the heatmaps [25]. One drawback of the heatmap-based approach is the potential loss of precision during the encoding and decoding process. Additionally, heatmap-based HPE is typically slower due to the need for complex post-processing. However, because heatmap-based HPE incorporates more supervision information, it achieves higher accuracy compared to

Regression-based HPE. The Convolutional Pose Machine proposed by Wei [25] can simultaneously learn image features and rely on spatial models dependent on the image. The Stacked Hourglass model constructed by Newell [12] effectively combines global and local information. In recent studies, some researchers have considered abandoning heatmaps due to its limitations. For example, SimCC [11] achieves state-of-the-art results on multiple datasets by treating keypoints detection as a coordinate classification task. In this paper, we adopt heatmap-based HPE, and after generating the heatmaps, a differentiable integration method [19] is applied to obtain the actual keypoints coordinates. In addition, the grayscale values corresponding to the coordinates in the original image are concatenated with the coordinates themselves and used as input to the GCN Refine module.

GCN for Pose Modelling

The human body shows a natural graph structure, so that some advanced work constructed graph networks to address human pose related problems, such as action recognition [27], motion prediction [9], 3D pose regression [1,28]. These works intuitively form the natural human pose as a graph and apply convolutional layers on it. Compared to other approaches, Graph Convolutional Networks demonstrate one compelling advantage when deal with human pose modeling problem: they are more effective in capturing dependency relationships between joints. Therefore, in paper [14], an Image-Guided Progressive GCN module is appended after the traditional CNN model to fine-tune the Initial Pose obtained from the CNN outputs, resulting in the Final Pose. This method yields significant improvements, especially on occluded joints. Inspired by [14], this study employs a similar approach by appending a GCN Refine Module after the CNN model, integrating hand skeletal structural information using the GCN Refine Module to enhance the model's accuracy

3 The Overall Structure of TBSA-Net

In this paper, we use HRNet [18] as the Backbone, where the output of HRNet serves as the input to Part-based branching network head (PBNHead) [20]. After passing through PBNHead, heatmaps are obtained. We then apply an integral method [19] to obtain the initial pose, marked as Hand1. Additionally, PBNHead generates two feature matrices, F_1 and F_2. By performing a grid sample operation on Hand1 with F_1 and F_2, we obtain feature vectors J_1 and J_2. Hand1, along with the grayscale values corresponding to the coordinates in the original image, J_1, and J_2, are used as inputs to the GCN refine module [14]. The GCN refine module fine-tunes Hand1 to generate Hand2 and Hand3. During the training phase, the joint loss of the three generated hands is used as the model's loss function. In the testing phase, TB-HJM is utilized to select the final hand output from the three generated hands that best matches the temperature distribution. The overall architecture of the model is shown in Fig. 1.

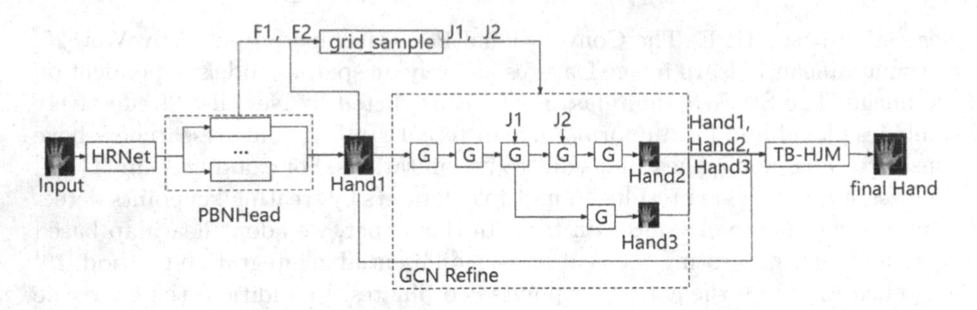

Fig. 1. The overall architecture of the model.

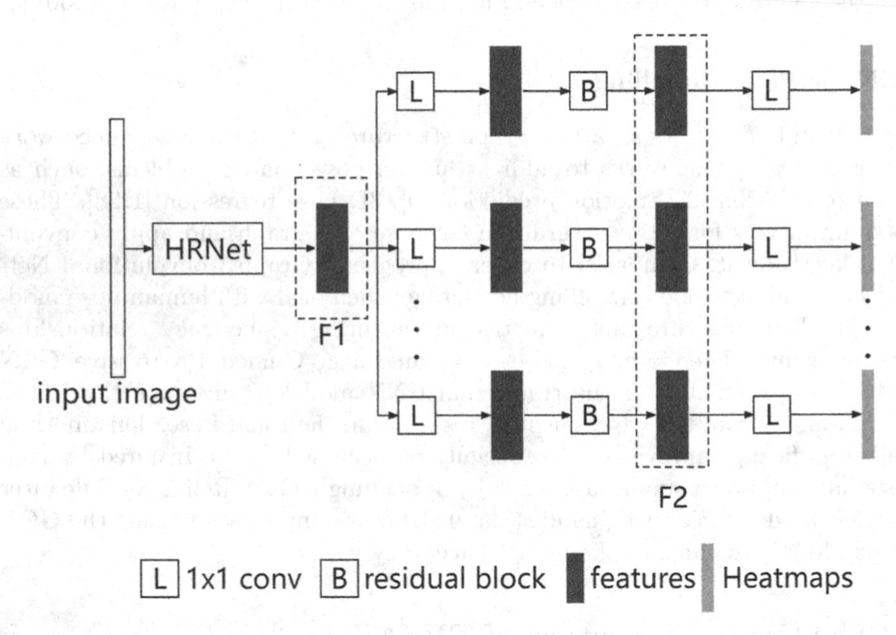

Fig. 2. The architecture of PBNHead, basically followed [20]

3.1 PBNHead

The multi-head design in this paper is mainly inspired by [20]. HPE is inherently a homogeneous multi-task learning (MTL) problem [16], with the localization of each part as a different task. Sharing a representation among related tasks can result in a more compact model and better generalization ability [16,30]. Following [20], this paper groups the keypoints accordingly. Keypoints with high correlation are grouped together, while keypoints with low correlation are assigned to different groups. Each group is predicted using a dedicated branch. By grouping the keypoints, it allows keypoints with high correlation to share all features, while keypoints with low correlation share only some features. The architecture of the PBNHead can be seen in Fig. 2.

After feeding the images into the backbone network to extract shared features, these features are then inputted into different branches to predict heatmaps for different key points. In this paper, HRNet [18], which performs well in the field of human pose estimation, is chosen as the backbone network. The hand key points are divided into five groups in this paper, and each branch predicts the heatmap for one group of key points. Each branch consists of a 1×1 convolutional layer and a residual module, and finally, a 1×1 convolutional layer is concatenated to obtain the heatmap. The grouping diagram of hand key points can be seen in Fig. 3, where (0–4), (5–8), (9–12), (13–16), (17, 20) are divided into separate groups, totaling five groups.

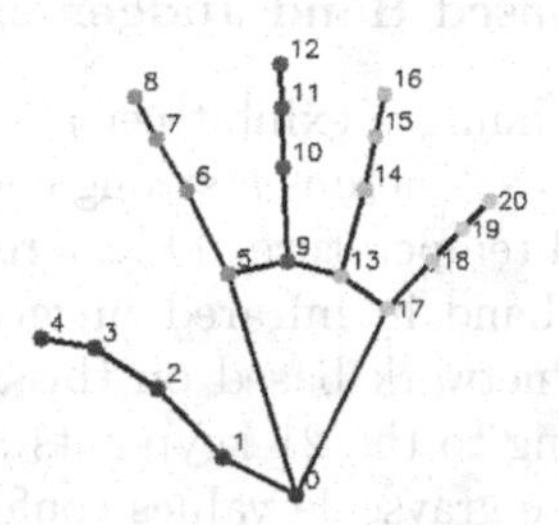

Fig. 3. Hand grouping diagram, each color represents a group, divided into a total of five groups.

3.2 GCN Refine Module

The design of the GCN Refine module in this paper is mainly inspired by the paper [14]. Following the approach presented in that paper, in this paper we incorporate the outputs of a traditional model into the GCN Refine module to fine-tune the results. Unlike [14], the GCN Refine module in this paper outputs three Hand Proposals, and the input is concatenated with the corresponding grayscale values of the key points. During the training stage, the loss is obtained from the generated three Hand Proposals. The designed GCN Refine module in this paper can be seen in Fig. 4.

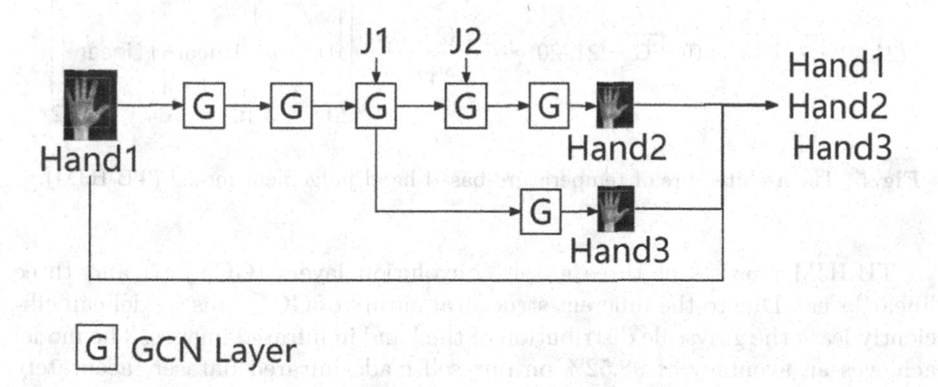

Fig. 4. The architecture of GCN refine module. Hand1, J_1 and J_2 are come from PBNHead, where Hand1 is come from the output heatmap of PBNHead.

The dimensions of input of the GCN Refine module is [N, 3], where N represents the number of key points, which is 21 in this paper. The 3 dimensions correspond to the x and y coordinates of the key point, as well as the grayscale value at that position. GCN Refine module consists of 4 GCN (Graph Convolutional Network) Layers. In the third and fourth layers, the features J_1 and J_2, which correspond to the key point's position in F_1 and F_2 respectively, are concatenated. Here, F_1 and F_2 represent the features outputted by the PBNHead.

3.3 The Temperature Based Hand Judgement Model

As warm-blooded creatures, humans exhibit certain regularity in the temperature distribution of their hands. There is a strong correlation between grayscale values in infrared images and temperature. Therefore, the grayscale distribution of the 21 keypoints of the hand in infrared images also follows certain patterns. This paper designs a network based on this regularity, which takes the grayscale values corresponding to the 21 keypoints in infrared images as input and determines whether these grayscale values conform to the grayscale distribution of the hand in infrared images. More specifically, the temperature based hand judgement model (TB-HJM) takes normalized grayscale values as input and outputs a one-dimensional vector with a length of 2 indicating whether the 21 grayscale values conform to the grayscale distribution of the hand keypoints in infrared images. It is important to note that this approach is only effective in infrared images since the numerical values in RGB images represent the brightness levels of the red, green, and blue channels and have no direct relation to the hand. Infrared image values, on the other hand, can reflect the temperature of objects, and the temperature of the hand exhibits strong regularity. Therefore, by inputting the grayscale values corresponding to the keypoints from the original image, the generated hand can be evaluated for its reasonableness. The structural of TB-HJM shows in Fig. 5.

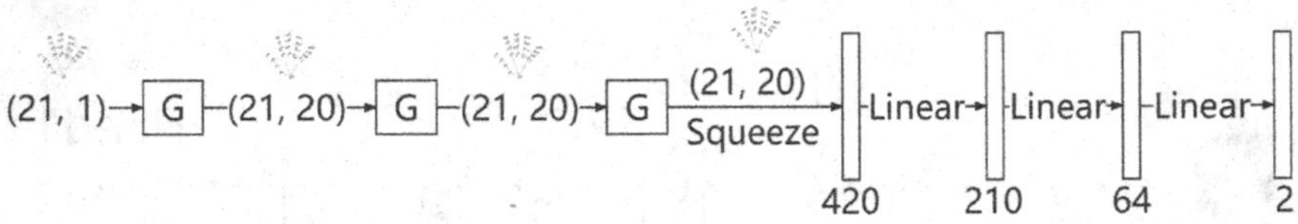

(21, 1) → G → (21, 20) → G → (21, 20) → G → (21, 20) Squeeze → 420 → Linear → 210 → Linear → 64 → Linear → 2

Fig. 5. The architecture of temperature-based hand judgement model (TB-HJM).

TB-HJM consists of three graph convolution layers (GCN) [7] and three linear layers. Due to the inherent structural nature of GCN, this model can efficiently learn the grayscale distribution of the hand in infrared images. The model achieves an accuracy of 98.52% on our self-made infrared dataset, accurately reflecting whether the input of 21 grayscale values conforms to the grayscale distribution of the hand in infrared images. During the training phase, the loss value can be dynamically adjusted by using TB-HJM to evaluate the model's output. Generated hands that conform to the grayscale distribution are assigned a lower loss, while hands that do not conform to the grayscale distribution are assigned a higher loss. By adjusting the loss value, the model is encouraged to generate hands that better align with the temperature distribution. During the testing phase, TB-HJM is utilized to select a hand that more closely matches the temperature distribution from multiple generated hands as the final output.

3.4 Loss Function

The loss function is an important component in machine learning, as it determines the optimization direction of the model. In this model, a multi-task learning optimization strategy is employed to learn the loss of three poses. During the training phase, TB-HJM is utilized to measure whether the output satisfies the grayscale distribution of the 21 hand key points in the infrared image. If the grayscale distribution does not comply, a higher loss should be assigned. The loss function used in this paper is as follows:

$$Loss = \sum_{i=0}^{2} \lambda_i (2 - P_i) Loss_{Hand_i} \tag{1}$$

The λ_i in Eq. (1) is used to control the weight of different Hand. P_i represents the degree to which $Hand_i$ conforms to the grayscale distribution (or it can be understood as the probability that $hand_i$ is a hand based solely on the grayscale distribution). Which $2 - p_i$ means that if $hand_i$ does not satisfy the grayscale distribution (i.e., p_i is close to 0), the loss will be amplified by a factor of two. If $hand_i$ satisfies the grayscale distribution (i.e., p_i is close to 1), the loss will remain unchanged. $Loss_{Hand_i}$ is MSE loss following this equation:

$$loss(\hat{y}, y) = \frac{1}{n} \sum (\hat{y}_i - y_i)^2 \tag{2}$$

4 Experiments

4.1 Infrared Hand Keypoints Dataset

The data collection device for the dataset is infiRay T2. The dataset was collected by six experimenters who performed various hand gestures in front of the camera to capture video data, which was then segmented into frames. Every fifth frame was selected and fed into an existing hand pose estimation model for automatic annotation, resulting in annotated data. Afterward, a manual selection process was used to remove incorrect and inaccurate annotations, resulting in 3,211 infrared images with annotations. Additionally, 177 infrared images were manually annotated as supplementary data. In the end, the dataset comprises a total of 3,388 images with a size of 455×346 pixels.

4.2 Experiments Setting

For training, we set the parameters $\lambda_1 = 1$, $\lambda_2 = 0.3$, $\lambda_3 = 1$ and $epochs = 210$. The batch size is set to 16. The initial learning rate is set to $5e{-}4$, and it decreases to one-tenth of its original size at steps 170 and 200. The input image size are 256×256. We utilize the AdamOptimizer for parameter optimization through backpropagation. In this paper, we utilize three widely used evaluation metrics in the Hand Pose Estimation field: AUC (Area Under the Curve), EPE (Euclidean Pixel Error) and PCK (Percentage of Correct Keypoints). We implement our model in PyTorch [13] and mmpose [2] and conduct experiments on one Nvidia GTX 1080Ti with 11GB memory. Additional details include the usage of torch-1.8.1, torchvision-0.9.1, cuda-11.1, and cudann-8.4.1.

For TB-HJM, we use the grayscale values corresponding to the ground truth hand coordinates as positive samples during training. Additionally, we randomly apply offsets of 0 to k pixels in distance and 0 to θ degrees in angle to the 21 key points of the positive samples. We consider the grayscale values corresponding to the offsetted key point coordinates as negative samples, simulating inaccurate hand coordinates generated by the model. The positive-to-negative sample ratio is set to $1:1$. The test set consists of 5136 samples, while the training set consists of 1286 samples. We set $k = 25$ and $\theta = 360°$. The final accuracy of the model is 98.52%. Therefore, the model can effectively determine whether the 21 grayscale values inputted satisfy the hand's grayscale distribution in infrared images.

We will provide a detailed description of the model's details. First, the input image is fed into HRNet. The output of HRNet serves as the input for PBNHead. PBNHead generates an N-dimensional heatmap and also outputs two features of different sizes F_1 and F_2. Using the grid sample method in PyTorch, we obtain the features J_1 and J_2 corresponding to the keypoints. To refine the coordinates obtained in the previous stage, we concatenate them with the corresponding temperature values from the original image. This concatenated input is then fed into the GCN Refine module, where fine-tuning is performed on the input data to generate two proposals. Using TB-HJM, the proposal that best matches the potential temperature distribution is selected as the final output. During the training phase, the joint loss of the three proposals is used. TB-HJM dynamically adjusts the weight of the loss to guide the model towards generating outputs that better match the potential temperature distribution.

4.3 Comparison Experiments

To validate the effectiveness and superiority of the proposed TBSA-Net for hand pose estimation in infrared images, we compare our model with SimpleBaseLine [26], RTMPose [5], and HRNet [18], as shown in Table 1. Our proposed model exhibits superior accuracy in the task of hand pose estimation in infrared images when compared to existing models. Specifically, our model outperforms HRNet by 1.72% in terms of AUC and achieves an improvement of 0.6448 by reducing the EPE from 3.02 to 2.38. And outperforms HRNet by 0.10% in terms of PCK@0.2, PCK@0.2 stands for Percentage of Correct Keypoints at a threshold of 0.2, which is a metric used to evaluate the performance of hand pose estimation models. It represents the proportion of predictions where the distance between the predicted keypoint and the ground truth keypoint is less than or equal to 20% of the image size in normalized image coordinates. These results demonstrate the efficacy of our approach in addressing the challenges posed by hand pose estimation in infrared images. This demonstrates the significant advantage of our proposed model in hand keypoint detection tasks specifically for infrared images.

Table 1. The result of comparison experiments.

	PCK@0.2	AUC	EPE
SimpleBaseline	99.90%	87.37%	2.8936
RTMPose	99.85%	89.21%	2.5584
HRNet	99.82%	87.92%	3.0204
TBSA-Net(Ours)	**99.92%**	**89.64%**	**2.3756**

4.4 Ablation Studies

Ablation Studies for TBSA-Net. To further validate the necessity and superiority of each module's modifications, we conducted ablation experiments on the Infrared Hand Keypoints Dataset. We progressively added modules to the baseline model to obtain the ablation experiment results. HRNet was used as the baseline model, and the effectiveness of each module was verified by adding them one by one. The experimental results are presented in Table 2. According to the experimental results, we observed a certain degree of accuracy improvement after adding each module, thus demonstrating the necessity and effectiveness of the modifications made to each module. The final experimental results confirm that the proposed algorithm effectively enhances the accuracy of hand pose estimation in infrared images. After adding the PBNHead to HENet, our model achieved an AUC improvement of 0.82%, increasing from 87.92% to 88.74%. Furthermore, after adding the GCN module, our model achieved an AUC improvement of 0.90%, increasing from 88.74% to 89.64%. The results of the ablation experiments demonstrate the effectiveness of each of our improvements.

Table 2. The result of ablation studies.

	PCK@0.2	AUC	EPE
HRNet	99.82%	87.92%	3.0204
HRNet+PBNHead	99.86%	88.74%	2.7085
HRNet+PBNHead+GCN(final)	**99.92%**	**89.64%**	**2.3756**

Ablation Studies for TB-HJM. We conducted some ablation experiments on TB-HJM and found that using three fully connected layers resulted in low accuracy. We then attempted to improve the TB-HJM model's accuracy by adding GCN layers to the model. We used the grayscale values corresponding to the 21 key points in the ground truth as positive samples and offset these key points to some extent and used the grayscale values corresponding to the offset key points as negative samples, with a positive-to-negative sample ratio of 1:1. The experimental results showed that the model's accuracy was higher after introducing the GCN layer. We conducted ablation experiments on the number of GCN layers and the dimensions and found that the best results were achieved with three layers and a dimension of 20. The experimental results are presented in Table 3.

Table 3. The impact of the number of GCN layers and dimensions on accuracy.

the number of GCN layers	dimensions	accuracy
3	10	98.29%
3	20	**98.52%**
3	30	98.44%
3	40	98.37%
4	10	97.90%
4	20	98.52%
4	30	98.44%
2	10	98.44%
2	20	98.29%
2	30	98.29%

5 Conclusion

Existing hand pose estimation (HPE) models are typically designed for RGB images and do not perform optimally in infrared images. In this paper, we proposed a model for the task of hand pose estimation in infrared images. The core idea is to incorporate hand structural information into the network by introducing PBNHead and GCN, aiming to enhance the performance of hand keypoints localization. Additionally, due to the strong regularity in the grayscale distribution of the 21 hand keypoints in infrared images, we design a TB-HJM model which can modify loss function to guide the model's descent direction, aiming to generate results that better conform to the grayscale distribution and improve the accuracy of the model. The model designed in this paper allows for the replacement of the Backbone, achieving a balance between accuracy and speed by using different Backbones.

Acknowledgements. This research was supported by the National Key Research and Development Program of China under Grant No. 2022ZD0115403, National Natural Science Foundation of China under Grant No. 62072236, and the Fundamental Research Funds for the Central Universities under Grant No. 56XCA2205404.

References

1. Ci, H., Wang, C., Ma, X., Wang, Y.: Optimizing network structure for 3D human pose estimation. In: Proceedings of the IEEE/CVF International Conference on Computer Vision, pp. 2262–2271 (2019)
2. MMPose Contributors: Openmmlab pose estimation toolbox and benchmark (2020). https://github.com/open-mmlab/mmpose
3. Geng, Z., Sun, K., Xiao, B., Zhang, Z., Wang, J.: Bottom-up human pose estimation via disentangled keypoint regression. In: Proceedings of the IEEE/CVF Conference on Computer Vision and Pattern Recognition, pp. 14676–14686 (2021)
4. Iqbal, U., Molchanov, P., Gall, T.B.J., Kautz, J.: Hand pose estimation via latent 2.5 D heatmap regression. In: Proceedings of the European Conference on Computer Vision (ECCV), pp. 118–134 (2018)
5. Jiang, T., et al.: RTMPose: real-time multi-person pose estimation based on MMPose. arXiv preprint arXiv:2303.07399 (2023)
6. Jin, S., et al.: Whole-body human pose estimation in the wild. In: Vedaldi, A., Bischof, H., Brox, T., Frahm, J.-M. (eds.) ECCV 2020. LNCS, vol. 12354, pp. 196–214. Springer, Cham (2020). https://doi.org/10.1007/978-3-030-58545-7_12
7. Li, G., Muller, M., Thabet, A., Ghanem, B.: DeepGCNs: can GCNs go as deep as CNNs? In: Proceedings of the IEEE/CVF International Conference on Computer Vision, pp. 9267–9276 (2019)
8. Li, J., Bian, S., Zeng, A., Wang, C., Pang, B., Liu, W., Lu, C.: Human pose regression with residual log-likelihood estimation. In: Proceedings of the IEEE/CVF International Conference on Computer Vision, pp. 11025–11034 (2021)
9. Li, M., Chen, S., Chen, X., Zhang, Y., Wang, Y., Tian, Q.: Actional-structural graph convolutional networks for skeleton-based action recognition. In: Proceedings of the IEEE/CVF Conference on Computer Vision and Pattern Recognition, pp. 3595–3603 (2019)
10. Li, W., et al.: Rethinking on multi-stage networks for human pose estimation. arXiv preprint arXiv:1901.00148 (2019)
11. Li, Y., et al.: SimCC: a simple coordinate classification perspective for human pose estimation. In: Avidan, S., Brostow, G., Cissé, M., Farinella, G.M., Hassner, T. (eds.) ECCV 2022, Part VI, pp. 89–106. Springer, Cham (2022). https://doi.org/10.1007/978-3-031-20068-7_6
12. Newell, A., Yang, K., Deng, J.: Stacked hourglass networks for human pose estimation. In: Leibe, B., Matas, J., Sebe, N., Welling, M. (eds.) ECCV 2016. LNCS, vol. 9912, pp. 483–499. Springer, Cham (2016). https://doi.org/10.1007/978-3-319-46484-8_29
13. Paszke, A., et al.: Pytorch: an imperative style, high-performance deep learning library. In: Advances in Neural Information Processing Systems, vol. 32, pp. 8024–8035. Curran Associates, Inc. (2019). http://papers.neurips.cc/paper/9015-pytorch-an-imperative-style-high-performance-deep-learning-library.pdf

14. Qiu, L., et al.: Peeking into occluded joints: a novel framework for crowd pose estimation. In: Vedaldi, A., Bischof, H., Brox, T., Frahm, J.-M. (eds.) ECCV 2020. LNCS, vol. 12364, pp. 488–504. Springer, Cham (2020). https://doi.org/10.1007/978-3-030-58529-7_29
15. Rogalski, A., Chrzanowski, K.: Infrared devices and techniques. Optoelectron. Rev. **10**(2), 111–136 (2002)
16. Ruder, S.: An overview of multi-task learning in deep neural networks. arXiv preprint arXiv:1706.05098 (2017)
17. Santavas, N., Kansizoglou, I., Bampis, L., Karakasis, E., Gasteratos, A.: Attention! a lightweight 2D hand pose estimation approach. IEEE Sens. J. **21**(10), 11488–11496 (2020)
18. Sun, K., Xiao, B., Liu, D., Wang, J.: Deep high-resolution representation learning for human pose estimation. In: Proceedings of the IEEE/CVF Conference on Computer Vision and Pattern Recognition, pp. 5693–5703 (2019)
19. Sun, X., Xiao, B., Wei, F., Liang, S., Wei, Y.: Integral human pose regression. In: Proceedings of the European Conference on Computer Vision (ECCV), pp. 529–545 (2018)
20. Tang, W., Wu, Y.: Does learning specific features for related parts help human pose estimation? In: Proceedings of the IEEE/CVF Conference on Computer Vision and Pattern Recognition, pp. 1107–1116 (2019)
21. Toshev, A., Szegedy, C.: Deeppose: human pose estimation via deep neural networks. In: Proceedings of the IEEE Conference on Computer Vision and Pattern Recognition, pp. 1653–1660 (2014)
22. Wang, D., Zhang, S.: Contextual instance decoupling for robust multi-person pose estimation. In: Proceedings of the IEEE/CVF Conference on Computer Vision and Pattern Recognition (CVPR), pp. 11060–11068 (2022)
23. Wang, Y., Peng, C., Liu, Y.: Mask-pose cascaded CNN for 2D hand pose estimation from single color image. IEEE Trans. Circuits Syst. Video Technol. **29**(11), 3258–3268 (2018)
24. Wang, Y., Zhang, B., Peng, C.: SRHandNet: real-time 2D hand pose estimation with simultaneous region localization. IEEE Trans. Image Process. **29**, 2977–2986 (2019)
25. Wei, S.E., Ramakrishna, V., Kanade, T., Sheikh, Y.: Convolutional pose machines. In: Proceedings of the IEEE Conference on Computer Vision and Pattern Recognition, pp. 4724–4732 (2016)
26. Xiao, B., Wu, H., Wei, Y.: Simple baselines for human pose estimation and tracking. In: Proceedings of the European Conference on Computer Vision (ECCV), pp. 466–481 (2018)
27. Yan, S., Xiong, Y., Lin, D.: Spatial temporal graph convolutional networks for skeleton-based action recognition. In: Proceedings of the AAAI Conference on Artificial Intelligence, vol. 32 (2018)
28. Zhao, L., Peng, X., Tian, Y., Kapadia, M., Metaxas, D.N.: Semantic graph convolutional networks for 3D human pose regression. In: Proceedings of the IEEE/CVF Conference on Computer Vision and Pattern Recognition, pp. 3425–3435 (2019)
29. Zheng, C., et al.: Deep learning-based human pose estimation: a survey. arXiv preprint arXiv:2012.13392 (2020)
30. Zhou, Z.H.: Machine Learning. Springer, Cham (2021)

Chaotic Particle Swarm Algorithm for QoS Optimization in Smart Communities

Jiaju Wang and Baochuan Fu(✉)

Suzhou University of Science and Technology, No. 99 Xuefu Road, Huqiu District, Suzhou, Jiangsu, China
fubc@163.com

Abstract. In smart communities, computer networks carry a large number of real-time computing tasks such as smart property, smart parking, smart home, etc., and these services are characterized by large data transmission and high concurrency, which require high real-time performance of the network. How to ensure the real-time network to reduce the task scheduling delay is the key problem to be solved when QoS optimization of smart community network. To this end, this paper deeply analyzes the characteristics of smart community task scheduling, firstly, establishes a community computing task scheduling model with computation time and computation cost as the optimization goal; then, proposes the optimization strategy of Chaotic Particle Swarm Algorithm for the stochastic nature of highly concurrent tasks, i.e., based on the basic algorithm of Particle Swarm to add the chaotic strategy in the initialization of the population and the optimization means of the adaptive factor in order to avoid falling into the local optimal and improve the optimization speed; finally, the time and cost overheads under different number of tasks are compared through simulation experiments, and the simulation results verify the effectiveness of the improved algorithm proposed in this paper in network QoS optimization.

Keywords: Time Delay · Task Scheduling · Network QoS · Smart Community · Chaotic Strategy

1 Introduction

With the development of new-generation information technology such as mobile Internet and big data, smart community, as a basic component of smart city, carries real-time computing tasks for diverse business scenarios such as smart property, smart parking, smart home, etc. These tasks are characterized by big data transmission and high concurrency [1], which require high real-time performance of the network. Therefore, how to ensure the real-time performance of the smart community network and reduce the delay and computation cost of task scheduling becomes a key issue in network QoS optimization.

Task scheduling is to assign mutually independent computing tasks in a certain time period to appropriate servers for processing according to different objectives. How

H. Jin et al. (Eds.): GPC 2023, LNCS 14504, pp. 251–265, 2024.
https://doi.org/10.1007/978-981-99-9896-8_17

to further reduce the network scheduling delay and lower the computation cost while ensuring QoS is the focus of this paper. The multi-task scheduling problem for smart community network computing resources is essentially a NP-C problem, i.e., a non-deterministic problem of polynomial complexity [2]. In order to solve the NP-C problem efficiently, on the one hand, holistic evolutionary algorithms are needed, such as PSO algorithms based on feedback mechanism (FBE), group intelligence algorithms based on competition mechanism (CSO), etc., which are mainly applied to multi-objective optimization, multi-peak optimization and dynamic optimization [3–5]; on the other hand, new co-evolutionary algorithms based on partition algorithms are needed, the main idea of which is to transform the NP-C in which the high-dimensional problem is transformed into multiple low-dimensional problems, and then each low-dimensional problem is regarded as a separate modular sub-problem, such as Random Grouping (RG), Multilayer Randomized Grouping (MLCC), and Differential Evolutionary Grouping (DG), etc. [6–8]. And although these algorithms have achieved better optimization results on high-dimensional problems to a certain extent, they perform poorly on multi-peak problems because they are affected by a large number of locally optimal solutions, and are prone to fall into locally optimal solutions, which do not allow them to globally optimize the QoS of smart community networks [19]. In the community network resource scheduling task, the following key dimensions of network QoS need to be considered: (1) bandwidth allocation, task scheduling needs to be reasonably allocated bandwidth; (2) fault tolerance, need to consider and ensure the reliability and fault tolerance of the whole system, to avoid affecting the normal work of the whole system because of a single point of failure; (3) reduce the delay, improve the efficiency of task execution and so on. Based on the above considerations, the smart community NP-C problem can be transformed into a network QoS optimization problem with multiple constraints and solved by such evolutionary algorithms as chaotic particle swarm.

Facing the QoS problem with multiple constraints, this paper proposes a strategy to use particle swarm optimization algorithm for task scheduling in the community on the basis of the above research results. Firstly, we establish a network computing scheduling model based on time and cost as the optimization objectives; secondly, for the stochastic nature of the highly concurrent tasks, we add and use chaotic strategies and adaptive factors to optimize and improve the particle swarm algorithm, to enhance the algorithm's global search ability and exploratory while avoiding falling into the local optimal solution, in order to improve the algorithm's performance and ensure the strategy's real-time reliability [20]; lastly, we selected appropriate benchmark test functions, and through simulation experiments on the time and cost overheads under different tasks and compared with the traditional methods, we found that the improved particle swarm algorithm achieved better time and cost results in community task scheduling, and improved the efficiency and performance of network task scheduling.

2 Community Network Computing Tasks and Their Scheduling Models

Task scheduling in community networks mainly involves three types of objects: user demanders, service providers and platform operators [12]. Assuming that the set of tasks in the community is $Task = \{task_1, task_2, task_3, \ldots, task_n\}$, m is the total number of tasks, and the set of servers is $Serv = \{serv_1, serv_2, serv_3, \ldots, serv_n\}$, n is the total number of servers. n is the total number of servers, so the correspondence between tasks and servers can be expressed by matrix TS, as shown in Eq. (1), ts_{ij} is 1 means that task i is assigned to server j, ts_{ij} is 0 means that task is not assigned to the corresponding server [9].

$$TS = \begin{bmatrix} ts_{11} & \cdots & ts_{1n} \\ \vdots & \ddots & \vdots \\ ts_{m1} & \cdots & ts_{mn} \end{bmatrix} \tag{1}$$

2.1 Task Calculation Model

There is a close relationship between the task computation model and the task completion time and the total cost of completing the task. The task computation model determines the time required to perform the task and the completion cost; the task completion time and cost are in turn affected by the task model, so these factors need to be considered comprehensively in task scheduling and optimization to achieve the best latency and cost.

(1) Task completion time

Task completion time is the entire time period from the start of submission of a task until the task is processed on a certain server, i.e., the sum of the task transmission time, the task queue waiting time, and the execution time of the task on the server. The transmission time is shown in Eq. (2), the task queue waiting time is shown in Eq. (3), and the task execution time is shown in Eq. (4). And the final total time of task completion for a certain time period should be the maximum completion time in each server, and the maximum completion time of each server is the maximum time spent for each task completion that it carries by itself.

$$TT = \begin{bmatrix} tt_{11} & \cdots & tt_{1n} \\ \vdots & \ddots & \vdots \\ tt_{m1} & \cdots & tt_{mn} \end{bmatrix} \tag{2}$$

$$QT = \begin{bmatrix} qt_{11} & \cdots & qt_{1n} \\ \vdots & \ddots & \vdots \\ qt_{m1} & \cdots & qt_{mn} \end{bmatrix} \tag{3}$$

$$FT = \begin{bmatrix} ft_{11} & \cdots & ft_{1n} \\ \vdots & \ddots & \vdots \\ ft_{m1} & \cdots & ft_{mn} \end{bmatrix} \tag{4}$$

$$TotalTime = \max\left\{\max\{tt_{ij} + qt_{ij} + ft_{ij}\}\right\} \tag{5}$$

(2) The total cost of completing the task
The cost of task scheduling in smart management and service in the community is mainly considered as the cost required for the task transmission in the network in the community (Eq. (6)) and the task completion on the server (Eq. (7)), and the total cost of the task completion is the combination of the two parts, as shown in Eq. (8).

$$CostFinish = ft_{ij} * serv_{j\_singleTime\_cost} \tag{6}$$

$$CostTrans = tt_{ij} * serv_{j\_bw} * serv_{j\_bw\_cost} \tag{7}$$

$$TotalCost = \sum\nolimits_{i=1}^{n} \sum\nolimits_{j=1}^{m} (CostFinish + CostTrans) \tag{8}$$

In Eq. (6), $serv_{j\_single\_cost}$ denotes the cost of server j per unit of time, i.e., $serv_{j\_singleTime\_cost} = serv_{j\_compute\_ability} * serv_{j\_core\_num} * serv_{singleCore\_cost}$ where, $serv_{j\_conpute\_ability}$ is the computational capability of a single CPU, $serv_{j\_core\_num}$ denotes the number of CPUs in the server, and $serv_{singleCore\_cost}$ is the cost per unit computational capability. Cost per unit computing power. In Eq. (7), $serv_{j\_bw}$ is the size of communication bandwidth of the server and $serv_{j\_bw\_cost}$ is the cost per unit bandwidth of the server [13, 14].

2.2 Optimal Scheduling Model Based on QoS Constraints

The task scheduling model mainly consists of a time affiliation function and a cost affiliation function, the time affiliation function is shown in Eq. (9) and the cost affiliation function is shown in Eq. (10).

$$Time(K) = \begin{cases} 0.9 * \frac{TotalTime - Time_{deadline}}{(TotalCost)^2} > Time_{deadline} \\ 0.1 * \frac{Time_{deadline} - TotalTime}{TotalCost * Time_{deadline}} \leq Time_{deadline} \end{cases} \tag{9}$$

$$Cost(K) = \begin{cases} 0.9 * \frac{TotalCost - Cost_{expect}}{(TotalCost)^2} > Cost_{expect} \\ 0.1 * \frac{Cost_{expert} - TotalCost}{TotalCost * Cost_{expect}} \leq Cost_{expect} \end{cases} \tag{10}$$

In Eq. (9)(10), $Time_{deadline}$ denotes the final deadline of all tasks, i.e., the upper limit of task completion time, and $Cost_{expect}$ denotes the ex-ante expected cost of all tasks. Therefore, this paper uses QoS with time and cost as the main evaluation index, which is also the main optimization goal of this paper, so the QoS profit and loss model G(K) can be expressed as:

$$G(K) = \alpha * Time(K) + \beta * Cost(K) \tag{11}$$

In Eq. (11), α and β are the weight factors of the two affiliation functions respectively, and $\alpha + \beta = 1$. And the weight factors α and β represent the degree of contribution of the two affiliation functions, namely, task completion time and computational resource cost, respectively, to the QoS scheduling model, whereas the maximum latency and the minimum bandwidth allowed by the QoS of the network are taken as constraints for the task optimization scheduling. By changing and adjusting the weights, the task scheduling model can dynamically allocate tasks and resources, such as the bandwidth, throughput and other QoS parameters therein, so as to improve the QoS of the whole system in order to better meet the community management and service demands.

3 Chaotic Strategies and Their Algorithms for QoS Optimization in Smart Community Networks

In the smart community, network task scheduling has the following characteristics: 1. Diversity, involving a variety of types of tasks; 2. Non-determinism, the scheduling process between the tasks interact with each other; 3. Real-time, some tasks for the high requirements of delay accuracy; 4. Multi-dimensional goals, in considering the timely completion of the task at the same time, but also pay attention to the delay, energy consumption and other multi-dimensional indicators.

Particle swarm algorithm is an evolutionary algorithm based on swarm intelligence, in which each "particle" represents a set of parameters to find the best solution by optimizing the fitness function. Compared with other evolutionary algorithms, the particle swarm algorithm has better robustness and search efficiency, and is less dependent on the initial conditions. When dealing with the smart community task scheduling problem, the particle swarm algorithm can help to find the global optimal solution and reduce the computational resources and time required by the traditional exhaustive search methods, so it becomes the choice of this paper.

However, the basic particle swarm algorithm has a local optimal solution problem in the task scheduling process of community network, which leads to the delay and energy consumption can't reach the global optimal; in addition, the task scheduling in the network has the characteristics of randomness and high concurrency, which makes the scheduling process ambiguous. Therefore, in order to optimize the characteristics of ambiguity and chaos, this paper introduces the chaotic strategy.

3.1 Elementary Particle Swarm Algorithm

Particle swarm algorithm is an optimization algorithm based on group intelligence, in which each "particle" represents a set of parameters, and the optimal solution is found by optimizing the fitness function. The algorithm has the characteristics of fast convergence, and easy to achieve parallel processing, is widely used in task scheduling and other fields. Compared with other evolutionary algorithms, the particle swarm algorithm is less dependent on the initial conditions and has better robustness and search efficiency. When dealing with task scheduling problems in community networks, particle swarm algorithms can help to find the global optimal solution and reduce the computational resources and time required by traditional exhaustive search methods, so they are a feasible choice. Now, the number of particles is set to be I and the spatial search dimension is d. The set of positions of the particles is $X_i = \{x_{i1}, x_{i2}, x_{i3}, \ldots, x_{id}\}$, , and the best position (i.e., the best fitness value) experienced by a particle is denoted as $P_i = \{p_{i1}, p_{i2}, p_{i3}, \ldots, p_{id}\}$. Also known as pbest, the index of the position of the best adapted value is generally denoted by g, i.e., P_g, also known as gbest, and the velocity of particle I can be represented by the set $V_i = \{v_{i1}, v_{i2}, v_{i3}, \ldots, v_{id}\}$. Then the velocity and position update formula of the particle swarm can be expressed as in Eq. (12) (13) as:

$$v_{id} = w * v_{id} + c_1 * rand * (p_{id} - x_{id}) + c_2 * rand * (p_{gd} - x_{id}) \tag{12}$$

$$x_{id} = x_{id} + v_{id} \tag{13}$$

where w is the inertia weight, c1 and c2 are acceleration constants, and rand is a random value varying in the range [0,1]. In addition, the velocity v_i of the particle should be limited by a maximum velocity v_{max}, i.e., the velocity on the dimension d of the particle should not be larger than the maximum velocity supported by the dimension, with the constraints shown in Eq. (14):

$$v_{i,d} \leq v_{max,d} \tag{14}$$

3.2 Chaotic Particle Swarm Algorithm

The optimization of this part is mainly reflected in two aspects. Firstly, chaotic sequences are added in the initialization stage of the population to make its initialization richer and the distribution space broader; secondly, in the local search process, in order to avoid falling into the local optimal solution quickly, the weights are updated through the adaptive algorithm to dynamically adjust the size and direction of the position and speed updates. The algorithm flow and optimization process are shown in Fig. 1:

(1) Chaotic initialization

In this paper, we explore the impact of the initialization phase of particle swarm algorithms on the evolutionary process and propose a particle swarm initialization method based on the chaotic formula. Compared with the traditional random initialization, we adopt the chaotic formula to map the particle swarm into the search space in order to achieve a uniform and random distribution of particles and search a wider search space. Considering the strong randomness and complexity of the chaotic algorithm, this paper uses the nonlinear characteristics of the chaotic algorithm to deal with the multi-constraint problem and helps to avoid falling into the local optimum, so as to search for the global optimal solution more efficiently and to improve the stability and robustness of the system.

According to the literature [21], a one-dimensional Logistic formula is used to generate chaotic variables, which is applied to the initialization of the algorithm as well as the chaotic updating of the parameters during the iteration process, and the chaotic formula is shown in Eq. (15):

$$y(n+1) = r * x(n) * (1 - x(n)) \tag{15}$$

where y(n + 1) ∈ [0, 1]; r ∈ [0, 4] is the logistic parameter.

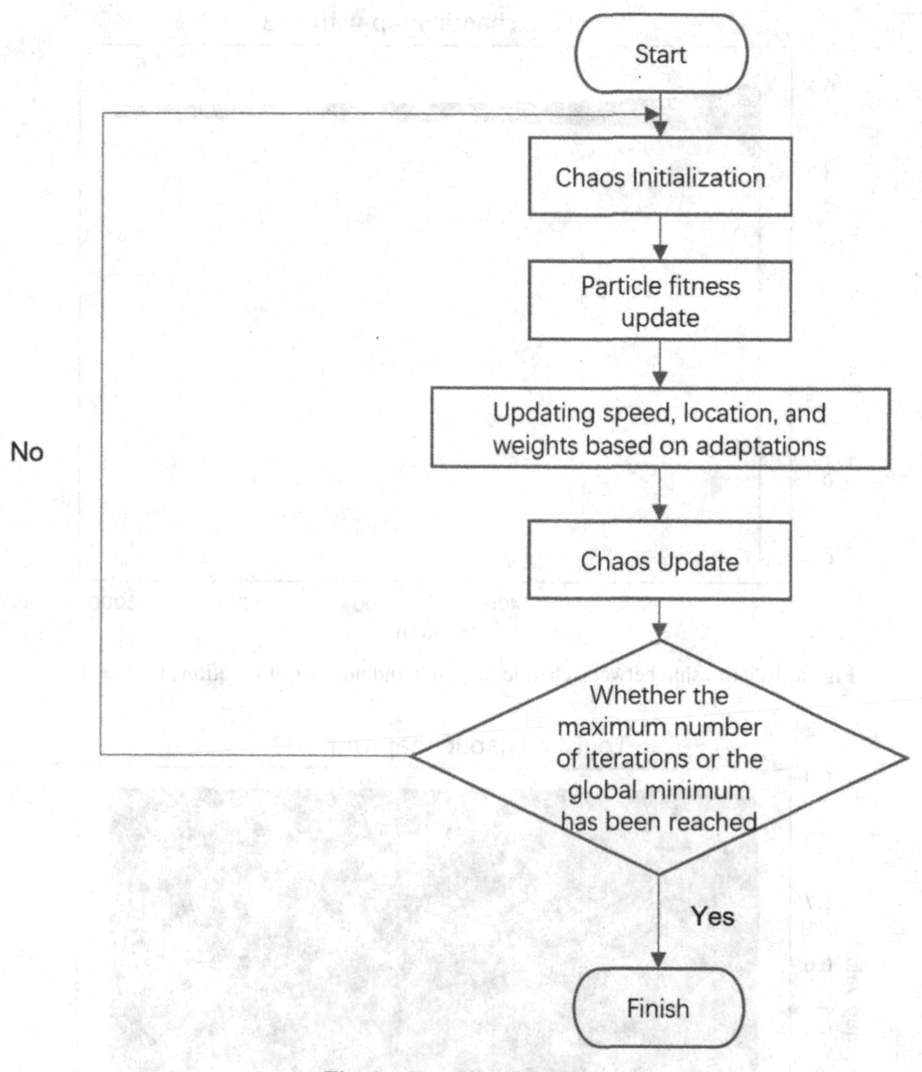

Fig. 1. Algorithm flowchart

The experimental results show that when r is closer to 4, the value range of y is closer to the average distribution to the whole region, when r is taken as 4, the system is in a completely chaotic state, when the uniformity of the mapping distribution reaches an extreme state [22], so in this paper, the value of r will be taken as 4. The relationship between the logistic chaotic mapping and the corresponding number of iterations in different cases of the value of r is shown in Figs. 1, 2 and 3 (Fig. 4).

(2) Local search optimization

In the traditional search algorithm, the inertia weight w is generally fixed to some constant value, which is easy to fall into the local optimal solution. Considering that

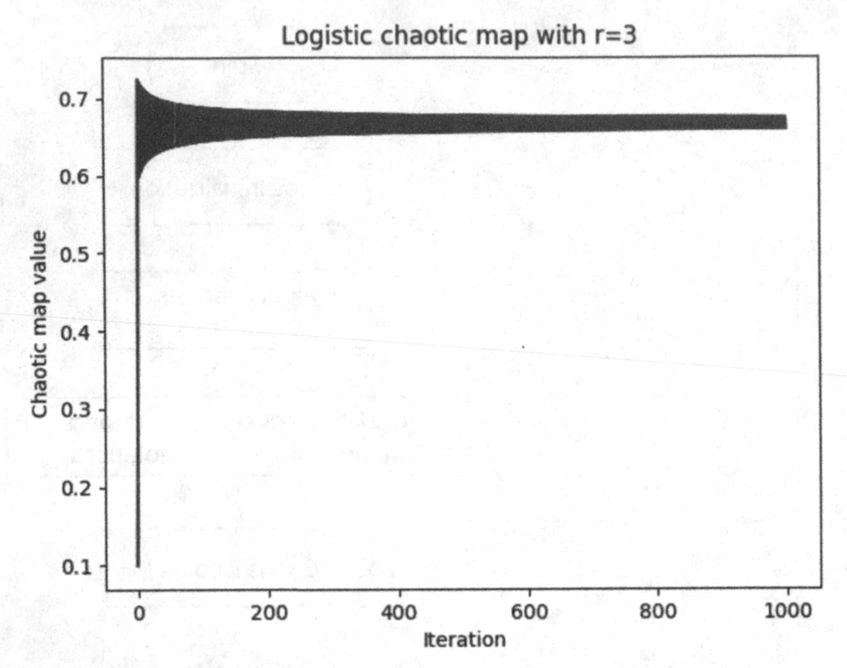

Fig. 2. Relationship between chaotic mapping and number of iterations for r = 3

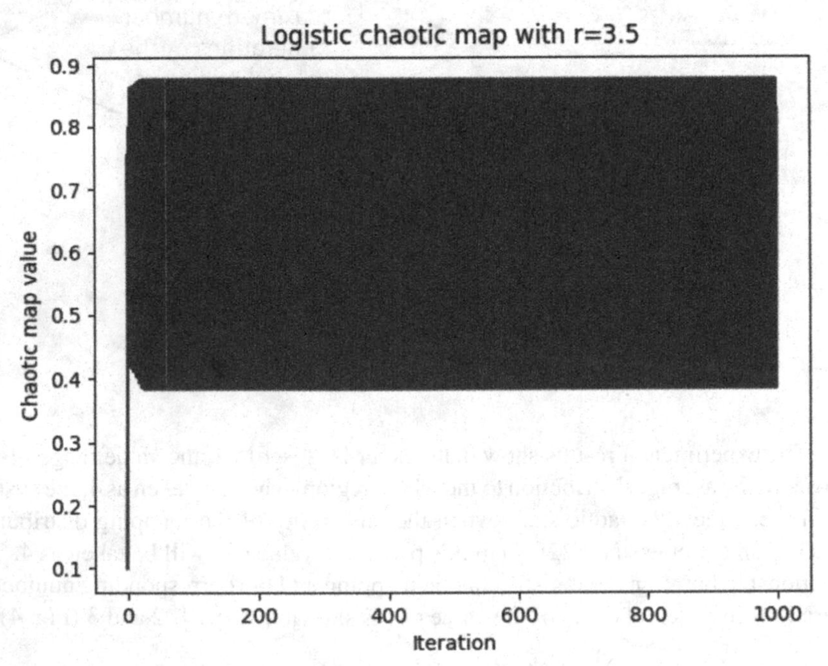

Fig. 3. Relationship between chaotic mapping and number of iterations for r = 3.5

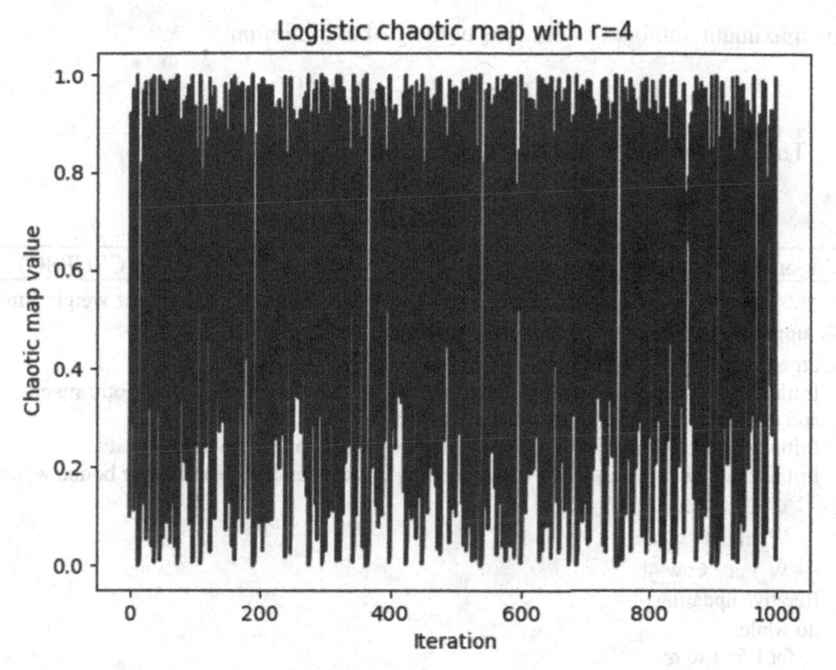

Fig. 4. Relationship between chaotic mapping and number of iterations for r = 4

the model developed in this paper is NP-C and has a nonlinear time-varying dynamic model, in order to optimize the local search, a nonlinear decreasing inertia weight is introduced. This weight reflects the inheritance of the previous speed capability and accelerates the global iterative search speed by a constant term setting as shown in Eq. (16):

$$w_d = w_{start} - (w_{start} - w_{end}) * [\frac{2d}{k} - (\frac{d}{k})^2] \tag{16}$$

where w_{start}, w_{end} are the upper and lower limit values of inertia weight factor, which generally take the values of 0.9 and 0.4 respectively.

(3) Objective function

In the particle swarm algorithm, the fitness value as a condition for selecting an individual, the smaller the fitness value is the better the individual, and vice versa, the worse the individual is, so in this paper, we need to select the individual with smaller fitness value for optimizing the community QoS-based task scheduling, and set l_i^t to be the position of the ith particle after t iterations, then the original QoS profit and loss model can be updated as in Eq. (17) as:

$$F\left(l_i^t\right) = \alpha * Time\left(l_i^t\right) + \beta Cost(l_i^t) \tag{17}$$

Since the optimization of QoS-based task scheduling sought in this paper is to obtain the minimum value, the objective function of the model in this paper is as in Eq. (18), and the result of the objective function is the return value after reaching the maximum number of iterations or the global minimum:

$$Fitness(l_i^t) = \min F(l_i^t) \tag{18}$$

The pseudo-code of the improved algorithm is as follows:

Algorithm Chaotic and Self-Update Particle Swarm Optimization (CSUPSO)

```
Inputs: number of particles N, dimension D, learning factors C1, C2, lower weight limit
w_lb, upper weight limit w_ub, maximum number of iterations max_iter
Output: particle swarm historical optimal solution gbest
1.  Initialize the particle swarm, generate the initial population through chaotic mapping,
    and calculate the current fitness value of each particle
2.  Initializing the optimal solution Pest[] and the global optimal solution gbest
3.  Initialization weights and their corresponding lower bound w_lb and upper bound w_ub:
      W ← w_max
      W_min ← w_ub
      W_max ← w_lb
4.  Iterative updating:
5.  do while:
6.     for I ← 1 to n:
7.        for J ← 1 to Dd
8.           v[i][j] ← w*v[i][j]+c1*r1*(Pbest[i][j]-x[i][j])+c2*r2*(Gbest[j]-x[i][j])
9.           x[i][j] ← x[i][j]+v[i][j]
10.          If current x[i] is better then
11.             Pbest[i]←x[i]
12.          If Pbest[i] is better then
13.             Gbest←Pbest[i]
14.       End for
15.       Updating weight W:
16.       If current Gbest don't update then
17.          W ← W*1.2
18.          If W > W_max then
19.             W ← W_max
20.       Else
21.          W ← W*0.8
22.          If W ← W_min then
23.             W ← W_min
24.       End if
25.    End for
26. End do
27. Output: Return the particle swarm historical optimal solution Gbest
```

4 Algorithm Testing and Simulation

In order to further verify the effectiveness of the algorithms in this paper for QoS optimization and task scheduling of computing resources, the following tests and simulations are prepared.

In the first part, it is the performance test of the optimized algorithm, which tests the performance of the algorithm's metrics through the benchmark test function. The test metrics are the dispersion of the results of different algorithms and the mean and standard values of the test function in fixed dimensions (dimension D = {2, 3, 5}) and intervals; the algorithms used as comparisons are the standard particle swarm algorithm, PSO, and the ant colony algorithm, ACO, respectively.

The second part of the simulation experiment is to optimize the QoS task scheduling model proposed in this paper, as shown in Eq. (18), by comparing the three algorithms in the case of large data tasks, to analysis the advantages and disadvantages of the algorithms proposed in this paper in terms of both time and resource consumption.

The test and simulation platform of this paper is Windows 11, 16 GHz RAM, 1 TB hard drive, and the simulation software is PyCharm equipped with python version 3.11. The population size is set to 100 and the number of iterations is 1000.

4.1 Algorithm Performance Test

Compare this paper with ACO, PSO in (2-dimensional, 5-dimensional, 10-dimensional) under three benchmark test functions (Table 1), the number of iterations is set to 1000, and the results of the data metrics are shown in Table 2, and the test metrics are the average and standard values.

Table 1. Benchmarking functions

Function Name	Basis functions
F1	$\sum_{i=1}^{n-1}[100(x_{i+1}-x_i^2)^2+(x_i-1)^2]$
F2	$10n+\sum_{i=1}^{n}[x_i^2-10\cos(2\pi x_i)]$
F3	$-20\exp\left(-0.2\sqrt{\frac{1}{n}\sum_{i=1}^{n}x_i^2}\right)-\exp\left(\frac{1}{n}\sum_{i=1}^{n}\cos(2\pi x_i)\right)+20+$ exp (1)

From Table 2, it can be found that the CPSO algorithm has better results in terms of the average and standard values of the three test functions, compared to the ACO, the traditional PSO numerical results already have a clear advantage, and compared to the traditional PSO, the advantage of the CPSO after chaotic initialization using chaotic formulas is further expanded. The algorithm in this paper illustrates, by means of two metrics, that after chaotic initialization, the local search optimization and the effect of chaotic update optimization.

Figure 5 shows the comparison results of three algorithms for smart community computing task fitness function, from the results in the figure, it can be seen that the three algorithms show a decreasing trend with the increase in the number of iterations, and from the comparison of fitness function, the CPSO algorithm is significantly lower than the other two algorithms, which indicates that the CPSO algorithm can be well applied to the high concurrency of the task scenarios and task scheduling in the smart community.

Table 2. F1–F3 function indicator results

Algorithm	Dim	F1		F2		F3	
		Average	Standard	Average	Standard	Average	Standard
ACO	2	985.7	983.5	16.2	25.7	39.8	89.7
	5	470.6	650.6	789.2	897.1	22.6	29.9
	10	720.6	748.7	106.4	495.5	286.8	822.8
PSO	2	0.12	0.1	1	1	1.9E−05	1.3E−06
	5	1.13	1.29	1.22	1.26	1.1	1
	10	149.5	108.6	5.45	2.92	1	1
CPSO	2	0.1	0.1	0.9	1	2E−05	1.6E−06
	5	1	0.9	1.11	1.04	0.002	0.002
	10	0.27	0.54	0.36	0.49	0.02	0.016

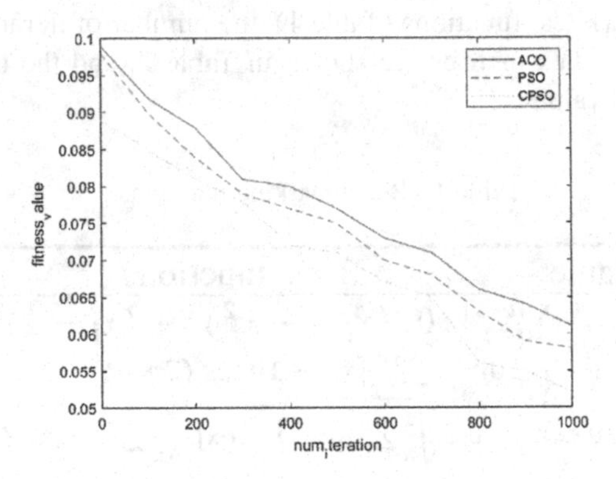

Fig. 5. Variation curves of task fitness function values for the three algorithms

4.2 Simulation Comparison of QoS Optimization Algorithms

Considering the complexity, variability and diversity of community tasks, the number of designed tasks are considered to be a large number of tasks, in order to observe the comparison of the completion time and cost consumption of the three algorithms under the large number of tasks in the smart community.

(1) Comparison of Completion Time

Figure 6 shows the results of the three algorithms' completion time comparison under different task numbers. In the case of large task numbers, the completion time of all three algorithms is more, but CPSO obviously rises more gently compared to ACO and PSO, which indicates that although there is an increase in the completion time of the tasks, the overall stability is better, and the delay and bandwidth consumed will be smaller, which are all advantages of CPSO in the mixing of the tasks. These advantages are due to the improvement of CPSO in chaos initialization, local search and chaos update CPSO saves 23% and 19% of time compared to ACO and PSO respectively, which indicates that CPSO can achieve better time completion results in the environment of large number of tasks in the smart community.

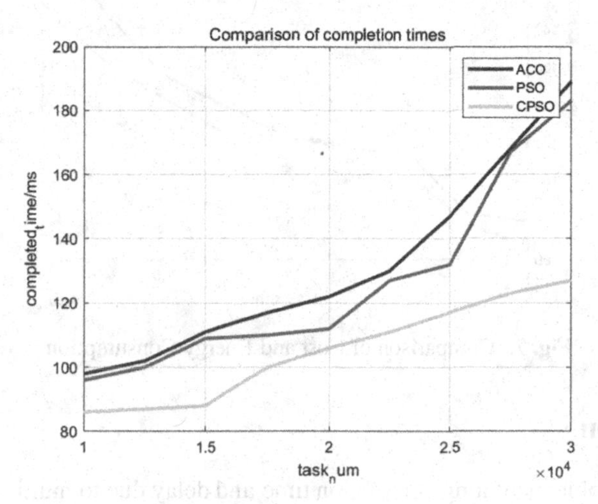

Fig. 6. Comparison of completion times

(2) Cost-energy consumption comparison

Figure 7 demonstrates the comparison results of the cost-energy consumption of the three algorithms under different number of tasks. In the case of large number of tasks, the CPSO cost-energy consumption has an obvious advantage compared with the other two algorithms, and CPSO saves 24% and 10% in terms of the cost-energy consumption compared with ACO and PSO, which indicates that CPSO has an obvious advantage in terms of the cost-energy consumption under the number of tasks.

From the synthesis and comparison of the results of (1) (2), it can be found that the algorithm of this paper is optimized to be used in the high concurrency task scenario of the smart community with better results, and it has plus good optimization results for the optimization of latency as well as the control of cost and energy consumption in the smart community.

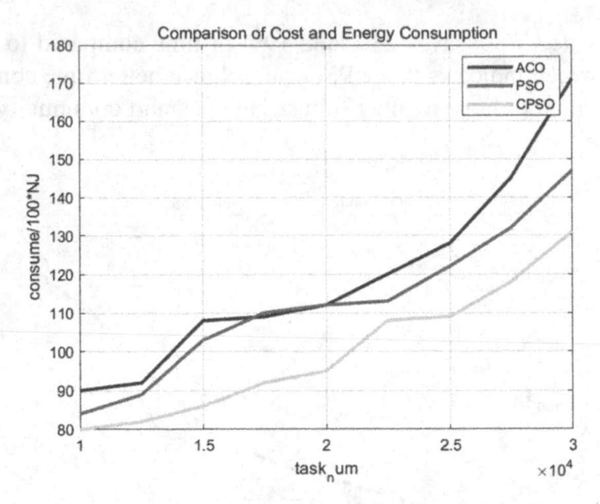

Fig. 7. Comparison of Cost and Energy Consumption

5 Conclusion

Aiming at the problems of long completion time and delay due to multi-task scheduling and high concurrency in smart communities, as well as high cost consumption, this paper proposes the optimization of particle swarm algorithm based on chaos update, establishes a QoS multi-task scheduling model based on time and cost in smart community scenarios, and carries out the initialization optimization of the particle swarm algorithm, local search optimization, and chaos update to improve the performance of the particle swarm algorithm. The local search optimization and chaotic update improve the performance of the particle swarm algorithm. Simulation experiments show that the optimized particle swarm algorithm can improve the efficiency of task scheduling and reduce the completion time and cost and energy consumption, and in the next study, the addition of neural networks and other techniques for dynamic planning and dynamic modelling will be further considered to further reduce the impact of delay and energy consumption in the smart community.

References

1. Arunarani, A.R., Manjula, D., Sugumaran, V.: Task scheduling techniques in cloud computing: a literature survey. Futur. Gener. Comput. Syst. **91**, 407–415 (2019)
2. Singh, B., Dhawan, S., Arora, A., et al.: A view of cloud computing. Int. J. Comput. Technol. **4**(2b1), 50–58 (2013)
3. Yuen, S.Y., Chow, C.K.: A genetic algorithm that adaptively mutates and never revisits. IEEE Trans. Evol. Comput. **13**(2), 454–472 (2008)
4. Tan, Y., Yu, C., Zheng, S., et al.: Introduction to fireworks algorithm. Int. J. Swarm Intell. Res. (IJSIR) **4**(4), 39–70 (2013)
5. Zhang, F.: A parallel collaborative large-scale differential evolutionary algorithm for solving NP-hard problems. J. Cebu Acad. **33**(11), 101–107 (2018)
6. Liu, Z.G., Wang, J.K.: Node scheduling strategy for large-scale sensor networks based on particle swarm optimization. Control Decis. **27**(12), 1903–1906 (2012)

7. Shami, T.M., El-Saleh, A.A., Alswaitti, M., et al.: Particle swarm optimization: a comprehensive survey. IEEE Access **10**, 10031–10061 (2022)
8. Montemayor, J.J.M., Crisostomo, R.V.: Feature selection in classification using binary max-min ant system with differential evolution. In: 2019 IEEE Congress on Evolutionary Computation (CEC), pp. 2559–2566. IEEE (2019)
9. Song, F.Q.: Research on improved pollen-based algorithm in cloud computing QoS. Sci. Technol. Bull. **38**(11), 46–52 (2022)
10. Zhang, C., Feng, F.: Whale optimization algorithm improved by chaotic strategy and simplex method. Chinese Sci. Technol. Paper **15**(03), 293–299 (2020)
11. Li, C., Zhang, X., Dong, X., et al.: The impact of smart cities on entrepreneurial activity: evidence from a quasi-natural experiment in China. Resour. Policy **81**, 103333 (2023)
12. Xiong, W., Lim, M.K., Tseng, M.L., et al.: An effective adaptive adjustment model of task scheduling and resource allocation based on multi-stakeholder interests in cloud manufacturing. Adv. Eng. Inform. **56**, 101937 (2023)
13. Du, J., Zhao, L., Feng, J., et al.: Computation offloading and resource allocation in mixed fog/cloud computing systems with min-max fairness guarantee. IEEE Trans. Commun. **66**(4), 1594–1608 (2017)
14. Yao, J., Ansari, N.: QoS-aware fog resource provisioning and mobile device power control in IoT networks. IEEE Trans. Netw. Serv. Manage. **16**(1), 167–175 (2018)
15. Eberhart, R., Kennedy, J.: Particle swarm optimization. In: Proceedings of the IEEE International Conference on Neural Networks, Australia (1942, 1948)
16. Dokeroglu, T., Sevinc, E., Kucukyilmaz, T., et al.: A survey on new generation metaheuristic algorithms. Comput. Ind. Eng. **137**, 106040 (2019)
17. Dooley, K.J., Van de Ven, A.H.: Explaining complex organizational dynamics. Organ. Sci. **10**(3), 358–372 (1999)
18. Zhang, S., Shi, L., Zhao, X.: Data fusion for community smart parking under scenario perspective. Surv. Map. Bull. No. **552**(03), 61–66 (2023). https://doi.org/10.13474/j.cnki.11-2246.2023.0073
19. Fu, Y., Song, X., Zhang, C., et al.: Simulation study of LRMC QoS routing algorithm based on Mininet. Comput. Simul. **39**(08), 212–217 (2022)
20. Ge, B., Wang, Z.-C., Xin, Y., Li, S., Yuan, Z.-Q.: Dynamic real-time reliability prediction of bridge structures based on Copula- BHDLM and measured stress data. Measurement **203** (2022)
21. Zhao, Z.-G., Chang, C.: Adaptive chaotic particle swarm optimization algorithm. Comput. Eng. **37**(15), 128–130 (2011)
22. Wu, D., Zhou, Q., Wen, L.: Improved sparrow algorithm based on logistic chaotic mapping. J. High. Educ. Sci. **41**(06), 10–15 (2021)

Resource Binding and Module Placement Algorithms for Continuous-Flow Microfluidic Biochip in Intelligent Digital Healthcare

Zhongliao Yang[1,3], Huichang Huang[1,3], Zeyi Liu[1,3], Chen Dong[1,3], and Li Xu[2(✉)]

[1] College of Computer and Data Science, Fuzhou University, Fuzhou 350116, China
[2] College of Computer and Cyber Security, Fujian Normal University, Fuzhou 350007, China
xuli@fjnu.edu.cn
[3] Fujian Key Laboratory of Network Computing and Intelligent Information Processing, Fuzhou University, Fuzhou, China

Abstract. Continuous-Flow Microfluidic Biochip (CFMB), with their integrated features, bring traditional biochemical experiments on a single chip to accomplish complex operations and reactions through precise control, efficient reactions and emerging ways of saving reagents. In the field of intelligent digital healthcare, CFMB have attracted a lot of attention. However, traditional manual design schemes can no longer meet the needs of increasingly complex chip architecture design. Therefore, this paper proposes an automated design method for resource binding and module placement of CFMB based on a list scheduling algorithm and an improved Simulated Annealing algorithm. Through the resource binding and scheduling design based on the list scheduling algorithm, an effective scheduling strategy is generated, which effectively improves the biochip execution efficiency. In addition, the improved Simulated Annealing algorithm solves the module placement problem in the biochip in a limited physical space. Compared with some benchmark algorithms, the experimental results demonstrate the effectiveness of the method in the biochip design process and provide a practical framework for further development of CFMB in the field of intelligent digital healthcare.

Keywords: Continuous-Flow Microfluidic Biochip · intelligent digital healthcare · Simulated Annealing algorithm · scheduling strategy

1 Introduction

Continuous-Flow Microfluidic Biochip (CFMB) have revolutionized the laboratory field [1]. Also known as Lab-on-a-chip or Microfluidic Total Analysis Systems (μTAS), these chips are capable of automating biochemical experiments

H. Jin et al. (Eds.): GPC 2023, LNCS 14504, pp. 266–281, 2024.
https://doi.org/10.1007/978-981-99-9896-8_18

by precisely manipulating microliters or even nanoliters of fluid on a chip [2]. At the same time, CFMB are cost-effective to manufacture and therefore have great economic benefits. Compared with traditional laboratories, CFMB eliminate the need for large amounts of sample reagents, slower chemical reaction times, complex and expensive specialized instruments, and human interference factors, greatly promoting the development of intelligent digital healthcare.

Continuous CFMB which are a kind of CFMB provide a novel way to perform complex operations and reactions with the advantages of precise control, efficient reactions, and reagent savings. CFMB can effectively replace some traditional laboratories including biological, chemical, and medical industries. With the advantages of high sensitivity, small reaction volume, and lower cost, CFMB have been widely used in recent years in a variety of bioassays, such as disease diagnosis [3], drug screening [4], real-time DNA sequencing [5] and Antigen-detection [6].

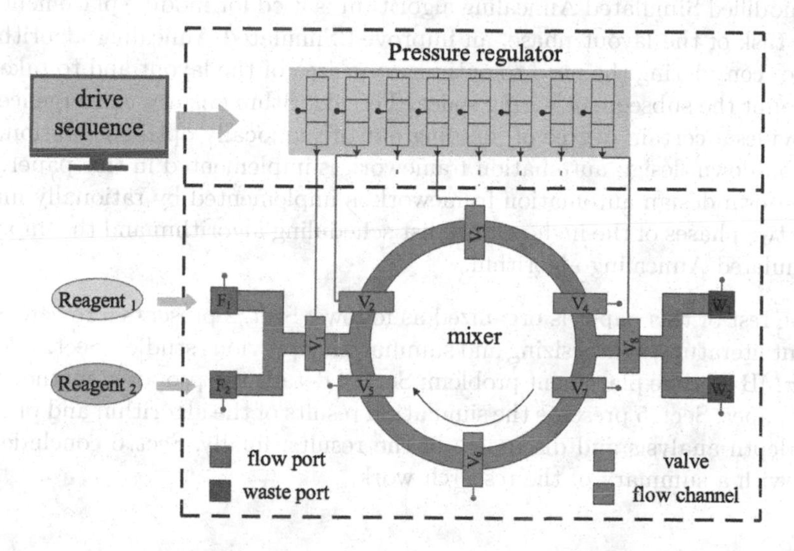

Fig. 1. Working diagram of Continuous-Flow Microfluidic Biochip.

As shown in Fig. 1, the most basic unit in CFMB is the valve, which is essentially an elastic membrane located at the junction of the flow and control layer pipelines. The control layer can be located above or below the flow layer, forming a "push-down" or "push-up" valve, respectively. External air pressure or hydraulic pressure will quickly reach the valve position through the control layer piping, causing it to deform, thus blocking the flow layer piping at the junction with it, and quickly recovering the deformation after the pressure disappears, thus enabling control of the fluid in the flow layer. In addition, the piping is the basic unit in CFMB, which can be used for both fluid storage and fluid transport and therefore needs to have good biocompatibility. Polydimethylsiloxane materials have exactly this property.

However, CFMB faces complex design issues. With the modern micrometers and nanometers process advancement, the integration scale of CFMB has increased. Currently, the design of CFMB is often performed manually using computer-aided design programs, where manual design requires designers with extensive interdisciplinary knowledge and manual involvement is prone to errors [7,8]. Traditional manual design solutions are no longer able to meet the requirements of increasingly complex chip architecture design. To solve the above problems, this paper proposes a top-down design flow based on the structural features and design flow of CFMB to automate the synthesis of CFMB architectures. The main contributions of this paper are as follows:

1. A high-priority list scheduling algorithm is proposed. A suitable priority evaluation method is designed for the high-level synthesis tasks in CFMB, considering that they are similar to process scheduling in operating systems, and based on the optimization goal of minimum time.
2. A modified Simulated Annealing algorithm is used for module placement. For the task of the layout phase, an improved Simulated Annealing algorithm is used, considering the need to optimize the area of the layout and to take into account the subsequent wiring space. The algorithm ensures convergence and provides a certain degree of jumping out of the locally optimal solution.
3. A top-down design automation framework is implemented in this paper. The top-down design automation framework is implemented by rationally linking the two phases of the high-priority list scheduling algorithm and the improved Simulated Annealing algorithm.

The rest of this paper is organized as follows: Sect. 2 presents a review of the relevant literature, synthesizing and summarizing previous studies; Sect. 3 defines the CFMB module placement problem; Sect. 4 details the proposed methodology of this paper; Sect. 5 presents the simulation results of the algorithm and provides an in-depth analysis and discussion of the results; finally, Sect. 6 concludes the paper with a summary of the research work.

2 Related Work

[9] presents a new microfluidic chip design using a three-layer PDMS valve structure instead of the commonly used two-layer structure. They demonstrated that these valves can reach a density of 1 million valves per square centimeter, which greatly exceeds current microfluidic large-scale integration. Therefore, they called this technique microfluidic ultra-large-scale integration. [10] proposes a new synthesis approach for microfluidic chip architectures that automates the design and implementation of microfluidic chips from a given biochemical application and library of microfluidic components. They evaluated the effectiveness and efficiency of the proposed approach through practical applications and synthetic benchmark tests. However, this type of chip also faces challenges such as high architectural complexity, difficult design, and low resource utilization.

[11] presents a novel method to transform the binding and scheduling problem into a maximum-cluster problem, which overcomes the drawback that previous heuristic algorithms cannot guarantee optimality. And an efficient branch delimitation algorithm is designed, which utilizes some pruning strategies and heuristic functions to accelerate the search process and reduce the time and space complexity. In [12], a component-based generic device concept is proposed, which can flexibly represent various types and functions of microfluidic components and the connection relationships between them, improving the abstraction level and scalability of the design. Combining the advantages of fixed and dynamic scheduling, it can make real-time decisions based on the characteristics of the operation and the state of the chip, improving the flexibility and efficiency of scheduling. [13] first proposes an integrated approach for flow-through microfluidic chips considering sieve valves and specific biological execution constraints, which can better meet the needs of biochemical experiments and bridge the gap between design automation and biology. In addition, the method integrates three common biological execution constraints, namely immediate execution, reciprocal execution, and parallel execution. However, these methods do not consider some practical constraints, such as valve failure, fluid mixing, and temperature variation, which may affect the performance and reliability of the chip.

Besides, there are many other practical problems faced in the CFMB design process. In [14], a method is proposed to transform the flow channel wiring problem into a right-angle Steiner minimum tree problem with obstacles, which can effectively handle obstacles such as preplaced components and valves on the chip. And using a heuristic function and a branch-and-bound strategy, it can find near-optimal or suboptimal solutions quickly, reducing the time and space complexity. [15] proposes a physically integrated approach considering distributed flow channel storage that is able to take full advantage of the dual function of the flow channel to improve the execution efficiency and resource utilization of the chip. In [16], an architecture synthesis approach considering fault tolerance is proposed to generate a chip architecture that can cope with valve failures with limited chip area and cost. However, these approaches still fail to provide a reliable top-down design automation framework in CFMB large-scale integration, which prevents CFMB from being integrated at scale in the intelligent digital healthcare domain.

3 Preliminaries

This subsection will describe in detail the problem of CFMB high-level integrated design and formulate the problem.

3.1 Problem Description

The flow layer design of CFMB mainly includes architecture synthesis and application mapping. Among them, the architecture synthesis is divided into two stages: high-level synthesis design and physical design. The high-level synthesis design aims to address the biochemical protocol representation, component assignment strategy, the binding relationship between components and operations, and operation scheduling. The physical design, on the other hand, is divided into two parts: component layout and pipeline wiring. Application mapping is to map biochemical protocols to the architecture based on architecture synthesis to meet the requirements of protocol execution.

High-level integrated design, also known as front-end design, is a process that serves as a guide for the subsequent physical design. Its main research includes three aspects: first, the analysis and interpretation of biochemical protocols; second, the determination of component assignment schemes; and third, the development of operation scheduling sequences and the binding relationships between operations and components. Based on the directed acyclic graph and component assignment strategy, the mapping relationships between different operations in the biochemical protocol and components in the component library need to be determined. The analysis and interpretation of biochemical protocols are mainly based on the given biochemical protocol specification book, which provides clear definitions for each operation, including execution time and type, dependencies between operations, and some special constraints.

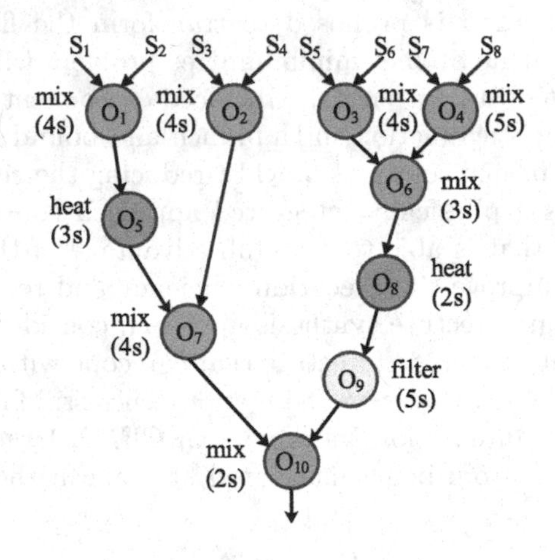

Fig. 2. EA Biochemical Protocol [10].

When performing the design of digital CFMB, the biochemical protocol can be abstracted as a directed acyclic graph $G(O, E)$ [17,18]. Here the set of nodes O represents operations, including type and execution time, while the set of edges E represents the dependencies between operations. As shown in Fig. 2, it is a biochemical protocol graph that satisfies the CFMB design requirements [10], where the main steps include mixing, heating, filtering and so on. In addition to this, after determining the representation of the biochemical protocol as a directed acyclic graph, a component library $L(T, H, N, P)$ needs to be given in the flow layer design of the CFMB. This component library contains information related to the type T, the size H, the available quantity N of the various components, and the driving phase P of the valves that make up the components. Table 1 shows the component library corresponding to the biochemical protocol of Fig. 2 [10] and one possible component assignment scheme Allocation [2].

Table 1. EA biochemical protocol component diagram [2].

T	H/150 μm	N	Allocation	P
mixer	30 * 30	3	3	Ip1/Ip2/Mix/Op1/Op2
filter	120 * 30	2	1	Ip/Filt/Op
heater	40 * 15	2	1	Ip/Heat/OP

3.2 Problem Formulation

The resource binding and module placement problem of CFMB can be defined as follows:

Input:

Bioassay sequencing graph, where G represents the set of operations, including start operation O and end operation E, as shown in Fig. 2.

The module library describes the resources and quantity N available for each operation type, with different module dimensions (i.e. width W, length L, and execution duration T), as shown in Table 1.

Design specifications, Chip size constraints $B_{W \times H}$; maximum duration T_{max}, etc.

Output:

N-to-1 binding relationship between operations and components

Execution time slot for each operation

A set of layout solutions, including the position and placement direction of each component

Objective:

Minimize the execution time of the entire biochemical protocol;Minimize chip area and consider subsequent wiring design.

Constraints:

Consider dependency constraints between operations :

$$start\,(O_j) \geq end(O_i), \forall (O_i, O_j) \in E \tag{1}$$

Satisfy the reuse constraint of component and operand numbers, that is, the demand for a module at the same time must not exceed the actual number that the module can provide:

$$!\exists C_j \in L \;\; C_j = Bind(O_i), \forall O_i \in O \tag{2}$$

where O_i represents the operation node in the biochemical protocol graph, and C_j represents the component in the component library

A certain space needs to be reserved between two components for subsequent wiring, that is, the distance between modules must be greater than 1:

$$|dist(m_i, m_j)| \geq 1, m_i, m_j \in module \tag{3}$$

where, dist represents the closest distance between two modules

4 Proposed Method

For the CFMB design flow, this subsection describes in detail the proposed top-down design flow to automate the synthesis of CFMB architectures in this paper.

4.1 Overall Introduction of the Algorithm

Based on the structural characteristics of CFMB, the designer goes through a series of design steps to finally generate a CFMB chip capable of executing a given biochemical protocol based on a given biochemical protocol model and component library.

This paper focuses on the first stage in architectural synthesis design. That is, in the high-level synthesis phase, the allocation strategy of components is determined based on the given component library, and the planning of the binding relationships between components and operations is carried out, while the location information of components and the routing information of the flow layer pipeline are determined. The architectural synthesis in this phase determines the basis of the entire CFMB design. In the binding and scheduling process of high-level synthesis, the generated binding relationships between components and operations have an impact on the routing between components.

Phase 1: High-level synthesis. In the first step, determine the type and quantity of components to be used. In the second step, check all ready operations and calculate the priority of each operation. In the third step, the operation of selecting the ready queue is added to the queue. The fourth step is to release the operation resources in the running queue after the fastest execution operation is completed. The fifth step is to check whether all operations have been bound and scheduled, if not, return to the second step, otherwise enter the second stage.

Phase 2: Module placement. In the first step, Randomly generate an initial layout solution based on the type and number of components used in Phase 1. In the second step, the current layout solution is perturbed (displacement, steering) to obtain a new layout solution. The third step is to update the local optimal solution according to the formula. The fourth step is to judge whether the number of disturbances at the current temperature reaches the set value, if not, return to the second step, otherwise enter the fifth step. The fifth step is to judge whether all the temperatures have dropped to the minimum, if not, return to the first step, otherwise output the layout solution.

4.2 Binding and Scheduling Design Based on List Scheduling Algorithm

Where component libraries are allowed, a component assignment policy needs to be defined to specify the correspondence between components and operations, and to determine the execution time of operations. This can be achieved through binding and scheduling policies.

Operation Priority Calculation. In this paper, a list scheduling algorithm based on a priority policy in operating systems to improve overall system efficiency is used to introduce an improved strategy that combines the characteristics of the binding and scheduling problem in the CFMB flow layer design. The algorithm aims at minimizing the overall scheduling time to solve the binding and scheduling problem in CFMB flow layer design. In a given component library, the number of components of each type is smaller than the number of corresponding operation types in the biochemical protocol. Therefore, we need to calculate the priority for each operation to select the operation that requires the most resources to be executed when resource competition occurs.

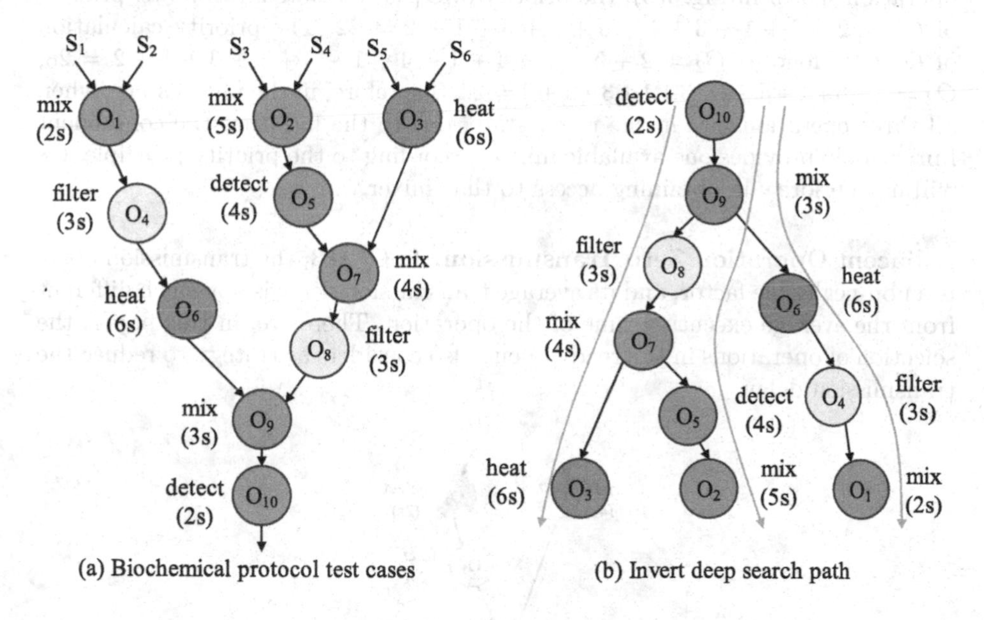

Fig. 3. Time-based strategy.

The goal of the optimization binding and scheduling algorithm is to minimize the execution time of the entire biochemical protocol. According to the observation of the biochemical protocol, as shown in Fig. 3(a), the execution of the protocol ends after the last operation O_{10} completes, because the execution of each operation depends on the full completion of its parent operation. Reversing the arrows of the biochemical protocol diagram, it can be considered as a tree structure containing three depth-first search paths. The weighted length Tl of each path l can be calculated by formula (4):

$$T_l = \sum_{i=1}^{m} Exec_i + \sum_{j=1}^{n} Trans_j \tag{4}$$

where m denotes the number of operations on the path, $Exec_i$ denotes the execution time of operation i; n denotes the number of channels on the path, and $Trans_j$ denotes the transmission delay of path j. According to the barrel theory, in order to minimize the execution time of the whole biochemical protocol, we need to complete the operation with the longest current weighted path length from the target node (i.e., the endpoint) as much as possible. Therefore, in this paper, the priority of each node is set as the weighted path length from that node to the endpoint.

Due to the unclear connection relationship between components at this stage, the transmission delay is also unknown. In order to calculate the priority of operations, we set the transmission delay between operations as unit time. Except for the input delay of the mixer, which is 2, the input delay of all other components is 1. For example, when calculating the priority of the biochemical protocol operation shown in Fig. 3(b), the priority of O_1 is calculated as follows: priority of $O_1 = 2 + 2 + 1 + 3 + 1 + 6 + 1 + 3 + 1 + 2 = 22$; The priority calculation of O_2 is as follows: $O_2 = 2 + 5 + 1 + 4 + 1 + 4 + 1 + 3 + 1 + 3 + 1 + 2 = 28$, $O_3 = 2+6+1+4+1+3+1+3+1+2 = 24$. Therefore, in the initial state, when all three operations are in the ready state, due to the fact that the component library only provides one available mixer, according to the priority principle, O_2 will have priority in obtaining access to that mixer.

Adjacent Operation Zero Transmission. In CFMB, the transmission delay is a non-negligible factor, and its average transmission delay is not much different from the average execution time of the operation. Therefore, in this paper, the selection of operations in the ready queue also considers a strategy to reduce the transmission delay.

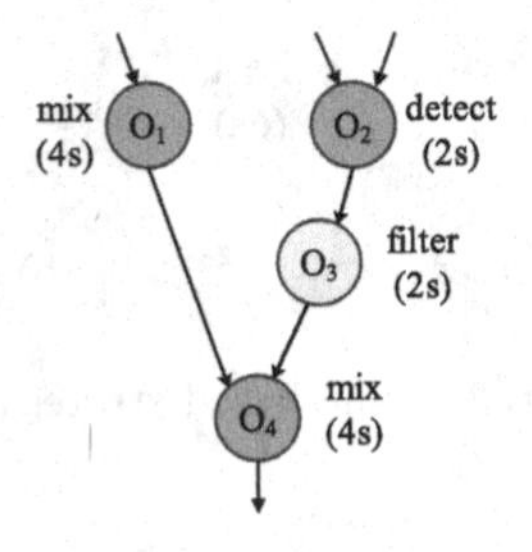

Fig. 4. Zero transfer policy.

For example, Fig. 4 illustrates a part of the biochemical protocol assuming that O_4's preorder input operations O_2 and O_3 complete their execution at the same time node. In this case, assume that there are two free Mixer resources, one of which is released by O_2 after its execution is completed. We bind the child operation O_4 to the same Mixer as its parent operation O_2 for the following reasons: first, the parent and child operations O_2 and O_4 have the same operation type; second, the resources occupied by the parent node are still free when the child node starts execution. By this binding method, the transfer time delay between parent and child operations can be eliminated and the time for droplet transfer from the O_2 operation to the O_4 Mixer can be reduced, thus decreasing the reaction time.

List Scheduling Algorithm. During the execution of the algorithm, operations undergo various state changes. Initializing all operations to the blocking queue is one of the key steps. Next, we need to check the operations in the blocking queue for dependency constraints to determine which operations can be added to the ready queue. The operations that meet the conditions are bound with the idle components according to the "parent-child" operation zero transfer and priority policy and are added to the running queue. At the same time, some of the operations in the ready queue may be rejoined to the blocking queue due to component reuse constraints. After that, the fastest-executing operation in the run queue will release its resources, complete the end operation, and wake up the blocking queue, thus entering the next loop. This process is repeated until all operations have been successfully executed.

After determining the priority calculation method and binding policy for each operation in the biochemical protocol, the following step-by-step list scheduling algorithm is used in this paper:

(1) Add the operations that can be ready to the ready queue, which includes operations without a parent node and operations with a parent node whose parent has already been executed. (2) Calculate the priority of the ready operations according to formula (4) and sort them in reverse order. (3) For each ready operation in turn, check whether the "parent-child" operation zero-transfer condition is met, and if so, bind it to the free component of the parent operation. (4) Bind the remaining operations in the ready queue to the idle component of the matching type in order of priority. (5) Determine the endpoint of the operation that finishes execution the fastest, release the other components that end at that point, and update the information of the related components and operations. (6) Determine whether each operation in the biochemical protocol has been executed, and exit the algorithm if it is finished, otherwise return to step (1) to continue execution.

4.3 Layout Design Based on Improved Simulated Annealing Algorithm

In the high-level integrated design, by obtaining the type and number of components used from the binding and scheduling solution, combined with the dimensional information in the component library, we can perform layout design. The goal of the layout design is to minimize the chip area while flowing out a certain routing space for the subsequent wiring design through the reasonable placement of components. In this paper, we analyze the characteristics of layout design and adopt an improved Simulated Annealing algorithm to accomplish this design task. The specific implementation process is as follows.

Module Placement Constraints. According to the Simulated Annealing algorithm with CFMB module placement characteristics, from a given input Comp, a random layout solution needs to be generated first, while satisfying the following arbitrary component location constraints and inter-component spacing constraints:

$$\begin{aligned} 0 \leq x_c - Hx_c, x_c + Hx_c \leq X_m \\ 0 \leq y_c - Hy_c, y_c + Hy_c \leq Y_m, \ \forall c \in Comp \end{aligned} \tag{5}$$

where, $x_(c)$ represents the center position of component c, $Hx_(c)$ $Hy_(c)$ represents the extension length of component c in the x-axis direction and the y-axis direction (i.e., half of the component side length), respectively,X_m and Y_m represents the maximum area length and width of a given chip.

The inter-component spacing constraint is shown in formula (6):

$$|x_i - x_j| + |y_i - y_j| \geq mids, \forall i, j \in Comp \tag{6}$$

where, indicates that the Manhattan distance between any two components i, $j$needs to be greater than or equal to $mids$. this value should vary dynamically. When the design size is small and the number of wiring is small, it can be used without reserving too much space, and vice versa, more space needs to be reserved for routing.

In this case,$mids$ are set as shown in formula (7):

$$mids = \varphi * Rtn \tag{7}$$

where Rtn is the number of wiring derived from the high-level integrated binding and scheduling solution to existing, and φ is a small number between 0 and 1.

After generating the initial random layout solution, we need to obtain the new layout solution by perturbation, which mainly includes the following three strategies: (i) arbitrarily swap the positions of two components; (ii) randomly change the position of one component; and (iii) randomly change the orientation of one component. After executing the strategies, the positions of the surrounding components need to be adjusted to satisfy the constraints expressed in formula (5) and formula (6).

Improved Simulated Annealing Module Placement Algorithm. In this paper, three improvement strategies are proposed in the traditional Simulated Annealing algorithm based on the specificity of the CFMB module placement task:

First, an initial solution Pc is randomly generated with a certain probability at each temperature T. Since in the traditional Simulated Annealing algorithm, if the initial solution is too extreme, it may result in falling into a local optimum even if the subsequent perturbations are very strong. Therefore, increasing the probability of regenerating a random solution during the algorithm allows the algorithm to search the solution space faster. Also, to implement the memory function, the global optimal solution is used to record the currently found optimal solution.

Second, a global state is added for remembering the global optimal solution that has been found. This can avoid losing the currently found optimal solution due to the acceptance link during the search process.

Finally, we introduced a random variable in updating the temperature to implement the process of rewarming, thus activating the acceptance probability of each state and avoiding the algorithm from wandering at the local optimum.

5 Experimental Results and Analysis

This section will present the description of the experimental setup, the simulated experimental protocol, and the analysis of the results. The results of the simulated experiments will also be presented.

5.1 Experimental Setup

In order to verify the effectiveness of the algorithm proposed in this paper, the architectural synthesis work will be performed in this section for the six standard biochemical protocol test cases listed in Table 2.

Table 2. Biochemical protocol information and number of allocated components [10].

	Biochemical protocol					
	PCR	IVD	EA	Syn1	Syn3	Syn4
operand	7	12	10	10	30	40
(Mixer, detector, filter, heater)	(3, 0, 0, 0)	(2, 3, 0, 0)	(3, 0, 1, 1)	(2, 1, 1, 1)	(5, 1, 2, 3)	(6, 3, 2, 2)
Edges	6	6	9	9	29	39

5.2 Protocol Scheduling Based on List Scheduling Algorithm

To verify the effectiveness of the list priority-based scheduling algorithm in this paper, a comparison is made with the benchmark experiment first-come, first-served strategy. Table 3 shows the results of binding and scheduling six test cases using two different scheduling strategies.

Since the transfer path between components is unknown, the transfer time is set to 1 unit of time by default. The Mixer requires 2 units of input time,

Table 3. Comparison of two scheduling strategies.

	PCR	IVD	EA	Syn1	Syn3	Syn4
FCFS scheduling	21	28	31	50	264	534
List scheduling	21	28	23	35	168	354
optimization rate	0	0	25.8	30	36.4	33.7

while the rest of the components have 1 unit of input time. In the PCR and IVD protocols, the priority scheduling and first-come, first-served scheduling policies yield the same binding and scheduling times due to the relatively single operation. However, in the other four biochemical experiments, the two strategies produced increasingly different results for scheduling the same biochemical protocol as the size of the biochemical protocol increased. Among these four biochemical protocols, the lowest improvement of 25.8% and the highest improvement of 36.4% are achieved by the list-based priority scheduling algorithm in this paper. The solutions obtained by the priority scheduling strategy are on average 31.5% lower than those of the first-come, first-served scheduling strategy. Therefore, it can be seen that the priority list scheduling algorithm is effective in CFMB.

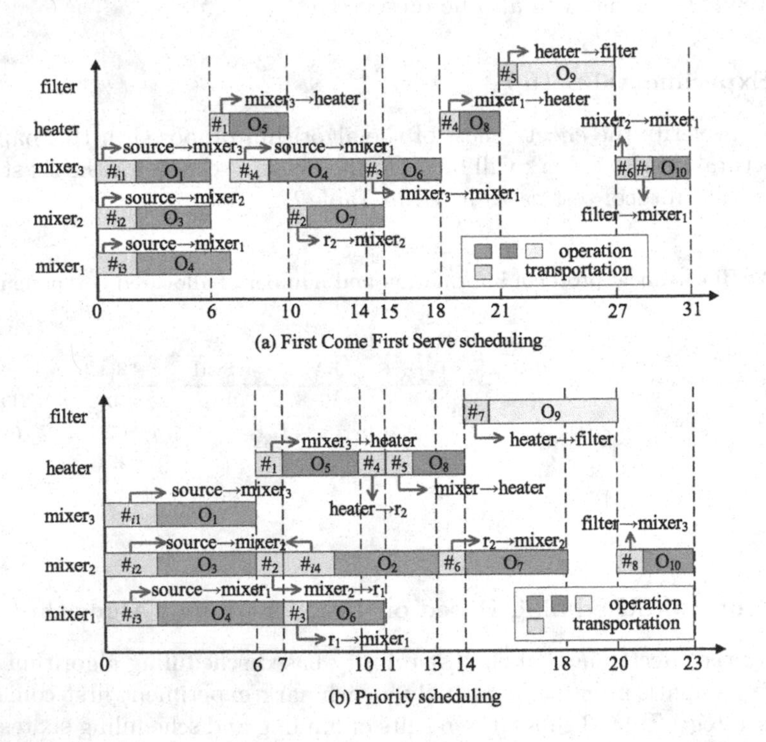

Fig. 5. Comparison chart of scheduling algorithms.

As can be seen in Fig. 5, the binding and scheduling results based on the priority scheduling algorithm outperform the first-come, first-served scheduling strategy in terms of resource utilization and overall execution time. In this example, the latter increases the execution time by 34.8% over the former. As the size of the biochemical protocol increases, this gap will become more significant.

5.3 Module Placement Results for Simulated Annealing Algorithm

In the layout design stage, a layout solution needs to be constructed based on the component types and numbers determined by high-level synthesis, and the location and orientation of each component need to be determined within the specified range. In addition, some space needs to be left between components because routing between components is required in the wiring phase. In this chapter, a Simulated Annealing algorithm incorporating an improved strategy is used for the optimization of the layout solution. The parameters of the algorithm are set as follows: the initial temperature is $T_{init} = 1000$, the lowest temperature is $T_{min} = 1$, the number of iterations at each temperature is $Ite = 10$, and the temperature decrease rate is $\alpha = 0.7$.

In the layout phase, the area size Area of the chip can be calculated using formula (8), which finds the boundaries of the four directions of the layout and thus determines the area of the rectangle. The overall line length *Splen*, which is predicted from the set of component connection relations rt obtained in the binding and scheduling phase, can be calculated using formula (9). In formula (9), each element in Rt represents the component coordinates of the starting and ending points, where $((x_0, y_0), (x_1, y_1))$ denotes the x and y coordinates of the starting and ending points. The optimization objectives of the layout stage are shown in formula (10), where ξ and κ are equilibrium parameters, which are set to 2 and 5, respectively, in this experiment.

$$\begin{aligned} Area =& (Max(x_a - Hx_a) - Min(x_b - Hx_b)) \\ & \times (Max(y_c - Hy_c) - Min(y_d - Hy_d)), \forall a, b, c, d \in Comp \end{aligned} \tag{8}$$

$$splen = \sum_{i=1}^{|Rt|} \left| x_0^i - x_1^i + y_0^i + y_1^i \right| \tag{9}$$

$$E = \xi \times Area + \kappa \times Splen \tag{10}$$

To verify the effectiveness of this algorithm, this section also uses the Hill-Climbing algorithm as the benchmark algorithm for comparative implementation in the layout design phase. The Hill-Climbing algorithm randomly generates an initial layout solution and then performs n variations on it, accepting only the optimization results and selecting the optimal solution among the n variations as the final result. Table 4 shows the standard test case solutions obtained using the Simulated Annealing algorithm and the Hill-Climbing algorithm for layout design with the same input conditions. These solutions are the results of an average of 30 layout designs to avoid errors that may be caused by a single experiment. The results show that the Simulated Annealing algorithm outperforms the Hill-Climbing algorithm in terms of area and line length, achieving an average decrease rate of 24.3% and 22.3%, respectively, which demonstrates the effectiveness of the Simulated Annealing algorithm in layout design.

Table 4. Module Placement Experiment Results.

biochemical protocol	simulated annealing algorithm		hill climbing algorithm		rising rate (%)	
	area	line length	area	line length	area	line length
PCR	35	19	42	25	16.7	24
IVD	183	51	234	76	21.8	32.9
EA	204	81	263	103	22.4	21.4
Syn1	195	79	271	114	28	30.7
Syn3	2034	652	2846	851	28.5	23.4
Syn4	3056	921	4261	1186	28.3	22.3
Avg	-	-	-	-	24.3	25.8

6 Conclusion

In this paper, a novel method for CFMB high-level synthesis design is proposed in response to the fact that the manual design scheme in the traditional CFMB design flow can no longer meet the increasingly complex design requirements of the chip architecture. The method firstly adopts a resource binding and scheduling design based on a list scheduling algorithm, and secondly, applies an improved Simulated Annealing algorithm for module layout design. The experimental results are compared with various benchmark algorithms to verify the effectiveness of the priority scheduling algorithm proposed in this paper in the binding and scheduling of high-level synthesis and the advantages of the Simulated Annealing algorithm using the improved strategy in the layout design process.

Acknowledgements. This work is supported by the fund of Fujian Province Digital Economy Alliance, the National Natural Science Foundation of China (No. U1905211), and the Natural Science Foundation of Fujian Province (No. 2020J01500).

References

1. Melin, J., Quake, S.R.: Microfluidic large-scale integration: the evolution of design rules for biological automation. Annu. Rev. Biophys. Biomol. Struct. **36**, 213–231 (2007)
2. McDaniel, J.M.: Design Automation of Continuous Flow-Based Microuidic Biochips. University of California, Riverside (2016)

3. Becker, H.: Microfluidics: a technology coming of age. Med. Dev. Technol. **19**(3), 21–24 (2008)
4. Guo, W., et al.: A survey on security of digital microfluidic biochips: technology, attack, and defense. ACM Trans. Des. Autom. Electron. Syst. (TODAES) **27**(4), 1–33 (2022)
5. Levenspiel, O.: Chemical Reaction Engineering. Wiley (1998)
6. Dong, C.: A survey of DMFBs security: state-of-the-art attack and defense. In: 21st International Symposium on Quality Electronic Design (ISQED), vol. 2020, pp. 14–20. IEEE (2020)
7. Stanford Microfluidics Foundry: Designing your own device: basic design rules (2019)
8. Ye, Y., et al.: A novel method on discrete particle swarm optimization for fixed-outline floorplanning. In: 2020 IEEE International Conference on Artificial Intelligence and Information Systems (ICAIIS), pp. 591–595. IEEE (2020)
9. Araci, I.E., Quake, S.R.: Microfluidic very large scale integration (mVLSI) with integrated micromechanical valves. Lab Chip **12**(16), 2803–2806 (2012)
10. Minhass, W.H., et al.: Architectural synthesis of flow-based microfluidic large-scale integration biochips. In: Proceedings of the 2012 International Conference on Compilers, Architectures and Synthesis for Embedded Systems, pp. 181–190 (2012)
11. Dinh, T.A.: A clique-based approach to find binding and scheduling result in flow-based microfluidic biochips. In: 18th Asia and South Pacific Design Automation Conference (ASP-DAC), vol. 2013, pp. 199–204. IEEE (2013)
12. Li, M., et al.: Component-oriented high-level synthesis for continuous-flow microfluidics considering hybrid-scheduling. In: Proceedings of the 54th Annual Design Automation Conference 2017, pp. 1–6 (2017)
13. Li, M.: Sieve-valve-aware synthesis of flow-based microfluidic biochips considering specific biological execution limitations. In: eDesign, Automation & Test in Europe Conference & Exhibition (DATE), vol. 2016, pp, 624–629. IEEE (2016)
14. Lin, C.-X., et al.: An efficient bi-criteria flow channel routing algorithm for flow-based microfluidic biochips. In: Proceedings of the 51st Annual Design Automation Conference, pp. 1–6 (2014)
15. Chen, Z.: Physical synthesis of flow-based microfluidic biochips considering distributed channel storage. In: Design, Automation & Test in Europe Conference & Exhibition (DATE), vol. 2019, pp. 1525–1530. IEEE (2019)
16. Huang, W.L., et al.: Fast architecture-level synthesis of fault-tolerant flow-based microfluidic biochips. In: Design, Automation & Test in Europe Conference & Exhibition (DATE), vol. 2017, pp. 1667–1672. IEEE (2017)
17. Su, F., Chakrabarty, K.: High-level synthesis of digital microfluidic biochips. ACM J. Emerg. Technol. Comput. Syst. (JETC) **3**(4), 1–32 (2008)
18. Su, F., Chakrabarty, K.: Architectural-level synthesis of digital microfluidics-based biochips. In: 2004 IEEE/ACM International Conference on Computer Aided Design, ICCAD-2004, vol. 2004, pp. 223–228. IEEE (2004)

Wireless and Ubiquitous Networking

Exploration and Application Based on Authentication, Authorization, Accounting in Home Broadband Scenario

Feng Tian[1(✉)], Haoshi Zhang[1], Pan Wang[2], and Xinkuan Wang[1]

[1] China Mobile Communication Group Jiangsu Co., LTD. Yangzhou Branch, Yangzhou, China
13952519669@139.com

[2] School of Modern Posts, Nanjing University of Posts and Telecommunications, Nanjing, China
wangpan@njupt.edu.cn

Abstract. With the rapid development of broadband market services, Internet Service Providers (ISPs) strive to improve the end-to-end network quality of home broadband. However, faced with the opaque and unknown indoor network environment, traditional monitoring methods cannot meet the requirements, and new methods should be explored to identify problems on the user-side network to improve the perception and experience of users. In this method, FreeRadius and DaloRadius are deployed on the Linux server, user AAA attributes are collected by the Radius server, poor-quality routers and non-direct connection TV Set-Top Boxes (STB) are monitored with the help of Python's data processing capability, and the home networking environment of users is optimized to improve the perception and experience of users.

Keywords: Home Broadband · FreeRadius · DaloRadius · SQL · Router Quality · Traffic Monitor

1 Introduction

With the rapid development of home broadband market services, gigabit high bandwidth and smart home devices gradually popularized, home users' requirements for high-quality broadband services continue to increase, and ISPs strive to improve the end-to-end network quality of home broadband. In the daily operation and maintenance of operators and network complaints, users reported slow network speed or TV viewing delays, nearly half of which were caused by the poor performance of user-side routers and unreasonable indoor networking layout [10]. At present, the user-side networking environment is becoming more and more complex, and the major router brands are uneven. In the face of the opaque indoor networking environment such as the selection of user routers, connection mode, and downlink devices, ISPs cannot judge the client-side problems through traditional methods.

By using open-source software such as FreeRadius, SQL, and DalaRadius, the server platform is built and the data of Internet access authentication and accounting attributes of users can be customized to provide basic data support for home networking research, it

H. Jin et al. (Eds.): GPC 2023, LNCS 14504, pp. 285–295, 2024.
https://doi.org/10.1007/978-981-99-9896-8_19

innovatively analyzes home gateway and STB data, which significantly helps to improve the user home network. Meanwhile, clients can be added or deleted according to monitoring requirements. SQL statements can be used to query user data on the server, and network engineers can also operate directly on the front page or report output; Python scripts automatically check the amount of data of 850,000 *400 per day, a one-click operation, without manual intervention in the process [6]. This paper explores and practices the optimization of home broadband networks and has achieved certain results.

2 RADIUS Protocol Introduction

RADIUS, which stands for "Remote Authentication Dial In User Service", is a network protocol that defines rules and conventions for communication between network devices. RADIUS serves AAA (Authentication, Authorization, Accounting) functions, it is commonly used by ISPs, corporate and educational networks [5], it is a client/server structure based on UDP, the client passes the user information to the server, the server stores the user's identity information, authorization information, and access records, and returns the user's authentication results and configuration information. In the home broadband internet access scenario, the RADIUS workflow [14] is shown in Fig. 1:

1) The user provides the user name and password for broadband dial-up to apply for access.
2) The broadband access device NAS (Network Access Server, as a RADIUS client) obtains the user name and password and sends an access request packet to the RADIUS server.
3) The RADIUS server compares the user information with the database information to determine whether the user name and password are valid. If the authentication is valid, the user's permission information is sent to the RADIUS client in an access-accept packet. If the authentication is invalid, the access-reject response packet is returned.
4) After user authentication succeeds, the RADIUS client sends an accounting-request packet to the RADIUS server. The status-type value is started.
5) The RADIUS server returns the accounting start response packet.
6) The user starts to access network resources.
7) The RADIUS client sends an accounting request to the RADIUS server. The status-type value is stopped.
8) The RADIUS server returns the accounting end response packet.
9) The RADIUS client notifies the user that network access ends and the user stops accessing network resources.

RADIUS packets are encapsulated in UDP packets. There are 16 types of RADIUS packets [1]. Data in each domain of the packets is transmitted from left to right. Take accounting request packets as an example in Fig. 2:

1) The Code field holds one byte, which identifies the RADIUS packet type. The valid values of the RADIUS Code field are assigned as follows:

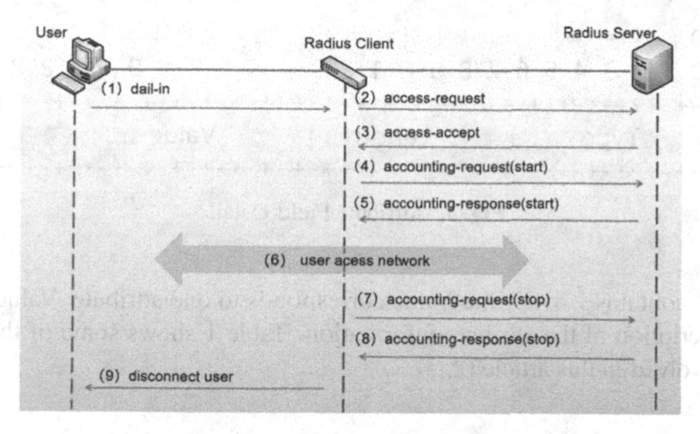

Fig. 1. Radius Workflow

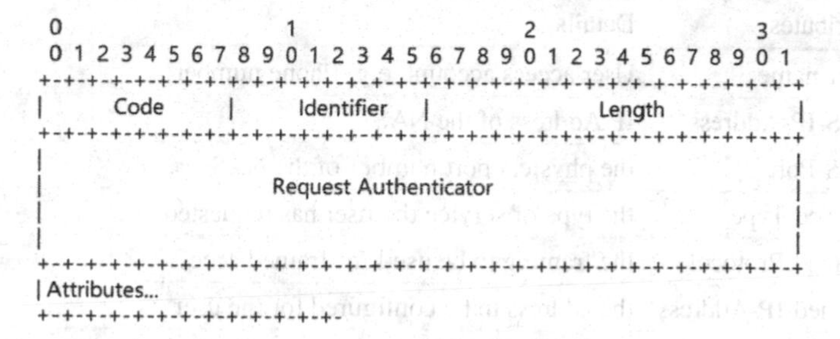

Fig. 2. Accounting Request Packet Format

2) The Code value 1–5 is common and indicates that the packet types are authentication request, authentication accept, authentication reject, accounting request, accounting response, and accounting success. If the Code field of the received packet is invalid, the packet is silently discarded.
3) Identifier: occupies 1 byte to match the request packet and response packet. If the request packet with the same source IP address, source UDP port, and the same Identifier field is received within a short period time, the server will consider it as a repeat request packet.
4) Length: contains 2 bytes, including the total length of the code field, identifier field, length field, authenticator field, and attribute field in the packet.
5) Request Authenticator: the request authentication key of the accounting request packet is a 16-byte MD5 hash value that is used to authenticate the messages between the client and server.
6) Attributes: the attribute field carries detailed information about authentication, accounting, authorization, and configuration through request packets and response packets. The TLV (Type-Length -Value) format is shown in Fig. 3 and is transmitted from left to right. All packet interactions are composed of different lengths of TLV triples.

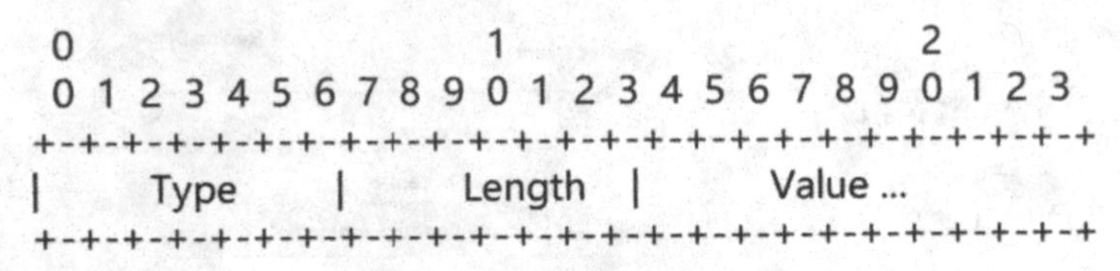

Fig. 3. Attribute Field Detail

The type contains 1 byte. One Value corresponds to one attribute. Value contains a detailed description of the attribute information. Table 1 shows some of the RADIUS attributes involved in this article [2, 3].

Table 1. RADIUS Attribute Types

Code	Attributes	Details
1	User-name	User access account, e.g. Phone number
4	NAS-IP-Address	IP Address of the NAS
5	NAS-Port	the physical port number of the NAS
6	Service-Type	the type of service the user has requested
7	Framed-Protocol	the framing to be used for framed access
8	Framed-IP-Address	the address to be configured for the user
31	Calling-Station-Id	the NAS to send in the Access-Request packet the phone number that the call came from
42	Acct-Input-Octets	Upload bytes
43	Acct-Output-Octets	Download bytes

3 Authentication and Accounting Based on FreeRadius

In this solution, a RADIUS server is constructed to implement user authentication and accounting. The server is CentOS 6.5 operating system and FreeRadius 2.2.6 is used to deploy the RADIUS authentication server. FreeRadius supports multiple databases, and MySQL is supported by default. In order to build a comprehensive environment suitable for the Web application platform, Apache, MySQL, and PHP are installed on the Linux system in advance to form a comprehensive LAMP (Linux, Apache, MySQL, PHP) environment and process architecture as shown in Fig. 4.

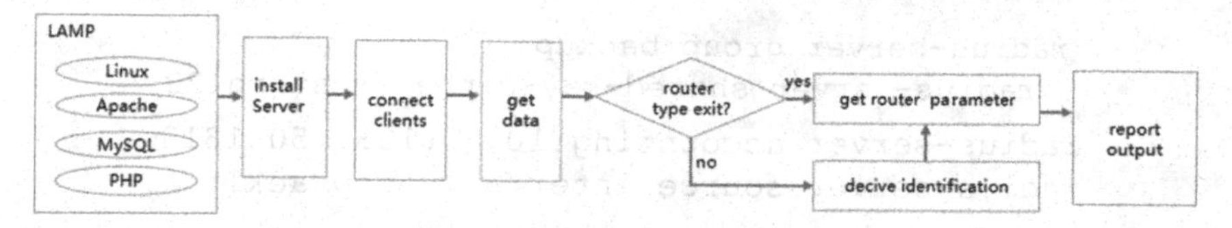

Fig. 4. LAMP and Process Architecture

3.1 Building FreeRadius Server

Install FreeRadius

FreeRadius is a developable and high-performance open-source RADIUS program. You are advised to install it as a binary installation package [4]. Run the yum command to install FreeRadius and its components in the CentOS system:

```
yum install -y freeradius freeradius-mysql freeradius-
utils
```

Here, Freeradius-utils is a toolkit for testing FreeRadius, and Freeradius-mysql is a database extension for FreeRadius.

Configure FreeRadius Server

The FreeRadius server configuration file is installed in the /etc./raddb directory. There are four configuration files: radiused.conf, clients.conf, user and sql.conf, radiusd.conf is the core configuration file of FreeRadius, which contains the basic information of the server, the environment variables of the configuration file and log file, the configuration of the module used for authentication and accounting, clients.conf is a client configuration file that stores authentication information about RADIUS clients. Use the VI editor to modify the client configuration file and add RADIUS clients to the server.

```
vi /etc/raddb/clients.conf
client x.x.x.x {
        ipaddr = x.x.x.x
        secret = testing123
        require_message_authenticator = no }
```

Configure FreeRadius Client

NAS (Network Access Server) is regarded as the client of RADIUS, it accepts user names and passwords submitted by users, then sends them to the RADIUS server for authentication [3]. On the corresponding NAS device, enter the AAA authentication interface, set the RADIUS service group, the address and port of the accounting server, and the loopback interface as the sending source interface.

```
radius-server group backup
  radius-server shared-key-cipher testing123
 radius-server accounting 10.35.133.150 1813
 radius-server source interface LoopBack1
```

Start Radius Service

Execute radius-X for the first boot, when “Ready to process requests” appears at the end of the code, it indicates a successful startup.

3.2 Configuring MySQL Database

In the LAMP environment above, MySQL service has been deployed and started [7], and then the FreeRadius database is configured.

Import DatabaseTables

Log into the database, enter the password of the administrator account root, create a database named radius, assign authorization to the user radius, and exit the database.

```
mysql -u root -p
create database radius;
grant all on radius.* to radius@"localhost" identified by
"radpass";
exit
```

Relog into the database, modify environment variables, import SQL databases, and run the following command:

```
mysql -u root -p
source /etc/raddb/sql/mysql/admin.sql
use radius;
source /etc/raddb/sql/mysql/schema.sql
```

Check Database Tables

After the import is complete, open the radius database, run the use radius command, and then show tables to view all tables in the radius database. The schema.sql database contains user authentication-related data tables [13]: radcheck (user check information table), radreply (user reply information table), radusergroup (user and group relationship table), redacct (accounting information table), and so on. Among them, raddact records the billing information of users and contains rich attributes in Table 1, which provides basic data support for subsequent collection and analysis.

Use SQL Language to query a single user and confirm that the server has collected user authentication data, such as the user internet access NASIP, NAS PORTID, billing start and end time, and gateway MAC.

At this point, the FreeRadius server has been set up, and the server can read the user authentication data on the client NAS device. However, FreeRadius does not have a WEB management interface, and it still needs to use the command line for maintenance, which is relatively complicated. Considering the convenience of network engineers, it is better to use the WEB interface for management.

4 DaloRadius Management

DaloRadius is an open-source third-party RADIUS WEB management tool with a graphical interface. It features user management, graphical reports, accounting, and authentication.

1) Download and decompress the DaloRadius source code, move the folder to the working directory of the Apache Web server, install the daloradius0.9–9 compressed package, decompress daloradius-0.9–9.tar.gz, and overwrite the daloradius source code [12].

```
wget http://download.pear.php.net/package/DB-
1.7.14RC2.tgz
wget
https://nchc.dl.sourceforge.net/project/daloradius/dalora
dius/daloradius0.9-9/daloradius-0.9-9.tar.gz
tar -zxvf daloradius-0.9-9.tar.gz
mv daloradius-0.9-9 daloradius
```

2) Set all files in the DaloRadius folder to be owned by the apache group and apache users, and set the permission of the daloraidus.conf.php file to rw-r--r--.
3) Create a DaloRadius data object in MySQL, import fr2-mysql-daloradius-and-freeradius.sql and daloradius.sql into the radius database [12], and modify the DaloRadius configuration. Set the sql connection with FreeRadius.

```
mysql -u root -p radius < mysql-daloradius.sql
mysql -u root -p radius < fr2-mysql-daloradius-and-
freeradius.sql
```

4) Set the connection mode of the daloradius database: Change the CONFIG_DB_PASS password in line 33 of the daloradius.conf.php file, save the password, and restart the httpd service.

```
vi /var/www/html/daloradius/library/daloradius.conf.php
$configValues['CONFIG_DB_PASS'] = '123456 '
```

5) Open a browser to http://10.35.133.150/daloradius, and enter DaloRadius interface, the default login user name and password are the administrator and the radius. The DaloRadius interface displays online user account information and supports CSV report export, which is simple, intuitive, and easy to operate.

5 Application Evaluation

The RADIUS database stores AAA information of the user, and the DaloRadius export report contains general attributes of the user, including Username,User IP Address,User MAC Address, Start Time, Total Time and NAS IP Address. These basic data provide accurate and reliable support for the analysis and exploration of home networking environment and user internet habits [11].

Scenario 1: The STB is not Directly Connected

To relieve the pressure of traditional unicast TV technology on servers, ISPs introduce more flexible and interactive multicast TV, and program resources can be pulled to the local server to achieve multicast data interaction. The more users, the more advantages of multicast technology [9]. Because multicast mode uses specialized channels, it is isolated from network data. The STB must be connected to the gateway's LAN port to obtain independent channel bandwidth. When a home user connects STB to the LAN port of the home gateway to watch TV, the MAC address numbers of each LAN port of the gateway are usually continuous, and the probe in STB can collect the information of the connected device, make a correlation comparison between the MAC of the connected device of STB and the MAC of the LAN port of the home gateway, and process the data with PYTHON, lambda x:str(x[0:-1]) compares the MAC digits of the both [17]. If they are consistent, it is considered that the STB is directly connected to the gateway; otherwise, it is not directly connected. Then, the maintenance staff will optimize user home networking and directly connect the STB to the LAN port of the gateway, thereby improving the user experience [16]. About 450 new TV subscribers are installed everyday, as shown in Fig. 5, since the implementation of this method in March this year, the proportion of newly installed TV direct connections increased from 81.29% to 90.23% daily, and the proportion of total user direct connections increased by 2.36PP.

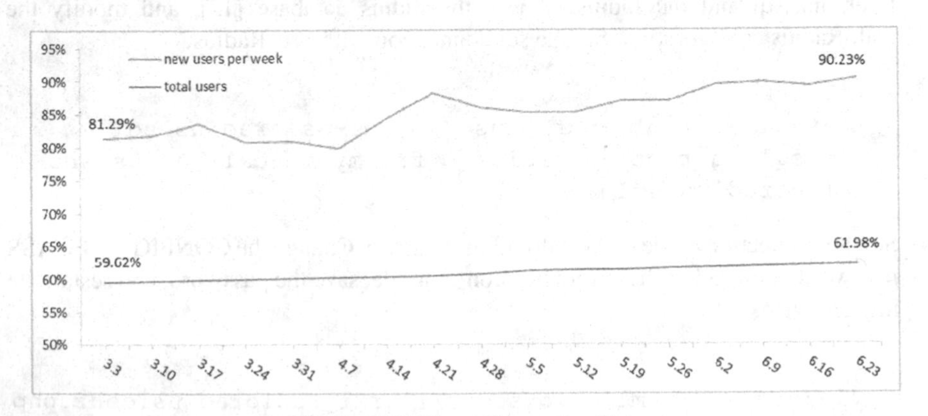

Fig. 5. STB Direct Connection Rate

Scenario 2: Router Quality Identification

At present, the common household router brands and types on the market are uneven, and the data forwarding capability and signal strength of various routers are different, and there are often disconnected or hung, which has become an important factor affecting the quality of indoor internet access [8]. DaloRadius reports the time for users to dial up and down each time, analyzes the parameter information of the router connected to the gateway for the user who dials frequently within a specified time, and classifies the router manufacturer and type, then puts them into the database of suspected poor-quality routers [15] as shown in Fig. 6. For users who use similar manufacturers and types of routers, it is actively recommended that users buy intelligent network equipment or replace high-quality routers. From a passive response to an active solution, ISPs innovate the maintenance method, accelerate the problem rectification and repair the perception of users further.

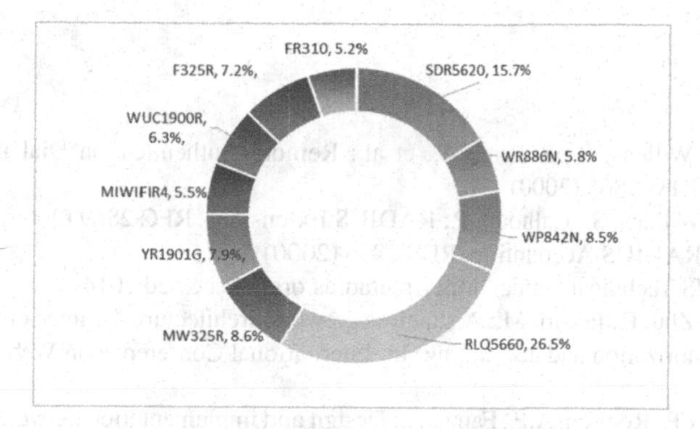

Fig. 6. Poor-quality Router Classification

We analyzed 82,000 user routers in a certain area, and 424 users were frequently online and offline. After taking some user samples and replacing them with high-quality routers, a lot of user network performances are significantly improved, and the effect of network quality improvement is shown in Fig. 7.

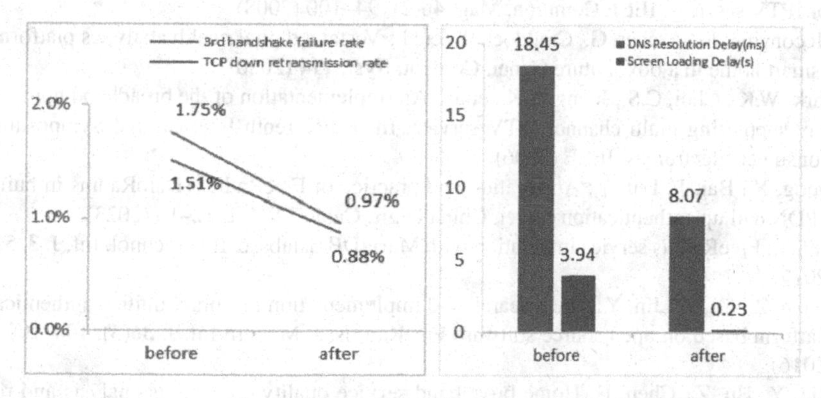

Fig. 7. Network Quality Improvement

6 Conclusion and Future Work

Using FreeRadius to build an authentication platform is a common practice in the industry. This scheme is closely related to daily production work, we use the RADIUS attributes collected by the RADIUS server to explore and practice the optimization of the home broadband network and have achieved certain results. At the same time, it meets the requirements of network engineers to simplify RADIUS management. In addition, DaloRadius reports can output data such as online and offline time points, upstream and downstream traffic of users each time they log in. We still have a lot to explore in future work with DaloRadius report data. For example, predicting the off-network trend of low-active users, upgrading high-value potential customers, and monitoring user abnormal traffic.

References

1. Rigney, C., Willens, S., Rubens, A., et al.: Remote Authentication Dial in User Service (RADIUS), RFC 2865 (2000)
2. Rigney, C., Willens, S., Calhoun, P.: RADIUS Extensions, RFC 2869 (2000)
3. Rigney, C.: RADIUS Accounting, RFC 2866 (2000)
4. FreeRADIUS Technical Guide. http://freeradius.org/. Accessed 2014
5. Kung, H.T., Zhu, F., Iansiti, M.: A stateless network architecture for inter-enterprise authentication, authorization and accounting. In: International Conference on Web Services DBLP (2003)
6. Paramitha, A.P., Rochim, A.F., Fauzi, A.: Design and implementation network administrators account management system based on authentication, authorization, and accounting based on TACACS and LDAP. IOP Conf. Ser. Mater. Sci. Eng. **803**, 012040 (2020)
7. Lorincz, J., Udovičić, G., et al.: Architecture of SQL databases for WLAN access control and accounting. In: Proceedings of 15th International Conference on Software, Telecommunications and Computer Networks – SoftCOM (2007)
8. Wang, P., Ye, F., Chen, X.: A smart home gateway platform for data collection and awareness. IEEE Commun. Mag. **56**(9), 87–93 (2018)
9. Degrande, N., Laevens, K., De Vleeschauwer, D., et al.: Increasing the user perceived quality for IPTV services. IEEE Commun. Mag. **46**(2), 94–100 (2008)
10. Mcconville, R., Archer, G., Craddock, I., et al.: Vesta: a digital health analytics platform for a smart home in a box. Future Gener. Comput. Syst. **114** (2020)
11. Park, W.K., Choi, C.S., Jeong, Y.K., et al.: An implementation of the broadband home gateway supporting multi-channel IPTV service. In: IEEE Tenth International Symposium on Consumer Electronics. IEEE (2006)
12. Dong, X., Bai, J., Liu, Y.: Application and practice of FreeRadius+DaloRadius in building VPDN dail up authentication server. China Digit. Cable TV J. **1**, 12–16 (2023)
13. Liu, T.: FreeRadius service integration with MariaDB database. Inf. Technol. Inf. J. **3**, 52–54 (2017)
14. Wang, Z., Yu, X., Jin, Y.: The research and implementation of library unified authentication platform based on open source software FreeRadius. J. Modern Inf. J. **36**(5), 104–109+127 (2016)
15. Xie, Y., Hu, Z., Chen, F.: Home broadband service quality complaints analysis and router solutions. Telecommun. Technol. J. **2**, 69–71 (2019)

16. Shkedi, R.: Smart TV detection of STB user-control actions related to STB-originated content presentation speed. US20210266615A1 (2021)
17. Xu, Y., Sun, C., Du, C.: Home potential customers identified based on DPI probe technology. Commun. Manage. Technol. J. 3, 29–32 (2022)

Autonomous Communication Decision Making Based on Graph Convolution Neural Network

Yun Zhang, Jiaqi Liu(✉), Haoyang Ren, Bin Guo, and Zhiwen Yu

School of Computer Science, Northwestern Polytechnical University, Xi'an 710072, China
zhang__yun@mail.nwpu.edu.cn, {jqliu,goub,zhiwenyu}@nwpu.edu.cn

Abstract. As a method of multi-agent system cooperation, multi-agent communication can help agents negotiate and adjust behavior decisions by exchanging information such as observation, intention, or experience during operation, improve the overall learning performance, and achieve their learning objectives. However, there are still some challenging problems in multi-agent communication. With the expansion of the multi-agent system scale, the global complete massive information will bring great resource overhead, and the introduction of redundant communication will lead to the difficulty of agent policy convergence, and affect the joint action and target completion. In addition, predefined communication structures have potential cooperation limitations in dynamic environments. In this paper, we introduce a dynamic communication model based on the graph convolution neural network called DCGN. Empirically, we show that DCGN can better cope with the dynamic update of tasks in the process of helping agents complete task information interaction, and can formulate more coordinated strategies than the existing methods.

Keywords: multi-agent reinforcement learning · multi-agent communication · graph neural convolution

1 Introduction

Reinforcement Learning (RL) is an important method in the domain of Machine Learning [1]. RL is an optimal behavior strategy learning method with environmental feedback as the input target. Evaluate the action and modify the strategy based on the reward. During this interaction, the agent learns strategies to maximize the return or achieve the goal. As the number of agents increases, RL evolves into Multi-Agent Reinforcement Learning (MARL). The environment in a MARL scenario is complex, and dynamic, and in general, agents cannot collect global information, they only perform tasks in their own subspace [2]. These characteristics make the learning process very difficult. In MARL, agents need to be related to each other to complete tasks without being completely knowledgeable in the environment. They can cooperate, compete with each other, or both cooperate and compete with each other simultaneously.

H. Jin et al. (Eds.): GPC 2023, LNCS 14504, pp. 296–311, 2024.
https://doi.org/10.1007/978-981-99-9896-8_20

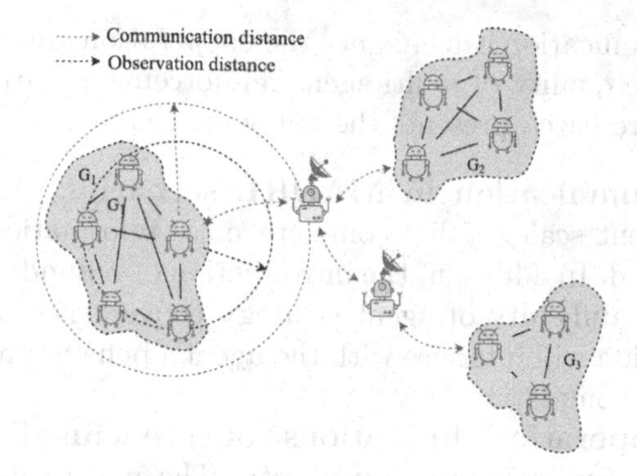

Fig. 1. Agents with limited local observation expand their receptive domain through communication.

Multi-agent cooperation is a very important research field, through the research of intelligent agent cooperation, we can solve many problems in reality. At present, the methods to promote multi-agent cooperation can be roughly divided into the following categories: 1) The multi-agent system is regarded as a whole. This kind of method is based on the perspective of a single agent, but it has the problem of dimension explosion and is not efficient. 2) Design of learning framework. This type of work is intended to explore a feasible learning framework to integrate existing RL technologies into multi-agent environments. 3) Communication between agents. That is, agents coordinate their strategies by sending and receiving abstract communication information to analyze the situation of other agents in the environment. Compared with other approaches, communication-based approaches are easier to implement and can be independent of centralized learning styles, preferring real-world systems and accomplishing collaborative tasks when the environment is partially observable [3], as shown in Fig. 1.

There are three effective ways for multi-agent communication to promote cooperation: 1) Multi-agent communication based on the gate mechanism. Information vectors from other agents can be taken as part of their own input, and these vectors contain some communication content that can be learned synchronously with gradient descent and pass information between agents [4]. 2) Multi-agent communication based on attention mechanism. This type of approach allows each agent to actively choose which agents to send messages to through the attention mechanism [10]. 3) Multi-agent communication based on graph neural network. Such methods focus on modeling the relationships between agents and learning cooperation using the underlying features generated by the graph convolution layer from the increasing acceptance domain [13]. However, these methods do not solve both the problem of reducing communication redundancy and the problem of adaptive multi-agent systems to dynamic environments.

Effective communication can enhance the cooperation among agents and ultimately improve the quality of multi-agent reinforcement learning strategies [4]. In this paper, our research presents the following challenges:

1. **Efficient communication in MADRL scenarios**. With the expansion of the multi-agent scale, global complete mass information will bring great resource overhead. In addition, the introduction of redundant communication will lead to the difficulty of agent strategy convergence, and inaccurate or wrong information will interfere with the agent's behavior and affect the joint action and goal completion.
2. **Potential cooperation limitations of predefined communication structures in dynamic environments**. The multi-agent environment is highly dynamic and the strategies of other agents change at run time. Whether agents need to communicate, which agents to communicate with, what kind of information to communicate, and how to effectively utilize information from other agents are all problems that need to be analyzed and solved [5].

According to these problems, this work introduces a **Dynamic Communication based on Graph Neural Network (DCGN)** based on graph convolution. First of all, DCGN proposes an agent relationship-building module for a dynamic environment to help agents build a decentralized self-organizing communication topology based on situation and task space. The model can better depict the relationship between agents and the influence of individual situation changes on information interaction. Secondly, we expand the action space of the agent, and the agent decides whether the agent needs to communicate when acting, to guide the next action. Finally, the time relationship regularization module is used to strengthen the consistency of the attention matrix between actions, which helps stable cooperation in the growth period of the intelligent body. The main contributions are as follows:

1. We introduce a Dynamic Communication based on Graph Neural Network (DCGN) based on graph convolution, which establishes the correlation between agent interaction and topology in a dynamic environment at all times so that agents can learn communication strategies together. Graph convolution communication helps agents expand their perceptive domain and improve the quality of strategies. At the same time, the empirical replay neural network is used to reduce the number of interactions with the environment and efficiently utilize the diversified training data.
2. We expand the action space of the agent and take communication as a kind of action of the agent itself, so that the agent can decide whether to communicate independently. The action of agents deciding whether to communicate is associated with the communication topology of a multi-agent system. The communication structure of the agent will change dynamically with the change in the agent's behavior and environment.
3. We have experimentally proved that DCGN can better cope with dynamic updating of tasks and help real-time decision-making in the process of helping agents complete task information interaction, and has strong adaptability

and practicability in the field of complex agent network communication. Compared to the advanced baseline approach, the DCGN increased rewards by 160% and reduced communication volume by 57%.

2 Related Works

In this section, we introduce the existing mainstream multi-agent reinforcement learning methods for multi-agent communication. From the perspective of methods, they can be roughly divided into algorithms based on gate mechanism, algorithms based on attention mechanism, and algorithms based on graph neural network.

2.1 Multi-agent Communication Based on Gate Mechanism

The function of the gate mechanism is to enable the agent to receive vectors from other agents as part of the input. These vectors contain some communication content, which can learn synchronously with the gradient descent, and transfer information between agents. Sukhbaatar et al. [6] designed a classic communication network called CommNet, which averages all the agent states at the previous moment and aggregates them with the agent states at the current moment through LSTM to predict the action. Jiang et al. [4] proposed ATOC reduce useless communication and improve the communication mode between agents to improve communication efficiency. Singh et al. introduce the IC3Net [8], which also aims to reduce useless communication and can control whether the agent needs information from other agents. Hu et al. [15] introduced ETC-Net, which is superior to other methods in reducing bandwidth usage and still preserves the cooperative performance of multi-agent systems to a maximum extent. Ding et al. [16] designed I2C to standardize agent strategies to better utilize communication information and improve the performance of various multi-agent collaboration scenarios.

2.2 Multi-agent Communication Based on Attention Mechanism

Attention mechanism as a screening and decision mechanism, is used to decide who to communicate with and how to effectively aggregate hidden information. Vain et al. [9] introduced the attention mechanism to calculate the attention vector of each agent's output communication information. It uses the multi-head attention mechanism to realize the sum of the weight contributions of other agents from different perspectives. Das et al. introduced TarMAC [10] which not only solves the problem of when to communicate but also the problem of who to communicate with. At the same time, it uses an attention mechanism to aggregate information. Liu et al. [17] proposed a new game abstraction mechanism based on two-stage attention networks called G2ANET, which can indicate whether there is interaction and the importance of interaction between two agents. Malysheva et al. [18] proposed a novel approach, called MAGnet, to multi-agent reinforcement learning that utilizes a relevance graph representation of the environment obtained by a self-attention mechanism.

2.3 Multi-agent Communication Based on Graph Neural Network

The communication model based on a graph neural network focuses on modeling the relationship between agents and uses the potential features generated by the graph convolution layer from the increasing acceptance domain to learn cooperation. As a method of establishing communication topology, a graph Neural Network is more suitable for this information propagation environment in dynamic node communication, for complex interaction and communication capture [11]. Jiang et al. [12] introduced a model called DGN, which is based on a graph network and uses multi-head dot product attention as a convolution kernel to calculate the interaction between agents [13]. Yuan et al. [14] proposed that GraphComm can provide explicit relationships through some knowledge background, to better model the relationship between agents. Liu et al. [17] introduced GA-COMM and GA-AC to simplify the learning process while achieving better asymptotic performance. Kim et al. [19] proposed a framework called SchedNet, in which agents learn how to arrange themselves, how to encode messages, and how to select operations based on received messages. Niu et al. [20] proposed MAGIC, setting up a schedule to decide when to communicate and with whom, and using graph attention networks with dynamic graphs to process communication signals.

The methods mentioned above either focus on reducing communication redundancy or focus on decentralization to achieve distributed communication, without balancing the two. Moreover, a fixed communication structure is not conducive to the construction of cooperative relationships between agents. Therefore, we introduce a Dynamic Graph Neural Network (DCGN), which comes from the adaptive response to the change of environment, and the agent itself decides whether to communicate.

3 Problem Formulation

In this section, we model the multi-agent communication decision-making problem. First, we give the communication topology diagram and the definition of the multi-agent communication decision. Then we define the environment of the multi-agent system, which is represented by tuple M, and we summarize what each element represents.

3.1 Communication Topology

We use communication topology to represent communication relationships between agents. The communication between agents can be established by default. For a given agent collection $V = \{v_1, v_2, v_3, ..., v_n\}$, where n represents the number of agents. We define $E = \{(v_i, v_j, w) \mid v_i, v_j \in V, w > 0\}$ to be the set of edges with weights, where w is the weight. The communication topology graph can be divided into directed or undirected graphs according to whether there is the direction of edge e_{ij}, and can be divided into powerless or weighted graphs according to weight, which can be expressed as resource cost or distance. The communication topology graph $G = (V, E)$ is constructed, and the adjacency matrix is represented by $C = (e_{ij})_{n \times n}$. For $(v_i, v_i, w) \in E$ the formulaic representation of e_{ij} is as follows:

$$e_{ij} = \begin{cases} 1, & if\ v_i \text{ chooses communication} \\ 0, & otherwise \end{cases} . \tag{1}$$

3.2 Multi-agent Communication Decision

We hope that the agent can independently decide whether to communicate, so we define the multi-agent communication decision.

In the environment, at each time step t, the agent selects an action $a_t \in A$ from its own action set, and the action space is $A = \{a_1, a_2, \ldots, a_K\}$. At the same time, the agent has its local observation $o_t \in O$ and the current state $s_t \in S$. In addition, the agent will also receive a reward $r_t \in R$ to mark the good or bad actions it has taken in the environment. Often, the reward depends on the entire preceding sequence of actions and observations. Feedback about an action may only be received after thousands of time steps have passed. The most important thing is to consider the communication strategy $C_\phi = f(S, O, R, C)$, which is determined by their states, observations, rewards, and the communication topology adjacency matrix between multiple agents. The subscript ϕ is used to distinguish from adjacency matrix C. Of course, these are not necessary and can be extended, even to add the relevant information of other agents at the last moment.

3.3 Problem Define and Example

According to the limited observation of each communication agent, we formulate the multi-agent communication decision-making problem as a Partially Observable Markov Decision Process (POMDP). In particular, we define a tuple $M = (N, S, A, P, R, O, C)$ to represent a multi-agent communication system, respectively representing the number of agents, agent state space, agent action space, transition probability functions, reward functions, partially observations space and agent system adjacency matrix. Take the Schedule environment in the later experimental section as an example, which is an environment in which agents need to reach the task point to finish the tasks, and the movement of agents will have costs. The meanings of each element are as follows:

Agent: We assume that every agent is homogeneous, that is, has the same communication distance, observation distance, state space, and action space. In Schedule, the number of agents is $N = 5$.

State Space S**:** In Schedule, the state space is the location of the agent, and the next state is determined by the agent's action and probability function. The probability function depends on the environment.

Action Space A**:** We consider discrete action, the action set A_i of each agent is a finite set and contains the action extension of whether or not to communicate.

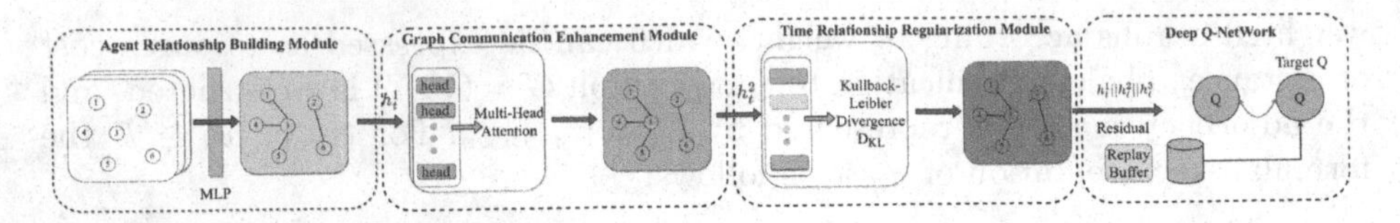

Fig. 2. The framework of communication autonomous decision making model DCGN.

In the Schedule, the action space of the agent is up, down, left, right, stay, and whether to communicate.
Reward R**:** The reward of the agent is related to the state and behavior of the agent. In the Schedule, each move loses a reward worth 1, while completing the task earns a reward worth 50.
Partially observation space O**:** The observation of each agent is limited, and it cannot observe the global information. For example, in the experimental part below, the observation of the agent is a 3×3 matrix, while the communication range is a 5×5 matrix. Agents can expand their receptive domain through communication.

4 The Proposed Method

In this section, we introduce the implementation of Dynamic Communication based on Graph Neural Network (DCGN). First, we provide an overview of the whole model. Then, we introduce the three modules of DCGN in detail.

4.1 The Overview of DCGN

The framework of DCGN is shown in Fig. 2. The DCGN mainly consists of three modules, which are illustrated below.

1. Agent Relationship Building Module (ARM): The function of this module is to divide the communication group according to the communication distance of the agent. The topology matrix represents the communication relationship of the agent.
2. Graph Communication Enhancement Module (GCM): The function of this module is to integrate information, extract potential features, and abstract the representation of the relationship between agents.
3. Time Relationship Regularization Module (TRM): The function of this module is to help agents form long-term consistent action strategies in a highly dynamic environment with a large number of mobile agents.

4.2 Agent Relationship Building Module

In the real environment, agents have a communication distance limit. We encourage agents to generate a decentralized self-organizing communication topology, which is more suitable for a dynamic environment. Agents within the communication distance can communicate with each other and become a communication group. When multiple groups select an agent at the same time, the agent acts as a bridge between different communication groups, allowing information to flow between different groups.

$$G_t^i = \{v_i, v_j |\ d_{ij} < rc,\ j \in N_{-i}\}, \tag{2}$$

where G_t^i is the group with agent v_i as the center at moment t, v_i is the ith agent, d_{ij} is the distance between agent v_i and v_j, rc is the communication distance. At the same time, the adjacency matrix C of the communication topology graph can be obtained, as shown in Eq. (1). When we consider whether the agent has communication behavior, the communication topology of the system is a directed graph; when we do not consider the communication behavior of the agent, the communication topology of the system is an undirected graph. We consider a directed graph, if edge e_{ij} between agent v_i and v_j is $e_{ij} = 1$ and $e_{ji} = 0$, then we say that the agent v_i only sends but does not receive information, other cases are similar.

4.3 Graph Communication Enhancement Module

The graph communication enhancement network module combines the features of a graph neural network, convolution kernel, and multi-head attention mechanism. By stacking more convolutional layers, the receptive domain of the agent gradually grows and more information can be collected, so the scope of cooperation can also increase. That is, the agent v_i can directly obtain the potential feature vector from the encoder of the one-hop node through a convolutional layer. By superimposing two layers, the agent can obtain the output of the first convolution layer of the single-hop node, which contains the information of the two-hop node.

In this module, the multi-head attention mechanism is used to build the hierarchical graph convolutional network, and multi-head dot product attention is used as the convolution kernel to calculate the interaction between agents, and each agent is regarded as an entity. For each agent v_i, there is a set of agents N in the local region. The input characteristics of each entity are projected onto the query, key, and value representations by each independent attention head. For agent observation o_i^t, MLP is used to encode it to generate low-dimensional feature vectors:

$$h_i^t = MLP\left(o_i^t\right) \tag{3}$$

For attention head m, the relation between $v_i, v_j \in N_i$ is calculated as:

$$\alpha_{ij}^m = \frac{\exp\left(\beta \cdot W_q^m h_i \cdot \left(W_k^m h_j\right)^T\right)}{\sum_{\epsilon \in N_i} \exp\left(\beta \cdot W_q^m h_i \cdot \left(W_k^m h_\epsilon\right)^T\right)}, \tag{4}$$

where β is a scaling factor, $W_q^m h_i, W_k^m h_j$ is parameter matrix. exp indicates exponential calculation. N_i indicates the neighbor of agent v_i. For each attention head, the value representations of all input features are weighted and added by relation. Then, the output of m attention heads of the agent v_i are connected in series and fed into the nonlinear single-layer MLP with ReLU to obtain the output of the convolutional layer:

$$h_i' = \sigma\left(\text{Concat}\left[\sum_{j\in N_i} \alpha_{ij}^m W_v^m h_j, \forall m \in M\right]\right). \tag{5}$$

The multi-head attention mechanism makes the relation kernel independent of the order of the input feature vectors and allows the relation kernel to jointly focus on different representation subspaces. More attention heads give more relational representations, making training more stable in experience. By using multiple convolutional layers, higher-order relational representations can be extracted and the interactions between agents can be captured effectively, which greatly helps cooperative decision-making.

4.4 Time Relationship Regularization Module

we adopt the paradigm of centralized training and distributed execution. In the training process, according to the empirical replay mechanism, in each time step, we store tuple (O, A, O', R, C) in the replay buffer B, where the observation set $O = \{o_1, o_2, \ldots, o_n\}$, action set$A = \{a_1, a_2, \ldots, a_n\}$, the observation after action $O' = \{o_1', o_2', \ldots, o_n'\}$, a reward set $R = \{r_1, r_2, \ldots, r_n\}$, the communication topology adjacency matrix $C = \{C_1, C_2, \ldots, C_n\}$. For simplicity, let's omit the time expression t. Then, small batch samples of $\mathcal{S}$ are randomly selected from B to minimize the loss. Let $M^k(O_i; \theta)$ represent the attention weight distribution represented by the relation of agent v_i in the convolution layer k, then:

$$\begin{aligned}\mathcal{L}(\theta) = \frac{1}{\mathcal{S}}\sum_{\mathcal{S}} \frac{1}{\mathrm{N}}\sum_{i=1}^{\mathrm{N}} \Big(& (y_i - Q(O_i, a_i; \theta))^2 \\ & + \alpha D_{\mathrm{KL}}\left(M^k(O_i; \theta) \| z_i\right),\end{aligned} \tag{6}$$

$$y_i = r_i + \beta \max Q(O_i', a_i'; \theta'), \tag{7}$$

where $z_i = M^k(O_i'; \theta)$, α is the regularization loss coefficient, D_{KL} is the KL divergence, the model is parameterized by θ. $\mathcal{S}$ is the state space of the agent. The discount factor has less influence on the state return value of the recent time and thus has more influence on the decision result so that the learning process is more stable. The target network is updated through $\theta' = \beta\theta + (1-\beta)\theta'$. During the calculation of *loss* in training, the communication adjacency matrix C is kept unchanged for two consecutive time steps. The *loss* gradient of all agents is accumulated to update the parameters. Each agent minimizes not only its own *loss*, but also the *loss* of other agents that it cooperates with. In the execution process, each agent only needs to obtain information from its neighbor agent through communication. The model is easy to scale and suitable for large-scale multi-agent deep reinforcement learning.

Regularization of time relationship helps agents to form long-term consistent action strategies in highly dynamic environments with a large number of mobile agents. This will further help agents form cooperative behavior, since many cooperative tasks require long-term, consistent actions of cooperative intelligence to obtain the ultimate reward. We will further analyze this point in the experiment. We use the deep Q network to simulate the expected state and actions taken by the agents under the strategy, and the residual network structure similar to DenseNet is adopted. For each agent, the features of all the previous layers are connected in series and input into the Q network, so that the observation representation and feature set from different receiving domains are reused. These observations and characteristics contribute differently to the consideration of strategies for cooperation at different scales. The interaction between agents can be better considered through residuals. From the point of view of individual agents, each agent receives the observations and intentions encoded by nearby agents, which makes the environment more stable.

5 Evaluation

For the experiment, two purely collaborative grid world agent environments, Schedule and Surviving, were used to validate the functionality of the methodology i.e., DCGN and baseline. Agents' rewards, communication volume, the number of steps, the number of deaths, and the number of communication agents are evaluated to explore the utility of autonomous communication from baseline. At the same time, we set up an ablation experiment to explore the role of TRM in our model. Finally, we explore the resource overhead in the experiment and confirm that our model can reduce the redundancy of communication.

5.1 Settings

To further simplify the problem, we use discrete action spaces that allow agents to advance or stay up, down, left, and right. In the environment, each agent corresponds to a grid and has a local view of a square view. There are an indefinite number of agents in the environment, and each agent corresponds to a grid. Each grid has a box representing the position of the agent, and the square view of a given observation distance represents the observation range of the agent, and the square view of the same given communication distance represents the communication range of the agent. Only agents within the communication range of the agent are allowed to communicate with each other. Each task or destination occupies a cell in the grid.

We take a 20×20 square network environment. With the agent as the center, the observation range is 3×3 grids. The agent can communicate with neighboring agents in the square area through a grid at a certain distance. The specific environment runs 4,000 rounds, every 20 rounds as a time step to save the results.

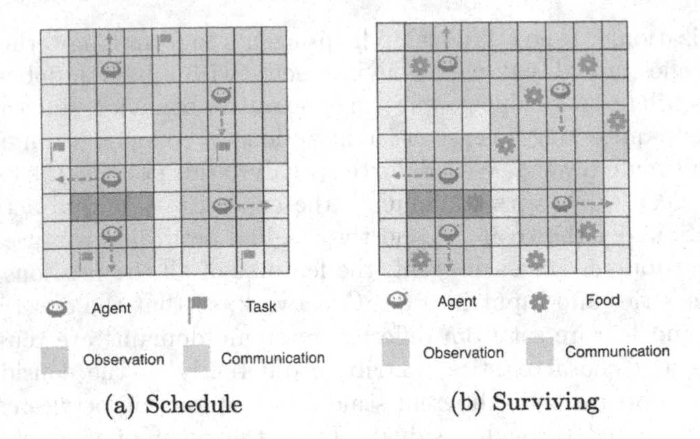

Fig. 3. The two environments used in the experiment, Schedule and Surviving.

Schedule environment is designed to test agents to perform reasonable task allocation through cooperation in a short period to complete tasks more efficiently, as shown in Fig. 3(a). A task period is 100 steps. The number of tasks is twice the number of agents and is randomly allocated in the environment. The agent loses 1 bonus for each move and gets 50 bonuses for completing the task. This is designed to make punishment and reward more immediate, as the agent must reasonably assign tasks to the nearest target point in a short time. Discrete actions of agents include moving and performing tasks, and communication can also be a part of the action space.

Surviving environment is designed to test the cooperation of agents in long-term tasks in a dynamic environment, as shown in Fig. 3(b). A task period is 500 steps. Agents start at 20 health and lose 1 health per step. The amount of food is kept at 100. The agent eating food will increase health and gain 10 health points. If the agent's health reaches 0, it deducts a reward worth 2, and if the agent's health does not reach 0, it gains a reward worth 5. The minimum health of the agent is 0. The agent needs food as soon as possible to survive. Similarly, the discrete actions of the agent include moving and eating, and communication can also be a part of the action space. During each time step, each agent can move to one of four adjacent grids or eat food at its location. Therefore, the agent must eat hard to maintain health and make the whole body gain high rewards. Once the agent eats the food, it completes the task.

5.2 Baselines

In this section, four baselines are used to illustrate the algorithm's solving effect in the fully cooperative scenario, which aims at a large amount of resource overhead caused by inter-agent communication, information redundancy caused by communication and incoherence between communication and behavior decisions, and is explained as follows:

1. **DQN**: Compared to Q-learning, DQN has a memory bank for learning previous experiences. In the implementation of this paper, all agents share the same DQN network.
2. **CommNet** [6]: The network realizes the definition of the communication process between agents through training, carries out the range of information interaction in each layer of the network, and the input and output between each layer can form an iterative relationship.
3. **IC3Net** [8]: IC3Net uses LSTM as the policy network. The communication objects between agents are not considered.
4. **DGN** [12]: DGN is an algorithm based on Q-learning. No judgment is made on whether the agent communicates.
5. **DCGN-KL:** The DCGN of the time relationship regularization module was removed to participate in the ablation experiment to demonstrate the effectiveness of the module.

5.3 Metrics

1. **Reward:** The average reward can reflect the reward obtained by the whole agent system and reflect the utility of the model from the side.
2. **Communication Volume:** The average communication volume is calculated according to the degree of the vertex of the communication topology graph and is a measurement standard of the communication volume in the agent system.
3. **The number of Finished tasks:** The average number of tasks completed by each agent in the agent system.
4. **The number of steps:** The average number of steps taken to complete the task reflects the efficiency of the agent to complete the task.
5. **The number of deaths:** The average number of deaths of agents per episode in a Surviving environment.
6. **The number of communication agents:** The average number of agents selected to communicate in each episode in both environments.

Through these metrics, we can see the situation of agents and cooperation, whether the communication is redundant, and better analyze the performance of the algorithm.

5.4 Results

Results on Autonomous Communication. In the Schedule environment, it can be seen from the data results in Fig. 4(a), Fig. 4(b), and Table 1 that: DCGN has the best performance in the index of the number of finished tasks and requires the least number of steps. According to this, we can see that: (1) Sharing state or communication can improve task performance in multi-agent reinforcement learning; (2) It is beneficial to the communication relationship

Table 1. Schedule results.

	DQN	CommNet	IC3Net	DGN	DCGN-KL	DCGN
Reward	−26.13	−24.15	−17.39	−9.50	−8.25	−0.86
Communication Volume	0	200	134	430	144	110
# Finished Tasks	1.5	1.6	1.6	1.6	1.6	1.7
# steps	71	68	66	72	74	64
# Communication agents	0	100	56	100	34	30

Table 2. Surviving results.

	DQN	CommNet	IC3Net	DGN	DCGN-KL	DCGN
Reward	−536.7	−783.7	−471	−317	−20	190
Communication Volume	0.0	1000.0	814.0	1826	970	770
# Finished Tasks	3.76	0.7	4.8	6.9	10.3	14.2
# deaths	433	469	442	400	364	330
# Communication agent	0	100	62	100	40	36

modeling and communication pruning of agents. It also shows that DCGN has better adaptability to dynamic environments and can reach cooperation in a short time.

In the Surviving environment, it can be seen from the data in Fig. 5(a), Fig. 5(b) and Table 2 that: DCGN had the best performance in reward and finished tasks metrics, and also had the lowest number of deaths. The result indicates that: (1) the method DCGN can form and maintain cooperation despite long-term dynamic change; (2) It is beneficial to give the agent the autonomy to decide whether to communicate or not, which can bring better results than the method of using decision modules. So. It is effective for DCGN to construct the autonomous communication network and make communication decisions through the agent, which can help the agent better choose whether to communicate or not and the communication object.

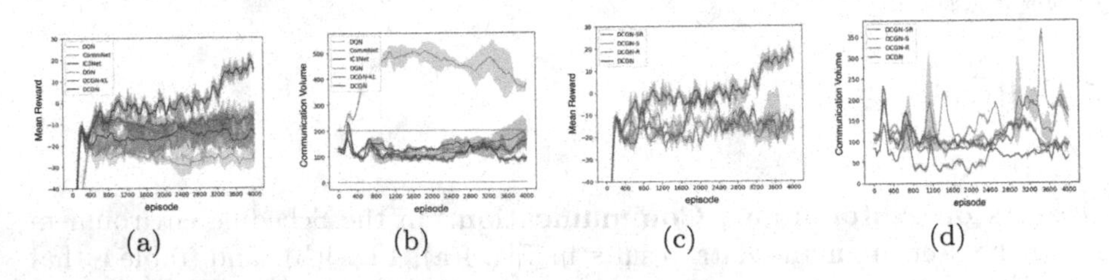

Fig. 4. Experimental results of Schedule environment. (a) Mean Reward. (b) Mean Volume. (c) Variant Mean Reward. (d) Variant Mean Volume

Ablation Experiment. The DCGN-KL we set is the DCGN that removes the time relation regularization module. Experimental results in the two environments show that the effect of the method of the time relation regularization module is slightly decreased, indicating that the time relation regularization module we proposed can help the cooperation between agents to maintain stability and consistency in a short time.

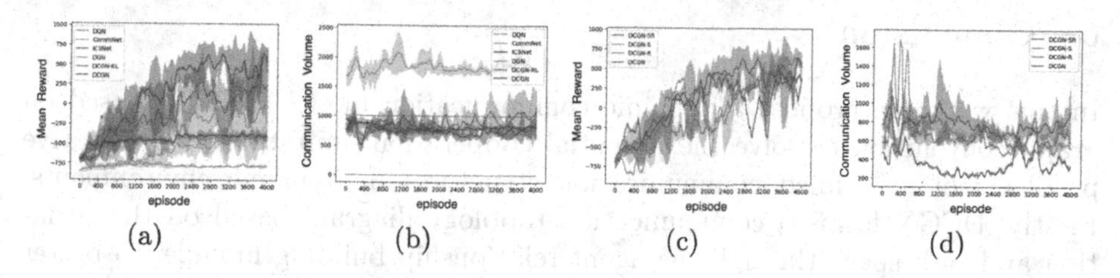

Fig. 5. Experimental results of Surviving environment. (a) Mean Reward. (b) Mean Volume. (c) Variant Mean Reward. (d) Variant Mean Volume

Resource Overhead. In terms of the communication volume index, the communication volume of DCGN is the smallest, followed by I3CNet, and the communication volume of DGN is the largest. We can conclude that: (1) In the case of communication, communication redundancy exists, that is, in some cases, more communication is not better; (2) In the case of universal communication, the methods of star topology use less communication, because they synthesize the state of all agents into a vector to calculate; (3) The method DCGN, can reduce the communication and improve the experimental results. (4) In the process of task execution, the number of agents selected for communication under the DCGN method is about half. Therefore, the autonomous decision-making communication of the agent can significantly reduce the communication. Moreover, compared with the single decision-making module, the method proposed is more robust to the communication cost, faster convergence, and more stable performance. The model can complete the task well on the basis of reducing the communication volume, autonomous communication brings more variable and simplified communication topology, information flow is more efficient, and redundant information is significantly reduced.

Agent Relationship Construction. Due to the judgment of whether agents communicate or not, the model does not receive or send data when all agents choose not to communicate, and there are three other variants correspondingly: When no communication is selected, the agent receives data but does not send data DCGN-R, sends data but does not receive data DCGN-S, and neither sends nor receives data DCGN-SR. As shown in Fig. 4(c), Fig. 4(d), Fig. 5(c) and Fig. 5(d), the results show that the complete effect of DCGN is improved compared with other variants, the reward is increased by 100% on average, and the overall training process is more stable. There is not much difference between DCGN-S, DCGN-R, and DCGN-SR. It can be concluded that when making a communication decision, the agent will measure the influence of others' information on itself, and the cooperation performance will be affected if the agent neither sends nor receives data.

6 Conclusion

In this work, we propose a dynamic communication model (DCGN) based on graph convolution to solve the potential cooperation limitations and adaptive problems of predefined communication structures in dynamic environments. Firstly, DCGN forms a communication topology diagram based on the situation and task space through the agent relationship building module, to better depict the interaction between agents and the influence of individual situation changes on information interaction. Secondly, the graph network communication enhancement module transforms the communication decision problem into an information propagation problem in the graph and proposes a graph convolution mechanism with multi-dot product attention for information exchange and aggregation among agents. Finally, the time relationship regularization module is used to strengthen the consistency of the attention matrix between actions, which helps stable cooperation in the growth period of the intelligent body. The experimental results show that DCGN can better cope with the dynamic update of tasks and help real-time decision-making in the process of helping agents complete the task information interaction. The agents can develop more coordinated and complex strategies than the existing methods. Moreover, it has strong adaptability and practicability in the field of complex agent network communication.

Acknowledgment. This work was partially supported by the National Science Fund for Distinguished Young Scholars (62025205), and the National Natural Science Foundation of China (No. 62002292, 62032020, 61960206008, 62102322).

References

1. Sutton, R.S., Barto, A.G.: Reinforcement Learning: An Introduction. MIT Press (2018)
2. Rizk, Y., Awad, M., Tunstel, E.W.: Cooperative heterogeneous multi-robot systems: a survey. ACM Comput. Surv. (CSUR) **52**(2), 1–31 (2019)
3. Foerster, J., Assael, I.A., De Freitas, N., Whiteson, S.: Learning to communicate with deep multi-agent reinforcement learning. In: Advances in Neural Information Processing Systems, vol. 29 (2016)
4. Jiang, J., Lu, Z.: Learning attentional communication for multi-agent cooperation. In: Advances in Neural Information Processing Systems, vol. 31 (2018)
5. Du, Y., et al.: Learning correlated communication topology in multi-agent reinforcement learning. In: Proceedings of the 20th International Conference on Autonomous Agents and MultiAgent Systems, pp. 456–464 (2021)

6. Sukhbaatar, S., Fergus, R., et al.: Learning multiagent communication with backpropagation. In: Advances in Neural Information Processing Systems, vol. 29 (2016)
7. Hochreiter, S., Schmidhuber, J.: Long short-term memory. Neural Comput. **9**(8), 1735–1780 (1997)
8. Singh, A., Jain, T., Sukhbaatar, S.: Learning when to communicate at scale in multiagent cooperative and competitive tasks. arXiv preprint arXiv:1812.09755 (2018)
9. Hoshen, Y.: Vain: attentional multi-agent predictive modeling. In: Advances in Neural Information Processing Systems, vol. 30 (2017)
10. Das, A., et al.: TarMAC: targeted multi-agent communication. In: International Conference on Machine Learning, PMLR 2019, pp. 1538–1546 (2019)
11. Kipf, T.N., Welling, M.: Semi-supervised classification with graph convolutional networks. arXiv preprint arXiv:1609.02907 (2016)
12. Jiang, J., Dun, C., Huang, T., Lu, Z.: Graph convolutional reinforcement learning. arXiv preprint arXiv:1810.09202 (2018)
13. Vaswani, A., et al.: Attention is all you need. In: Advances in Neural Information Processing Systems, vol. 30 (2017)
14. Yuan, Q., Fu, X., Li, Z., Luo, G., Li, J., Yang, F.: GraphComm: efficient graph convolutional communication for multiagent cooperation. IEEE IoT J. **8**(22), 16 359–16 369 (2021)
15. Hu, G., Zhu, Y., Zhao, D., Zhao, M., Hao, J.: Event-triggered multi-agent reinforcement learning with communication under limited-bandwidth constraint. arXiv preprint arXiv:2010.04978 (2020)
16. Ding, Z., Huang, T., Lu, Z.: Learning individually inferred communication for multi-agent cooperation. In: Advances in Neural Information Processing Systems, vol. 33, pp. 22 069–22 079 (2020)
17. Liu, Y., Wang, W., Hu, Y., Hao, J., Chen, X., Gao, Y.: Multi-agent game abstraction via graph attention neural network. Proc. AAAI Conf. Artif. Intell. **34**(05), 7211–7218 (2020)
18. Malysheva, A., Sung, T.T., Sohn, C.-B., Kudenko, D., Shpilman, A.: Deep multi-agent reinforcement learning with relevance graphs. arXiv preprint arXiv:1811.12557 (2018)
19. Kim, D., et al.: Learning to schedule communication in multi-agent reinforcement learning. arXiv preprint arXiv:1902.01554 (2019)
20. Niu, Y., Paleja, R.R., Gombolay, M.C.: Multi-agent graph-attention communication and teaming. In: AAMAS 2021, pp. 964–973 (2021)

Author Index

H. Jin et al. (Eds.): GPC 2023, LNCS 14504, pp. 313–314, 2024.
https://doi.org/10.1007/978-981-99-9896-8

Printed in the United States
by Baker & Taylor Publisher Services